ORAL AND MAXILLOFACIAL RADIOLOGY
A DIAGNOSTIC APPROACH

ORAL AND MAXILLOFACIAL RADIOLOGY
A DIAGNOSTIC APPROACH

David MacDonald, BDS, BSc(Hons.), LLB(Hons.), MSc, DDS(Edin.), DDRRCR, FDSRCPS, FRCD(C)

A John Wiley & Sons, Inc., Publication

This edition first published 2011
© 2011 David MacDonald

Blackwell Publishing was acquired by John Wiley & Sons in February 2007. Blackwell's publishing program has been merged with Wiley's global Scientific, Technical and Medical business to form Wiley-Blackwell.

Registered office: John Wiley & Sons Ltd, The Atrium, Southern Gate, Chichester, West Sussex, PO19 8SQ, UK

Editorial offices: 2121 State Avenue, Ames, Iowa 50014-8300, USA
The Atrium, Southern Gate, Chichester, West Sussex, PO19 8SQ, UK
9600 Garsington Road, Oxford, OX4 2DQ, UK

For details of our global editorial offices, for customer services and for information about how to apply for permission to reuse the copyright material in this book please see our website at www.wiley.com/wiley-blackwell.

Authorization to photocopy items for internal or personal use, or the internal or personal use of specific clients, is granted by Blackwell Publishing, provided that the base fee is paid directly to the Copyright Clearance Center, 222 Rosewood Drive, Danvers, MA 01923. For those organizations that have been granted a photocopy license by CCC, a separate system of payments has been arranged. The fee codes for users of the Transactional Reporting Service are ISBN-13: 978-0-8138-1414-8/2011.

Designations used by companies to distinguish their products are often claimed as trademarks. All brand names and product names used in this book are trade names, service marks, trademarks or registered trademarks of their respective owners. The publisher is not associated with any product or vendor mentioned in this book. This publication is designed to provide accurate and authoritative information in regard to the subject matter covered. It is sold on the understanding that the publisher is not engaged in rendering professional services. If professional advice or other expert assistance is required, the services of a competent professional should be sought.

Disclaimer
The contents of this work are intended to further general scientific research, understanding, and discussion only and are not intended and should not be relied upon as recommending or promoting a specific method, diagnosis, or treatment by practitioners for any particular patient. The publisher and the author make no representations or warranties with respect to the accuracy or completeness of the contents of this work and specifically disclaim all warranties, including without limitation any implied warranties of fitness for a particular purpose. In view of ongoing research, equipment modifications, changes in governmental regulations, and the constant flow of information relating to the use of medicines, equipment, and devices, the reader is urged to review and evaluate the information provided in the package insert or instructions for each medicine, equipment, or device for, among other things, any changes in the instructions or indication of usage and for added warnings and precautions. Readers should consult with a specialist where appropriate. The fact that an organization or Website is referred to in this work as a citation and/or a potential source of further information does not mean that the author or the publisher endorses the information the organization or Website may provide or recommendations it may make. Further, readers should be aware that Internet Websites listed in this work may have changed or disappeared between when this work was written and when it is read. No warranty may be created or extended by any promotional statements for this work. Neither the publisher nor the author shall be liable for any damages arising herefrom.

Front cover photo credits:
Top image: Courtesy of Dr. Montgomery Martin
Second image from top: Courtesy of Dr. Babak Chehroudi
Bottom image: Courtesy of Dr. Montgomery Martin

Library of Congress Cataloging-in-Publication Data

MacDonald, David, 1955-
  Oral and maxillofacial radiology : a diagnostic approach / David MacDonald.
       p. ; cm.
  Includes bibliographical references and index.
   ISBN 978-0-8138-1414-8 (hardcover : alk. paper)   1. Mouth–Radiography.   2. Maxilla–Radiography.   3. Face–Radiography.   I. Title.
   [DNLM: 1. Diagnostic Imaging.   2. Stomatognathic System–pathology.   3. Diagnosis, Oral. WN 230]
   RK309.M33 2011
   617.5'22075–dc22
                                      2010041339

A catalogue record for this book is available from the British Library.

This book is published in the following electronic formats: ePDF 9780470958797; ePub 9780470958803

Set in 10.5 on 12 pt ITC Slimbach by Toppan Best-set Premedia Limited
Printed and bound in Singapore by Fabulous Printers Pte Ltd

1   2011

To my mother, my daughter, Amy, and to my wife

Contents

Author and contributors		ix
Preface		xi
Part 1	**Introduction**	3
Chapter 1	Basics of radiological diagnosis *D. MacDonald*	5
Chapter 2	Viewing conditions *D. MacDonald*	37
Chapter 3	Physiological phenomena and radiological interpretation *D. MacDonald*	44
Part 2	**Advanced imaging modalities**	47
Chapter 4	Helical computed tomography *D. MacDonald*	49
Chapter 5	Cone-beam computed tomography *D. MacDonald*	59
Chapter 6	Magnetic resonance imaging *D. MacDonald*	67
Chapter 7	Positron emission tomography *D. MacDonald*	84
Chapter 8	Basics of ultrasound *D. MacDonald*	88
Part 3	**Radiological pathology of the jaws**	91
Chapter 9	Radiolucencies *D. MacDonald*	93
Chapter 10	Radiopacities *D. MacDonald*	151
Chapter 11	Maxillary antrum *D. MacDonald*	195
Chapter 12	Temporomandibular joint *D. MacDonald*	225
Chapter 13	Imaging of the salivary glands *D. MacDonald*	233

Chapter 14	Fractures of the face and jaws *D. MacDonald*	244
Chapter 15	Osseointegrated implants *T. Li and D. MacDonald*	249

Part 4 Radiological pathology of the extragnathic head and neck regions — 267

Chapter 16	Introduction *D. MacDonald and M. Martin*	269
Chapter 17	Benign lesions *M. Martin and D. MacDonald*	278
Chapter 18	Malignant lesions *M. Martin and D. MacDonald*	304

Index — 341

Author and Contributors

Author

Dr. David MacDonald, BDS, BSc(Hons.),
LLB(Hons.), MSc, DDS(Edin.), DDRRCR,
FDSRCPS, FRCD(C)
Associate Professor and Chairman, Division of
Oral and Maxillofacial Radiology
Faculty of Dentistry
The University of British Columbia
Vancouver, BC, Canada

With contributions by

Dr. Montgomery Martin, MD, FRCP(C)
Clinical Director, Department of Diagnostic
Imaging
British Columbia Cancer Agency
Faculty of Radiology
The University of British Columbia
Vancouver, BC, Canada

Dr. Thomas Li, BDS, MSc, DDRRCR,
DGDP(UK), FCDSHK, FHKAM
Former Head, Oral and Maxillofacial Radiology
Faculty of Dentistry
The University of Hong Kong
Currently in full-time Oral and Maxillofacial
Radiology specialist practice

Preface

The purpose of this textbook is to guide diagnosticians of all skill levels in generating a diagnosis for lesions affecting the face and jaws. Although its primary readership will be oral and maxillofacial and head and neck specialists, much of it is relevant to the general and specialist dentist and senior dental student, who, in service of the community at large, are most likely to encounter these lesions first. Therefore, the figures are appropriately detailed to facilitate comprehension and correlation with current standard textbooks with which the dentist is likely to be familiar.

This book focuses on new and/or important lesions and their appropriate imaging needs. These imaging needs include the modalities of helical and cone-beam computed tomography, magnetic resonance imaging, and positron emission tomography. Ultrasonography is introduced.

Over the last decade, imaging in dentistry has been substantially transformed by the advent of cone-beam computed tomography. The moderate-to-large fields of view of this modality display the base of the skull and the neck. Although these regions are the proper interpretative remit of the medical radiologist, the nonradiologist reader should be able to recognize any abnormality that may be displayed in these regions so that it can be appropriately referred for diagnosis by a radiologist. This book bridges the gap between current textbooks in oral and maxillofacial radiology and those of head and neck (medical) radiology by including Chapters 16, 17, and 18, cowritten with a medical radiologist and dedicated to the more common and important lesions likely to be imaged in the neck and base of the skull.

ORAL AND MAXILLOFACIAL RADIOLOGY
A DIAGNOSTIC APPROACH

Part 1
Introduction

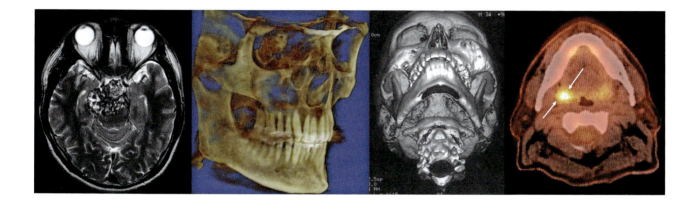

Chapter 1
Basics of radiological diagnosis

Introduction

The clinician should understand how the image is made and the normal anatomy and its variants in order to be able to identify artifacts, particularly those that can mimic the appearance of disease. Although these elements, as they present on *conventional radiography*, are addressed in detail by the wide range of dental radiology texts currently available, this textbook's figure legends note features caused by incorrect panoramic technique, artifacts, and variations of normal radiographic anatomy. Figure 1.1 outlines the main attributes of the imaging modalities that are featured in this textbook. These imaging modalities have been broadly divided into conventional radiography and advanced imaging.

Diagnosis in oral and maxillofacial radiology is most frequently based both on the clinical findings (including presenting complaint and history) and on the features observed on conventional radiographs. A definitive diagnosis is possible for a large proportion of lesion types that present to the primary care dentist. These lesions do not include just those lesions of inflammatory origin that present as periapical radiolucencies (on histological examination: *granuloma*, *periapical cyst*, or periapical abscess) and *condensing osteitis*, but also *dentigerous cysts* and *dense bone island* (also known as *idiopathic osteosclerosis*). They are not only the most frequently occurring lesions affecting the jaws, but a majority of them also have distinctive clinical and radiological presentations. Some other lesions such as *florid osseous dysplasia*, the *cementoblastoma*, the *compound odontoma*, and some cases of *odontogenic myxoma* can be definitively diagnosed solely on their radiological appearance. In those situations where a definitive diagnosis is not possible, a differential diagnosis should be developed. This will consist of two or more lesions. Such cases are frequently referred to a specialist as much for a diagnosis as for treatment. In order to assist the reader in his/her diagnosis this textbook is illustrated throughout with diagnostic flowcharts.

There is an expectation that the images created should adequately display the area of clinical interest with the purpose of addressing those clinical questions that indicated the need for the investigations. Thus the image or images should display the entire area of pathology and be free of artifacts. Therefore, an unerupted third molar should not only include the entire tooth and its follicle, but also at least a clear margin of 1 mm around them. This would allow the clinician to determine whether it is close to the mandibular canal or any other adjacent structure.

An example of inadequacy of the radiography resulted in a Canadian dental malpractice case that continued for 12 years through at least five courts before it was concluded, presumably settled.[1] The only positive result of this failure to include only 98% of a third molar was its not insignificant contribution to Canadian law specifically and common law in general. From reading the case it is abundantly clear that if an adequate radiograph or radiographs had been taken in the first instance this case would have had little grounds upon which to proceed, and the spilling of so much legal ink and personal and professional distress would have been avoided.

Radiographs are prescribed for three reasons, diagnosis, presurgical planning and follow-up. Those prescribed for the purpose of diagnosis and/or presurgical planning should be made prior to biopsy because this can change the radiology of the lesion appreciably. This is particularly so with regard to advanced imaging such as *helical computed tomography* (HCT) and *magnetic resonance imaging* (MRI). Two cases demonstrate the effects of biopsy prior to HCT.

The biopsy of an odontogenic myxoma, a locally invasive benign neoplasm, prior to HCT,

Oral and Maxillofacial Radiology: A Diagnostic Approach, David MacDonald. © 2011 David MacDonald

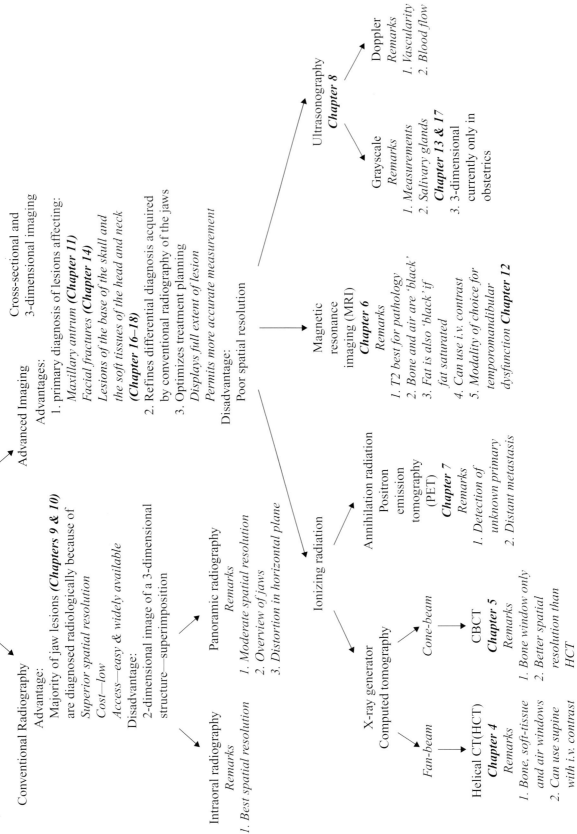

Figure 1.1. The modalities used in oral and maxillofacial radiology. This is an overview of the main imaging modalities, including remarks concerning their clearest clinical uses, relative advantages over other modalities, and limitations of use.

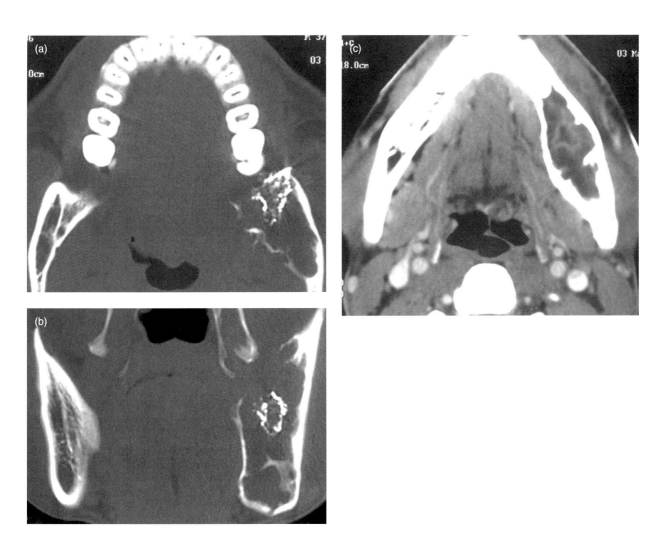

Figure 1.2. A computed tomograph of an *odontogenic myxoma* carried out after the lesion had been biopsied. The biopsy site still has its dressing in place (Figure 1.2a and 1.2b). As a result there was enhancement (Figure 1.2c) by the intravenous contrast at the site biopsied that is more likely to reflect hyperemia in response to the trauma of surgery. **Note:** All the major blood vessels including the facial and lingual arteries are enhanced in Figure 1.2c. Figure 1.2c reprinted with permission from MacDonald-Jankowski DS, Yeung R, Li TK, Lee KM. Computed tomography of odontogenic myxoma. *Clinical Radiology* 2004;59:281–287.

provoked an inflammatory response within the depth of the lesion, which was enhanced by the intravenous contrast (Figure 1.2). Contrast is recommended for lesions, which include a neoplasm or a vascular lesion in their differential diagnosis. This, with regard to neoplasms, is important to determine local invasion of adjacent soft tissues, which would need to be resected along with the rest of the neoplasm.

Figure 1.3 displays a case of *fibrous dysplasia*, which caused a substantial expansion of the affected mandible. When it was first seen by general surgeons unfamiliar with its manifestation in the jaws they performed multiple biopsies. These biopsies created their own artifacts on a subsequent HCT. These artifacts were loss of cortex and dysplastic tissue exuding through a biopsy site.

Conventional radiography will be the first imaging modality to be prescribed to investigate further a lesion occurring within the bony jaws obvious to or suspected by the clinical history and/or examination. For the majority of lesions affecting the jaws, conventional radiography is likely to be the sole imaging modality deemed clinically necessary. The principal advantages of conventional radiography are its superior spatial

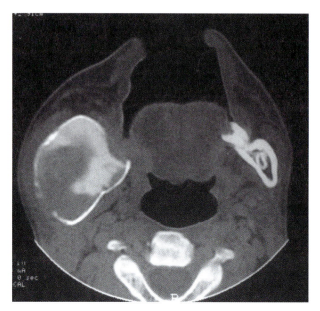

Figure 1.3. This is a bone-window axial computed tomography of *fibrous dysplasia* affecting the vertical ramus of the mandible. The cortical defects are the result of several biopsies performed prior to referral for computed tomography. Such operations can largely invalidate any clinically important radiological findings because these, if erroneous, could lead to a wrong diagnosis and inappropriate treatment. **Note:** Radiology is very central to the diagnosis of specific *fibro-osseous lesions*, discussed later.

resolution (especially of the intraoral technologies), low radiation dose, and low cost. It is also available in the dental office or surgery. It is most likely that this prescription will include a panoramic radiograph that may be accompanied by intraoral radiographs. These images may be in either analogue (film) or digital format. An overview of the various conventional radiographic technologies is set out in Table 2.1. The panoramic radiograph permits an overview of the jaws from condyle to condyle. It also permits comparison between sides. These premises can be valid only if the patient is properly positioned within the panoramic radiographic unit exposed by the most appropriate exposure factors and the image is properly developed. Finally it is also expected that the resultant image is properly reviewed (read) under optimal viewing conditions (see Chapter 2). To reiterate, all prescriptions for a radiological investigation must be based upon a thorough clinical examination. Although there is little, if no, place for routine radiographic screening in the modern practice of dentistry, every image should be carefully reviewed to identify any pathology that may be incidental to the patient's complaint and the results of the clinical examination.

The panoramic radiograph in addition to permitting determination of the specific features of the lesion or suspected lesion that prompted its making, can also reveal macroscopic abnormalities such as size differences and changes in a specific anatomical location (Figure 1.4) Furthermore, it can compliment the clinical examination by confirming defects in the dental development, such as the number, eruption, size, and even structure of the teeth (Figure 1.5). Because these features have been fully addressed in other texts and are generally well understood, space constraints preclude offering images of them here.

The various lesions, occurring within the face and jaws, often present with similar features at certain stages. Most will at some stage present as a radiolucency as they create space for further growth within the bony jaws. The borders of this radiolucency give a further clue as to their intrinsic behavior. Encapsulated benign neoplasms and many uninfected cysts grow at a moderate pace and are generally well defined. They may even have a cortex. Infected lesions and malignancies are generally associated by a poorly defined margin reflecting their more aggressive infiltrative expansion into previously normal bone. Sometimes, if the infected lesion becomes less virulent the adjacent bone may respond by laying down more bone on the trabeculae resulting in sclerosis.

Slow-growing lesions, such as most cysts and encapsulated neoplasms, can displace teeth and adjacent structures such as the mandibular canal and cortices. More aggressive lesions are more likely to resorb them. Some malignancies, such as a *squamous cell carcinoma*, will destroy structures with very little displacement, whereas others will provoke a periosteal reaction such as the onion layer typical of *osteogenic sarcoma* or *Ewing's tumor*. Such periosteal reaction can occur in *chronic osteomyelitis*. Such periosteal reactions are frequently seen in the extragnathic skeleton[2] but are infrequently seen in the jaws.

After the lesion has been properly imaged and reviewed the clinician reaches the point at which s/he wants to identify the lesion. Because the aim at this stage is to achieve, if possible, a definitive diagnosis it follows that this is best accomplished if the images of the lesion have been scrupulously reviewed. To this end I developed the rule of the

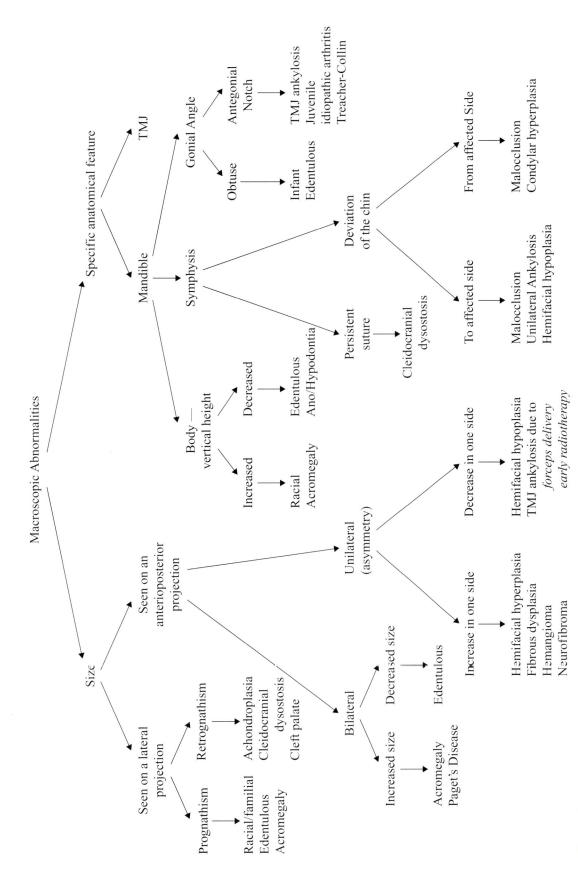

Figure 1.4. Classification of macroscopic abnormalities.

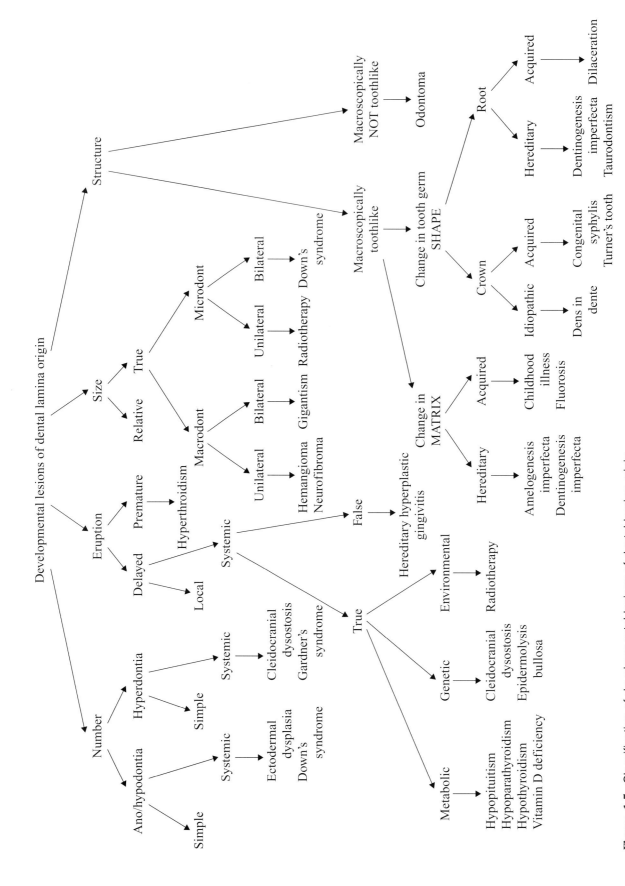

Figure 1.5. Classification of developmental lesions of dental lamina origin.

"Five S's" (shade, shape, site, size, and surroundings) and its ancillary "Three D's" (diameter, density, and displacement. There are many lesions that can be definitively diagnosed at this stage, but many others require further investigations, which could include advanced imaging.

In order to ensure that the most appropriate investigations are applied, the provisional diagnosis should be restricted to no more than 3 lesions if possible, placing the most likely in the first position so the most appropriate investigation can be performed to determine whether it is that lesion. An important exception to this "most likely" rule is potential seriousness of outcome of the lesions. Table 1.1 compares clinical outcomes according to a 10-step (0 through 9) hierarchy of seriousness of outcomes. The higher placed lesions have the more serious outcomes.

The selection of the lesions can vary among clinicians depending upon that particular lesion's presentation and frequency within a particular clinician's patient pool. The age, gender, and ethnic origin of the particular patient and site of predilection are perhaps overemphasized in most teaching programs. The main problem with this is that many lesions frequently present first outside their expected age ranges. Occasionally, this expected age range may simply be out of date. An example is fibrous dysplasia; the majority in a recent systematic review first presented in the third decade and older. If the predilection of a lesion is less than 80% for a particular feature, its value as a major diagnostic tool should be discounted unless it may hint at a serious lesion that should not be overlooked or inappropriately treated. One such lesion is the *ameloblastoma*, the most common odontogenic neoplasm globally. This 80% limit is reflected in the *receiver operating characteristics' (ROC) area under the curve (AOC)*.[3]

Another source for inaccuracy is that lesions are often superficially reported as *relative period prevalence* (RPP), which is not only dependent upon their proportion but on that of the other lesions within the same group of lesions, such as odontogenic neoplasms. The RPP not only varies between communities,[4] but it is also dependent upon the edition of the *World Health Organization* (WHO) classification of odontogenic neoplasms used. Many previously classified odontogenic neoplasms are no longer formally considered as such. An example is the cementifying fibroma (then later combined with the "ossifying fibroma", previously considered to be a separate lesion, as the *cemento-ossifying fibroma*),

once considered by the 1971 WHO edition[5] to be an odontogenic neoplasm is now considered to be a wholly osseous neoplasm, the ossifying fibroma. Some other lesions are reclassified as neoplasms. The *parakeratotic variant of the odontogenic keratocyst* is now, according to the 2005 WHO edition,[6] *keratocystic odontogenic tumor*, a neoplasm and thus no longer a cyst, whereas the orthokeratotic variant remains a cyst, the *orthokeratinized odontogenic cyst*. The same has also happened to the *calcifying odontogenic cyst*, which is now according to the 2005 edition the *calcifying cystic odontogenic tumor*. Such changes render RPP increasingly unreliable.

After a diagnosis has been made the clinician has a choice of three broad approaches to the lesion's management. These have been summarized in the rule of the 3 R's. *Refer* (to an appropriate colleague) and *review* are obvious, whereas *recipe* (treatment) requires an explanation. This is derived from the apothecary's "barred R," now often reduced to *Rx* derived from the Latin imperative *Recipe!* meaning *Take!* or *Receive!* This is still printed at the top-left corner of prescriptions for pharmaceuticals and/or other treatment.

The nomenclature used throughout will be, as far as possible, that used by the 2005 edition of the *World Health Organization Classification of Tumours*.[6] Common synonyms will appear in parentheses with the first appearance of each term in each chapter. As far as possible the morphology code of the *international classification of diseases for oncology* (ICD-O) will be provided along with the invaluable behavior codes ("/0" for benign; "/3" for malignant, and "/2" for uncertain). Although, the vast majority of lesions are diagnosed and treated in oral and dental practice solely on clinical and radiological criteria, the overwhelming majority of such lesions are sequelae of dental caries. There are many other lesions, such as cysts and neoplasms, in which a definitive diagnosis based on their histopathology is necessary.

Radiological Features

The radiological features central to the diagnosis of oral and maxillofacial lesions are encapsulated as the Five S's and Three D's rules. Although the use of these rules is most apposite for conventional radiography, they can also be applied when viewing HCT's "bone-windows" (Chapter 4) or *cone-beam computed tomographic (CBCT)* images (Chapter 5).

Table 1.1. Scale of severity of outcomes/potential severity of outcomes of oral maxillofacial radiology*

9. *Resection, but high likelihood of recurrence or metastasis*
   Poorly differentiated squamous cell carcinoma
   Osteosarcoma
   Fibrosarcoma
   Adenoid cystic carcinoma (neural spread)
8. *Resection and lower likelihood of recurrence or metastasis*
   Well-differentiated squamous cell carcinoma (qualified by site)
   Chondrosarcoma
   Ameloblastic carcinoma
   Mucoepidermoid carcinoma
7. *Resection and likelihood of recurrence or metastasis rare*
   Solid ameloblastoma
   Verrucous carcinoma
   Odontogenic myxoma
6. *Enucleation and cytotoxic treatment (Carnoy's solution)*
   Unicystic ameloblastoma (provided not affecting posterior maxilla)
   Keratocystic odontogenic tumor (KCOT formerly the parakeratotic variant of keratocyst)
5. *Simple enucleation and high chance of recurrence (recurrence rate of 10% and over)*
   Aneurysmal bone cyst (ABC)
   Ameloblastic fibroma
   Ossifying fibroma (OF)
   Glandular odontogenic cyst (GOC)
   Cementoblastoma
   Pleomorphic (salivary) adenoma (PSA)
   Calcifying epithelial odontogenic tumor (CEOT)
   Calcifying cystic odontogenic tumor (CCOT)
4. *Simple enucleation and little chance of recurrence*
   Adenomatoid odontogenic tumor (AOT)
   Ameloblastic fibro-odontoma
   Osteoblastoma/osteoid osteoma
   Orthokeratinized odontogenic cyst (formerly the orthokeratotic variant of keratocyst)
   Giant cell lesions, (large ones may need resection)
   Complex odontoma
   Squamous odontogenic tumor
   Warthin's tumor
3. *Simple enucleation and no chance of recurrence (in a neoplastic fashion)*
   Periapical radiolucencies of inflammatory origin (either nonresponsive to orthograde endodontics or too large)
   Nasopalatine duct cyst
   Dentigerous cyst
   Compound odontoma
2. *Conservative surgery may be required only to improve aesthetics*
   Fibrous dysplasia (surgery is not indicated unless compelled by appalling aesthetics or risk of blindness)
   Cherubism
   Condensing/sclerosing osteitis (no treatment required, but treatment of the affected tooth may result in regression)
1. *No treatment generally required*
   Lingual bone defect
   Osseous dysplasia (florid and focal, but NOT familial or spontaneous forms)
   Retention pseudocyst
   Osteoma—solitary; nonsyndromal (ivory type could be surgically difficult)
   Traumatic/simple bone cyst
   Idiopathic osteosclerosis/dense bone island

*This table was inspired by the Richter scale for earthquakes. The scale is based on the general current treatment paradigms for each lesion.

SHADE

Shade reflects the *radiodensity* of the lesion or feature under consideration and is its most obvious radiological attribute. This is readily reflected in the greatest frequency of radiodensity referred to in reports.

The radiodensity of a lesion observed by conventional radiography is usually described as one of three manifestations, *radiolucency*, *radiopaque*, and *mixed*. The radiolucency appears *black* and represents an absence of the bone type normal for that site (Figure 1.6).

The radiopacity appears *white* and represents an excess of mineralized tissue—frequently abnormal mineralized tissue (Figure 1.7). This abnormal tissue is usually laid down by cells (almost invariably abnormal bone cells and their variants) due to dysplastic or neoplastic processes and may show some sort of structure. It is not always possible to determine the process by histopathology; three very different lesions, fibrous dysplasia (Figure

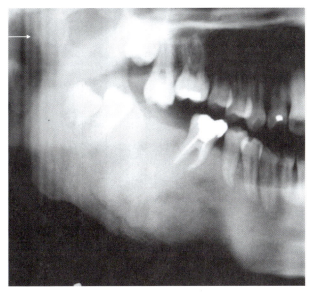

Figure 1.7. A panoramic radiograph displaying a generalized radiopacity of the posterior sextant. The mandibular canal has been reduced in thickness and displaced to the lower border of the mandible. Two unerupted molars are embedded within the vertical ramus. The lesion has expanded the body of the mandible vertically. This is *fibrous dysplasia*. **Note 1:** The mandibular canal is very obvious here as a radiolucent structure set against a background of abnormal (in this case dysplastic) bone. It has not only been displaced downward in this case, but also reduced in diameter and with a slightly irregular course. **Note 2:** The radiolucent presentation of the maxillary alveolus is a result of the superimposition of the air-filled oral cavity upon it. It may be prevented by instructing the patient to raise the tip of his/her tongue to contact the hard palate. **Note 3:** The secondary image of the contralateral mandible is superimposed upon the upper two-thirds of the vertical ramus. **Note 4:** The soft-tissue images of the soft palate and dorsum of the tongue are superimposed upon the upper third of the vertical ramus. The air space of the residual oral cavity between them presents as a radiolucent line, which has been mistaken to represent a fracture of the vertical ramus in other cases. Reprinted with permission from MacDonald-Jankowski DS. Fibrous dysplasia in the jaws of a Hong Kong population: radiographic presentation and systematic review. *Dentomaxillofacial Radiology* 1999;28:195–202.

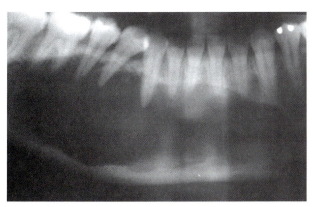

Figure 1.6. A panoramic radiograph displaying a well-defined unilocular radiolucency within the mandible extending from the right first molar's distal root to the junction between the contralateral canine and first premolar. The right lower border of the mandible has been eroded and displaced downward. The root of the right first premolar has been displaced distally. The root of the second premolar displays resorption. This is a *unicystic ameloblastoma*. **Note 1:** This panoramic radiograph had not been made using the optimal technique. It is in the head-down position. **Note 2:** The apparent root resorption or shortening of the teeth in the anterior sextant is most likely to be an artifact; due to its appearing outside the focal trough of the panoramic radiography. This happens particularly in the anterior sextant. **Note 3:** The horizontal band superimposed upon the roots of the right molars is the secondary image of the contralateral lower border of the mandible. Reprinted with permission from MacDonald-Jankowski DS, Yeung R, Lee KM, Li TK. Ameloblastoma in the Hong Kong Chinese. Part 2: systematic review and radiological presentation. *Dentomaxillofacial Radiology* 2004;33:141–151.

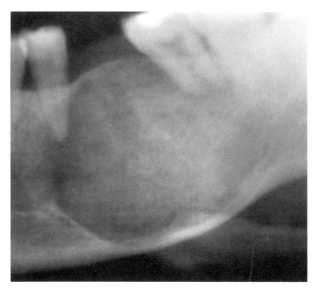

Figure 1.8. Panoramic radiograph displaying an *ossifying fibroma*. The lesion is well defined. It has a capsule of varying thickness. It has displaced downward the lower border of the mandible and displaced upward the alveolar crest. It has also displaced the mandibular canal toward the lower border of the mandible. It has displaced the root of the premolar forward and the roots of the molar distally. Its central radiodensity has a cotton wool pattern. **Note 1:** The partial superimposition of the hyoid bone on the lower border of the mandible is an indicator that the exposure had been made in the chin-down position. **Note 2:** The soft tissue of the gingival mucosa is observed in the edentulous space. Reprinted with permission from MacDonald–Jankowski DS. Cemento-ossifying fibromas in the jaws of the Hong Kong Chinese. *Dentomaxillofacial Radiology* 1998;27:298–304.

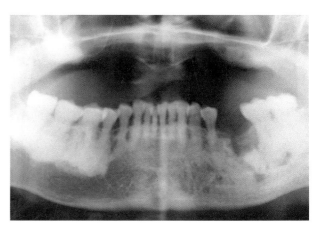

Figure 1.9. The panoramic radiograph exhibits radiopacities in all four posterior sextants. The mandibular lesions are confined to the alveolar process; that is, they are found above the mandibular canal, which can be seen in places. This is a case of *florid osseous dysplasia*. **Note:** The relative radiolucency of the anterior sextant of the maxilla is due to the superimposition of the residual oral cavity.

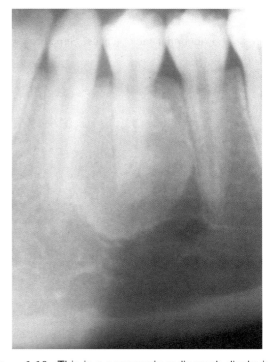

Figure 1.10. This is a panoramic radiograph displaying a well-defined radiopacity associated with the root of the first premolar. The periodontal ligament space is intact and of regular thickness separating it from the radiodense bone. This tooth displays an intact crown; there are no caries or restorations. There is also no periodontal bone loss. The radiopacity is in direct contact with the adjacent normal bone; there is no radiolucency space between them. *Idiopathic osteosclerosis* is also known as a *dense bone island*.

1.7), ossifying fibroma (Figure 1.8), and osseous dysplasia (formerly known as cemento-osseous dysplasia) (Figure 1.9) are entirely different lesions but display similar histopathological appearances, those of *fibro-osseous lesions*. This is discussed in detail in Chapter 10. Sometimes the bone is not *per se* abnormal but merely thickened trabeculae as found for idiopathic osteosclerosis (also known as dense bone islands) (Figure 1.10).

Occasionally mineralization can also be dystrophic; this is a deposition of mineral in soft-tissue lesions, such as calcification of lymph nodes (Figure 1.11), tonsils (Figure 1.11), sialoliths (Figure 13.6), antrolith acne scars, and so on. This is not laid down by bone cells but still may display some structure, usually as concentric layers of accretion (Figure 9.16).

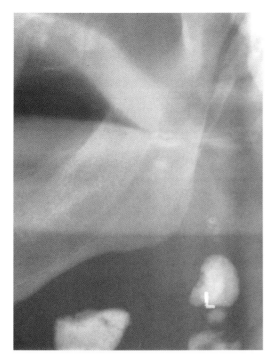

Figure 1.11. A panoramic radiograph displaying a number of normal and abnormal radiopacities. Structures, which are normally composed of soft tissue, can present as radiopacities either by being silhouetted against air, as already seen for the soft palate and tongue, or becoming calcified. The latter can occur secondary to an infection. Classically this infection was tuberculosis. The calcified structures are the lymph nodes (cervical jugulodigastric and submandibular nodes) and the palatine tonsil (small opacities superimposed upon the mandibular foramen). This calcification is dystrophic. Another calcified, but almost always normal structure, is the styloid process. **Note 1:** The soft palate and tongue are clearly visible. **Note 2:** The horizontal band of a smeared radiopacity occupying the superior two-thirds of this image represents the contralateral mandible.

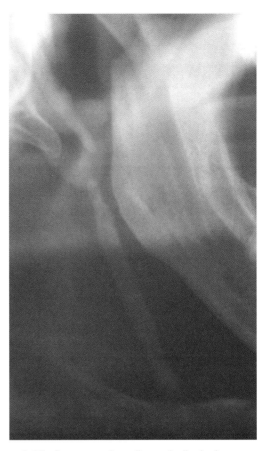

Figure 1.12. A panoramic radiograph displaying a normal-sized styloid process (extends no lower than the mandibular foramen; see Chapter 10 for more details) and a calcified stylohyoid ligament reaching the hyoid bone. The lesser horn is presented as a round radiopacity superimposed upon the superior margin of the hyoid bone. These are also normal features. **Note 1:** The pinna of the ear is superimposed upon the styoid process. **Note 2:** The condyle is fractured and displaced anteriorly. As it overlaps the superior vertical ramus, an increased radiopacity occurs at the site of this overlap. **Note 3:** The black line delineating the line of the fractured condyle represents the Mach band effect and is discussed further in Chapter 3. **Note 4:** The soft palate and dorsum of the tongue are in contact and the radiolucent line observed in Figure 1.3 is substantially absent. **Note 5:** The superior half of the image is occupied by the secondary image of the contralateral mandible.

Radiopacities can arise from variants of anatomy such as mineralization of the stylohyoid complex (Figure 1.12). The normally (not mineralized) soft-tissue structures can be present, of which the easiest to recognize are the tongue and soft palate, on panoramic radiographs and lateral cephalograms (Figure 1.12). The ear lobe (Figure 1.12) is also very frequently apparent. Fractures can result in opacities if the fractured ends overlap (Figure 1.12). Incorrect panoramic radiographic technique (head-down) can result in the superimposition of the body of the hyoid on the mandible, resulting in a radiopacity (Figure 1.13a) instead of its usual submandibular position (Figure 1.13b).

Mixed radiodensity describes a lesion presenting as a white area/s within a black area (Figure 1.14). This generally represents the deposition of mineralized tissue in an area where the bone type normal for that area had been previously removed to create space for the lesion, which subsequently undergoes mineralization.

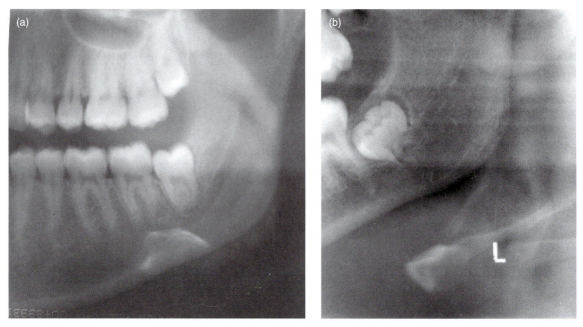

Figure 1.13. Panoramic radiographs displaying the hyoid bone. (a) Although this panoramic radiograph is correctly taken, the hyoid bone is superimposed upon the lower body of the mandible. This may be misinterpreted as a radiopaque lesion within the mandible. It is well defined and delineated by the black line of the Mach band effect. (b) The components of the hyoid bone, which are frequently apparent on panoramic radiographs and lateral cephalograms. The body and lesser and greater horns are observed as distinct entities. There are two depictions of the greater horn; the smaller and better detailed is the ipsilateral, whereas the longer and poorer detailed is the contralateral. The radiolucent area between the contralateral greater horn and the body represents the joint between them, which is frequently patent. **Note 1:** The secondary images of the contralateral mandible appear in both (a) and (b). **Note 2:** (a)There is a small air-filled space between the soft palate and the dorsum of the tongue, which is superimposed upon the mandibular foramen. **Note 3:** (b) The pinna of the ear.

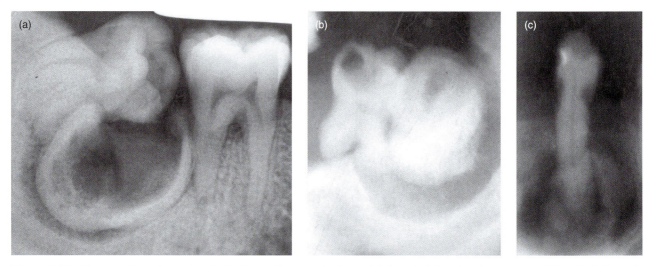

Figure 1.14. Panoramic radiographs displaying radiopaque lesions. (a) A well-defined radiolucency within which there is an annular (ringlike) radiopacity. This is an *annular odontoma*, which is a subset of the *complex odontoma*. (b) A well-defined radiolucency, associated with an almost wholly extruded molar tooth. Within the radiolucency and associated with the molar tooth is a well-defined radiopacity. This is a *complex odontoma*. (c) A radiolucency at the apex of an incisor. Within the radiolucency are several radiopacities. This is a case of *osseous dysplasia*.

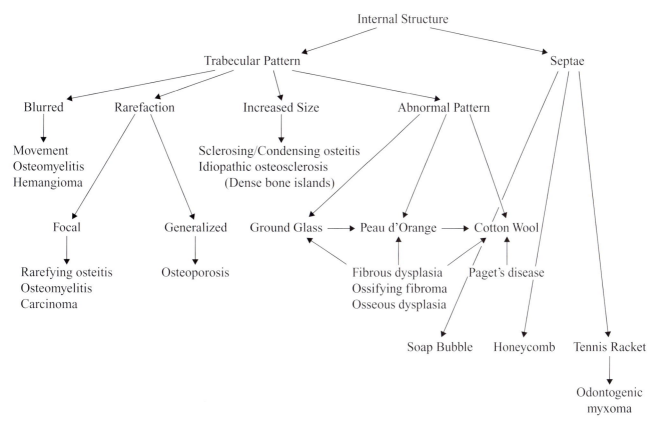

Figure 1.15. Internal structure of the lesion.

Those lesions that substantially present as radiolucencies are considered in Chapter 9, whereas those that most frequently present either complete radiopacities or as mixed lesions will be considered in the Chapter 10.

Having now determined that the lesion is radiolucent or at least partly radiopaque, consideration should then be given as to whether that radiopacity has an internal structure (Figure 1.15).

SHAPE

The shape of a lesion may give a clue to its broad behavior. If it has a smooth rounded shape, it is *unilocular*. Although this shape is typical of less serious lesions such as inflammatory cysts and dentigerous cysts, which can be readily enucleated with a minimal tendency to recur, it is frequently seen of *unicystic ameloblastomas* (Figure 1.16). Sometimes a generally rounded shape may present with an undulating or scalloped periphery (Figure 1.17) typical of *simple bone cysts*.

Those lesions whose outline has been broken into loculi by "septae" are *multilocular*. This shape is indicative of more serious lesions, which require more radical treatment such as resection because of their marked propensity to recur. Such lesions are the *solid (multilocular) ameloblastoma*, keratocystic odontogenic tumor, and odontogenic myxoma. The multilocular radiolucency can present with three basic patterns; soap-bubble, honeycomb (Figure 1.18), and tennis racket (Figure 1.19). With the exception of the tennis-racket pattern, which is virtually pathognomic for the odontogenic myxoma, the other two patterns have so far not shown a particular predilection for any specific lesion.

The clinician should not confuse multilocular with scalloping (Figure 1.20)!

It should be noted that for some lesions, particularly those cases observed in the younger patient

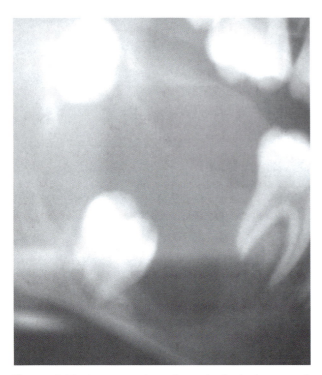

Figure 1.16. A panoramic radiograph displaying a well-defined radiolucency within the posterior sextant of the body of the mandible. Although there are 2 unerupted molars about its periphery, it is more intimately associated with the second molar, which has been displaced to the lower border of the mandible. The horizontally inclined third molar's follicular space is partially evident and is less likely to be contiguous with the larger lesion. The lesion almost wholly surrounds the second molar tooth, including its root. The absence of an attachment of the lesion at or within 1 mm of the cementoenamel junction (CEJ) and root resorption of the distal root of the first molar tooth suggest that the lesion is very unlikely to be a dentigerous cyst. This is a *unicystic ameloblastoma*. **Note 1:** The secondary image of the lower border of the mandible is partially superimposed upon the radiolucency. The inferior third displays the radiodensity that would have been obvious throughout the lesion if superimposition did not occur. The superior two-thirds displays a radiodensity, which is similar to the ground-glass appearance classically observed of fibrous dysplasia. This superimposition of the secondary image of the contralateral anatomy can be obviated on the panoramic reconstructions of computed tomography imaging. **Note 2**: The slightly more radiolucent superior third represents the substantial erosion or penetration of the cortex of the alveolar crest. The lesion immediately distal to the first molar is likely to be fluctuant.

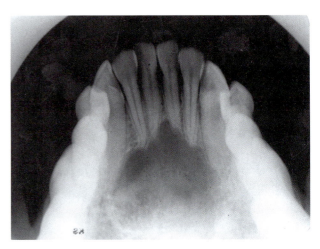

Figure 1.17. This standard anterior occlusal projection (of the anterior sextant) of the mandible displays a well-defined radiolucency, which exhibits scalloping between the roots of the anterior teeth. Their lamina dura is intact and they do not exhibit root resorption. The superior portion of the lesion appears to be more translucent than the inferior portion, because of the superimposition of the mention. This is a *simple bone cyst*. Reprinted with permission from MacDonald-Jankowski DS. Traumatic bone cysts in the jaws of a Hong Kong Chinese population. *Clinical Radiology* 1995;50:787–791.

and smaller (thus may themselves be at an early stage in their life history), are generally unilocular, whereas those cases observed in the older patient and larger may appear multilocular. Therefore, multilocularity may represent the maturity of a lesion rather than its tendency to recur if inappropriately (enucleated rather than resected) or inadequately treated.

Most cysts and a few neoplasms display hydrostatic expansion to assume a round (spherical in three dimensions) or oval shape, whereas others may assume a spindle or fusiform shape. Although the latter is classically associated with fibrous dysplasia (Figures 1.21 and 1.22), it can be observed for some neoplasms, such as the odontogenic myxoma and the keratocystic odontogenic tumor.

SITE

A solitary localized or single lesion suggests a local cause, whereas multiple lesions—particularly those affecting several sextants—suggest a systemic cause that could have general health implications. Although generally, if enough cases of a

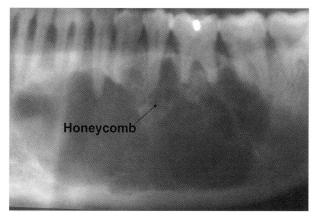

Figure 1.18. The panoramic radiograph shows a well-defined multilocular radiolucency extending from between the second molar and the junction between the contralateral canine and lateral incisor of the mandible. Many of the roots, particularly those of the first molar, display resorption, and those of the second molar are displaced distally. The lower border of the mandible has been both eroded and downwardly displaced. There are two multilocular patterns, the majority is of the soap-bubble pattern and a small area about the apex of the second premolar is of the honeycomb pattern. The latter is made up of multiple continuous cells of similar size, which together recall the appearance of a bee's honeycomb. This is a *solid ameloblastoma*. Reprinted with permission from MacDonald-Jankowski DS, Yeung R, Lee KM, Li TK. Ameloblastoma in the Hong Kong Chinese. Part 2: systematic review and radiological presentation. *Dentomaxillofacial Radiology* 2004;33:141–151.

lesion are reported they divide evenly between right and left, it is nevertheless important to record correctly this feature to avoid incorrect investigations or treatment for that particular patient. It is also clearly important to record correctly the jaw and sextants, not only for the above reason, but because some lesions have particular dispositions for a particular jaw and sextant.

Identify the affected jaw as the maxilla or mandible and the sextants as either anterior or posterior. The junction between the anterior (incisors and canines) and posterior (premolar and molars) sextants is arbitrarily defined by a vertical line passing between the canine and first premolar tooth.

Those lesions primarily affecting the maxillary antrum often present quite differently radiologically than they do in the mandible and anterior sextant of the maxilla. The lesions that affect the maxillary antrum will be considered separately in Chapter 11.

In order to determine between the alveolar and basal portions of the mandible, the relationship of the lesion to the mandibular canal should be reviewed. The equivalent feature for the maxilla is the hard palate. This is readily observed on panoramic radiographs or lateral cephalograms.

A lesion arising above the mandibular canal is in the alveolus and therefore likely to be an *odontogenic* lesion (see Figure 1.19), whereas a lesion below the mandibular canal is likely to be a *nonodontogenic* lesion (Figure 1.23). A lesion arising within the mandibular canal is likely to be a neural or vascular lesion. A lesion below the hard palate (esp. on panoramic radiographs) is likely to be an odontogenic lesion (Figure 1.24), whereas that arising above the hard palate is likely to be a nonodontogenic lesion (Figure 1.24).

If the lesion is in the alveolus, its relationship not only to teeth, but to a certain part of the tooth or teeth is important to refine further the differential diagnosis. If it is related to the crown of an unerupted tooth, this could suggest its origin within the follicle, whereas its relationship to the root of an erupted tooth with evidence of caries or periodontal disease could suggest an inflammatory cause and should provoke a testing of the pulp vitality of that tooth (pulp vitality testing is generally recommended for any tooth/teeth that are adjacent to a lesion). This clearly becomes less likely if the lesion is separated from the apex by a periodontal ligament space, which is represented by a near uniformly wide (0.2 mm) radiolucent line (Figure 1.25). The precise location of the lesion to the root is important; most inflammatory lesions are associated with the root apex, whereas this is less so if it is associated with the side of the root (Figure 1.26).

The *periodontium* is the overarching term for all tissues that surround and support the tooth. The *periodontal ligament space* is one of three components of the periodontium. The other two radiologically apparent components are the *lamina dura of the alveolar bone* and the *cementum of the root*. The main lesions that affect the periodontium have been set out in the flowchart in Figure 1.27. The length of the tooth directly affects the quality of the periodontium by determining the surface

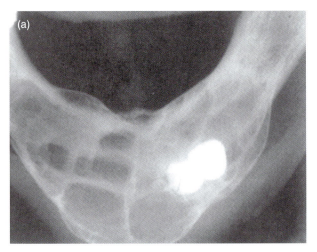
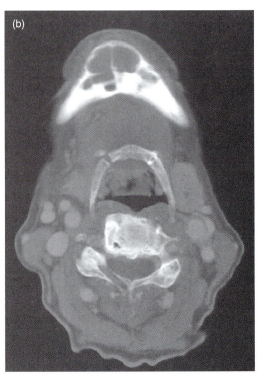
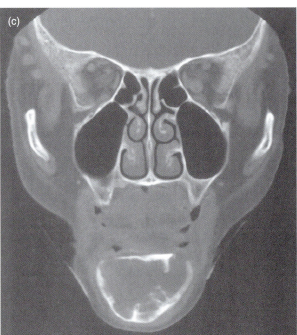

Figure 1.19. The true occlusal projection of the anterior sextant of the mandible (a) and the axial (b) and coronal (c) computed tomographic sections (bone window) display the tennis racket multilocular pattern, which is virtually pathognomonic of the *odontogenic myxoma*. (a) The "strings" of the tennis racket appear to completely transverse the entire anterior sextant. Images (b) and (c) instead display the septae confined to the periphery of the lesion, leaving a central "atrium" completely free of septae. **Note 1:** The shape of the lesion recalls the fusiform shape typically observed in fibrous dysplasia affecting the jaws. **Note 2:** (b) The patency of the synchondrosis of the hyoid bone with the lesser horn immediately adjacent to it is readily displayed. **Note 3:** (b) Enhancement of the major blood vessels, but none of the lesion. Reprinted with permission from MacDonald-Jankowski DS, Yeung R, Li TK, Lee KM. Computed tomography of odontogenic myxoma. *Clinical Radiology* 2004;59:281–287.

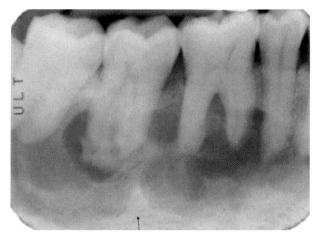

Figure 1.20. Periapical radiolucency of the mandibular molars displaying a well-defined radiolucency associated with the apices of all three molars and the second premolar tooth. The alveolar-facing margin appears scalloped, whereas there is a septum on the inferior margin. As a result this lesion is now considered to be multilocular. At the apex of the second molar tooth is a radiopacity. This appears to be dysplastic. This is a *simple bone cyst* containing an area of *osseous dysplasia*. **Note:** The more translucent area in the center of the lesion represents perforation or at least significant erosion of either the buccal or lingual cortex or both. Reprinted with permission from MacDonald-Jankowski DS. Traumatic bone cysts in the jaws of a Hong Kong Chinese population. *Clinical Radiology* 1995;50:787–791.

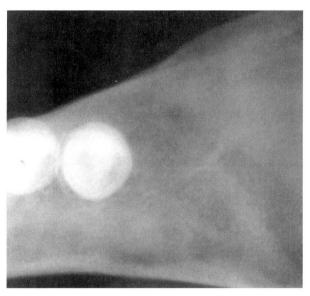

Figure 1.21. A true occlusal projection of the mandible displaying *fibrous dysplasia*. A well-defined margin between the dysplastic and adjacent normal bone is absent. The gradual expansion of the lesion from the adjacent normal bone is apparent. This pattern of expansion recalls the shape of a spindle, hence fusiform. Both the buccal and lingual cortices are greatly reduced in thickness in comparison to the normal cortex anteriorly. The reduction in cortical thickness is gradual and reflects the broad zone of transition typical of fibrous dysplasia. The radiodensity can be observed to vary in pattern, from ground glass, peau d'orange, and cotton wool. The first two are apparent here.

area available for periodontal fiber attachment. The size of the pulp in also entered both because the health of the root depends upon it, and it is just as easy to inspect it at the same time as the periodontium on the radiographs.

SIZE

The size of a lesion can be rendered in metric units (imperial units are still used but increasing less so) or according to their anatomical boundaries (Figure 1.28). The latter is particularly necessary if the lesion is displayed on a panoramic radiograph. Not only is this modality subject to substantial magnification but also distortion, particularly in the horizontal plane.[7]

Another method for determining size from a panoramic radiograph is using "the dental unit." Each tooth and the mesiodistal width of bone it spans is one unit, except for each lower incisor,

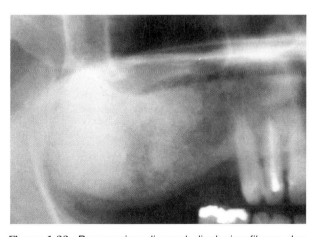

Figure 1.22. Panoramic radiograph displaying *fibrous dysplasia* affecting the right hemimaxilla. It exhibits similar fusiform expansion as is apparent in Figure 1.21. The dysplasia has involved the lower part of the posterior antral wall.

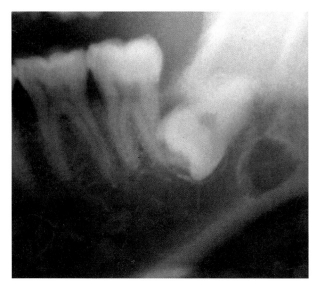

Figure 1.23. Panoramic radiograph exhibiting a well-defined radiolucency between the mandibular canal and the lower border of the mandible. This is the classical presentation of the *lingual bone defect*. The more radiolucent center represents the ostium on the lingual cortex, which is narrower than the larger defect mushrooming out within the basal process of the posterior mandible. **Note 1:** The semi-inverted unerupted third molar tooth has a normal follicle. **Note 2:** It is likely that there is no root resorption of the distal root immediately adjacent to the unerupted tooth. Persuasive evidence for this contention is derived from observation of the periodontal margin on the distal aspect of this root, through the crown of the third molar. **Note 3:** The horizontal "break" in the lower border of the mandible is caused by the Mach band effect enhancement of the superimposition of the hyoid body upon it.

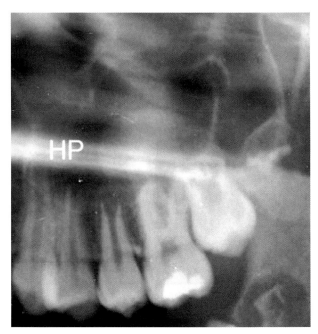

Figure 1.24. Panoramic radiograph displaying a soft-tissue opacity within the maxillary antrum. It is not associated with a carious or heavily restored tooth, which may suggest that the tooth's vitality has been compromised. This lesion is a *mucosal antral cyst,* also called a *pseudocyst.* **Note 1:** The hard palate (HP) presents as two images. The lower is its junction with the ipsilateral alveolus and the upper with the contralateral alveolus. **Note 2:** The soft tissue is visualized on a radiograph because it is silhouetted against the air-filled space of the maxillary antrum. This silhouetting is further enhanced by a black line around the mucosal antral cyst represents the Mach band effect. The same phenomena are associated with the visualization of the tongue, soft palate, and pharynx. **Note 3:** The root of the second premolar is still developing as evidenced by the presence of two "inverted chisels."

which counts for a half a unit. This can be extended into the ramus; the retromolar to the mandibular foramen, the mandibular foramen to the base of the condyle and coronoid processes, and the condyle and coronoid each account for one dental unit. This was recently used to compare the sizes of keratocystic odontogenic tumors as they appeared on a panoramic radiograph.[8] These give a reasonable estimate of the lesion's size, which may reflect an approach to surgery based on such units. Nevertheless, if surgery of a substantial lesion is contemplated, the use of advanced imaging, such as CT and MRI, permits very accurate measurements of lesions (Figure 1.29).

SURROUNDINGS

The lesion's effect on its surroundings is twofold, the degree of marginal definition and the effect on adjacent structures.

The degree of definition of the normal adjacent tissue-lesion zone of transition should be, as far as possible, objectively assessed. This is important because marginal definition is the most important radiological feature after radiodensity. Failure to use a standard objective parameter can result in significant differences of opinion between clinicians affecting the differential diagnosis. One such objective definition of margin definition was that pro-

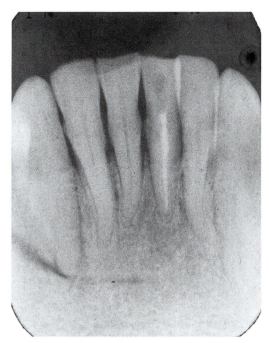

Figure 1.25. A standard anterior occlusal projection displaying an endodontically treated incisor. At its apex is a mature *osseous dysplastic* lesion. It may be surmised that the lesion presenting to the original clinician was that of a radiolucency. **Note:** It is not unusual to see such apical lesions associated with root-treated incisors. This suggests that the early radiolucent stage of this lesion had been mistaken for a periapical radiolucency of inflammatory origin. Pulp vitality is an essential investigation when the vitality of a tooth is questioned.

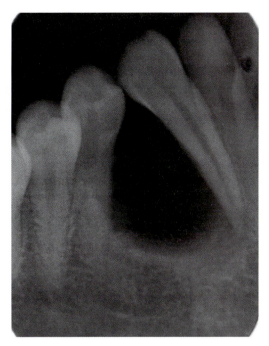

Figure 1.26. The periapical radiograph displays a well-defined radiolucency between the noncarious and pristine canine and first premolar tooth. A positive pulp vitality test ruled out a lateral radicular cyst. Other lesions that may give this presentation are remaining lesions in the differential diagnosis, the lateral periodontal cyst and the keratocystic odontogenic tumor (KCOT). At histopathology it was found to be the latter.

posed by Slootweg and Müller.[9] If s normal-adjacent tissue-lesion zone of transition is less than 1 mm, the lesion can be described as well defined and thus more representative of an uninfected cyst or benign neoplasm (Figure 1.30), whereas that which exceeds 1 mm is poorly defined suggestive of an inflamed lesion or a malignant neoplasm (Figure 1.31). This can be appreciated by running a 0.5 point pen around the periphery of the lesion displayed in an analogue format (film). If this can be achieved with ease then the margin is well defined.

Well-defined lesions may or may not have a cortex, which may assist in further refinement of the differential diagnosis (Figure 1.32). Although a cortex is, in the majority of cases, strongly suggestive of a benign lesion, be aware that multiple cortices resembling the layers of an onion (Figure 10.14) may suggest not only chronic inflammation but also some malignancies.

A cortex should be distinguished from sclerosis. A cortex is well defined with regard to both the lesion and the normal adjacent bone, whereas the sclerosis is poorly defined with regard to the latter (Figure 1.33).

The effect of the lesion on adjacent structures is expressed by the rule of the Three D's: diameter, density, and displacement; structures such as the mandibular canal can be affected by all three, whereas the cortex and the lamina dura are affected by only density and displacement.

Diameter

Changes in diameter are best seen in hollow structures such as the mandibular canal and mental and mandibular foramina. If their diameters are increased this suggests that there is a lesion within the structure, whereas if it is decreased the lesion is outside. See Figure 1.7, which displays a narrow mandibular canal invested by *fibrous dysplasia.*

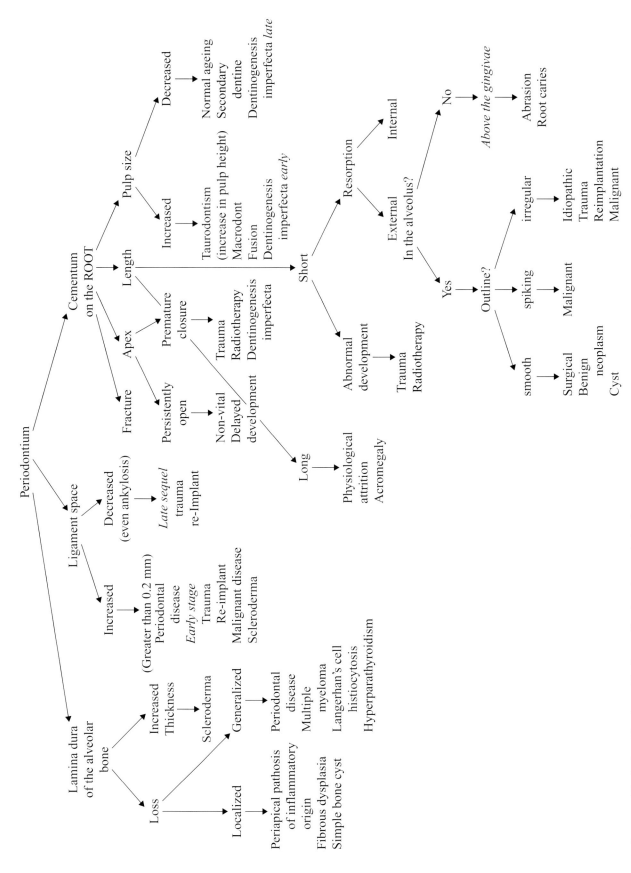

Figure 1.27. A classification of lesions affecting the periodontium.

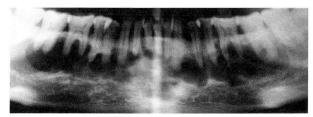

Figure 1.28. The panoramic radiograph exhibits a well-defined radiolucency, which occupies the entire length of the alveolus. This was a *simple bone cyst*, which arose from four original discrete lesions. Each of these original lesions recurred after surgery and eventually coalesced into one lesion. Reprinted with permission from MacDonald-Jankowski DS. Traumatic bone cysts in the jaws of a Hong Kong Chinese population. *Clinical Radiology* 1995;50: 787–791.

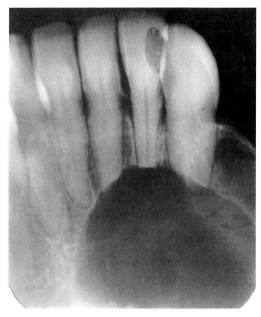

Figure 1.30. A periapical radiograph that displays a radiolucency with a well-defined periphery. The lesion has resorbed the roots in line with the bony outline of the lesion. The lesion is a *solid ameloblastoma*.

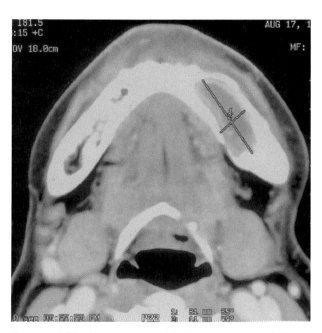

Figure 1.29. Axial computed tomograph (soft-tissue window) displaying a radiolucency within the mandible. The digital measurements are set out at the bottom of the frame. **Note:** Intravenous contrast media has enhanced the blood vessels. The tortuous outline of the lingual artery is observed near the midline anteriorly.

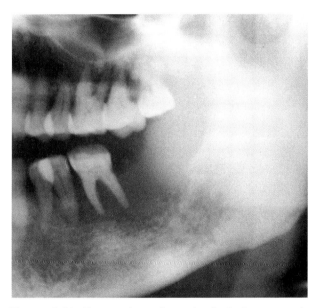

Figure 1.31. Panoramic radiograph displaying a poorly defined radiolucency occupying the posterior body of the mandible. There is almost no lamina dura associated with the first molar tooth. There appears to be a thick soft-tissue mass anterior to the vertical ramus. This is a *squamous cell carcinoma*. **Note 1:** The secondary image of the contralateral mandible is superimposed upon the vertical ramus. **Note 2:** The radiolucent region above the tongue represents the residual air-filled space of the oral cavity.

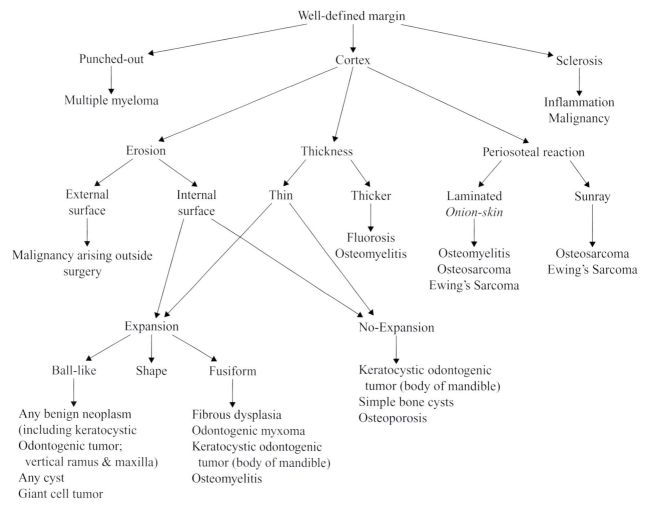

Figure 1.32. Assess the cortex for refinement of the differential diagnosis.

Density

Changes in density can be observed on teeth, cortices, and hollow structures. A reduction of density on part of a tooth root may suggest resorption either by the lesion or an anatomical structure such as the mandibular canal. Reduction in density of the cortices suggests erosion or even full perforation by the lesion (Figure 1.34). It should be appreciated that much of the radiolucency of a lesion is not derived from the absence of cancellous bone but also erosion, even perforation of either buccal or lingual cortex or both. When the last occurs the lesion's degree of radiolucency is higher and is usually associated with appreciable buccolingual expansion. Perforations of the cortex can occur in several places in the same lesion; if very large, these can give the illusion of multiloculation (Figure 1.35). Always look again for septae before arriving at this conclusion.

Changes in density of the mandibular canal (an increase in translucency-blackening) in association with a lesion or tooth suggest an intimate relationship between them, urging caution during surgery to minimize the risk of damage to the neurovascular bundle it contains. The mandibular canal can appear more translucent (blacker) and thus more conspicuous if the bone is abnormal as evident in the case of *fibrous dysplasia* in Figure 1.7.

Air-filled spaces such as the antrum and the pharynx are visible as radiolucent structure by virtue of their absence of any tissue that could attenuate the X-ray beam; in other words much of

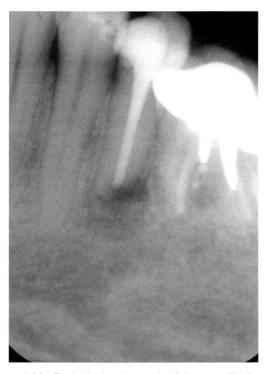

Figure 1.33. Periapical radiograph of the mandibular premolar region. The mental foramen, mandibular canal, and incisive canal are clearly obvious. Note the upward and backward bend of the canal toward the mental foramen. **Note 1:** The periapical radiolucency associated with the endodontically treated tooth displays root resorption. The radiolucency is well defined, but is not corticated. **Note 2:** The molar, which had also been endodontically treated, exhibits a radiolucency at the furcation. Within this radiolucency are radiopacities with a similar radiodensity to that of the root-filling material. Therefore, the former are likely to represent extrusion of cement though a perforation of the furcation into the tissues. The well-defined margin of the radiolucency has been enhanced by a zone of sclerosis apical to it.

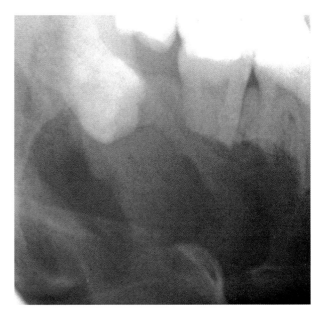

Figure 1.34. This is a oblique lateral projection of the posterior body of the mandible. Note the obliquely superimposed hyoid bone on the mandible and the contralateral angle of the mandible in the top-right corner. The radiolucency is well defined with a thin cortex. It is unilocular. Although the lesion is associated with the cementoenamel junction of the unerupted third molar, suggestive of a dentigerous cyst, the root resorption of the first and second molars is substantial. The last is more indicative of an ameloblastoma. This is a *unicystic ameloblastoma*. **Note:** The two vertical curved lines in the anterior half of the lesion arise from marked erosions or perforations of either the buccal or lingual cortex. Reprinted with permission from MacDonald-Jankowski DS, Yeung R, Lee KM, Li TK. Ameloblastoma in the Hong Kong Chinese. Part 2: systematic review and radiological presentation. *Dentomaxillofacial Radiology* 2004;33:141–151.

the beam passing though these structures is relatively unattentuated in comparison to the patient imaged. Density changes within are invariably increased densities. In the maxillary antrum this represents both discrete lesions and complete opacification by inflammatory fluid.

Displacement

The lesion can displace teeth, buccal and lingual cortices (Figure 1.36), the lower border of the mandible (Figures 1.8 and 1.37) and the antral floor, and the mandibular canal (see Figure 1.8). The types of lesions that most frequently displace adjacent structures are most benign neoplasms, particularly those with a capsule, and cysts.

ULTIMATE PURPOSES OF RADIOLOGICAL DIAGNOSIS

For the large majority of patients radiology is central in the treatment planning for caries, periodontal disease, and dentofacial disharmony (orthodontics and orthognathic surgery). In addition, radiology is important to

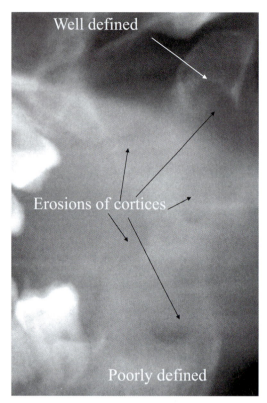

Figure 1.35. Panoramic radiograph displaying a radiolucency occupying the vertical ramus. The superior margin at the condyle is well defined, whereas that inferiorly appears poorly defined. The last was caused by the superimposition of the secondary image of the lower border of the contralateral margin upon the lower margin. The Swiss-cheese presentation of the affected vertical ramus reflects the occurrence of more erosions or perforations at certain sites rather than at others. This is a *unicystic ameloblastoma*. Reprinted with permission from MacDonald-Jankowski DS, Yeung R, Lee KM, Li TK. Ameloblastoma in the Hong Kong Chinese. Part 2: systematic review and radiological presentation. *Dentomaxillofacial Radiology* 2004;33:141–151.

1. Distinguish between a malignant and a benign lesion because early diagnosis enhances survival of the former
2. Prompt consideration of locally invasive benign neoplasms so that the most appropriate treatment can be provided to minimize recurrence
3. Prompt consideration of a hemangioma so that the most appropriate treatment plan can be formulated to avoid potential fatal exsanguination.

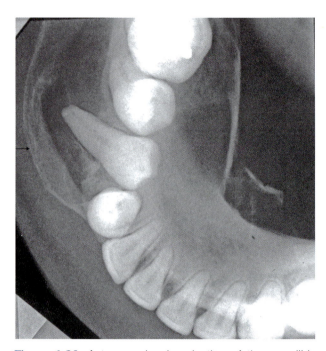

Figure 1.36. A true occlusal projection of the mandible already reviewed in Figure 1.6. There is substantial buccolingual expansion. Both the buccal and the lingual cortex are very thin, particularly buccally. In addition to the distal displacement observed in Figure 1.6. the root of the first premolar has also been displaced buccally. This is a *unicystic ameloblastoma*. Reprinted with permission from MacDonald-Jankowski DS, Yeung R , Lee KM, Li TK. Ameloblastoma in the Hong Kong Chinese. Part 2: systematic review and radiological presentation. *Dentomaxillofacial Radiology* 2004;33:141–151.

System of Evidence Used in This Textbook

Sackett et al.[10] defined a systematic review as a summary of the medical literature that uses explicit methods to search systematically, appraise critically, and synthesize the world literature on a specific issue. This means that unlike a traditional review the systematic review, like any other form of primary research, will have a "materials and methods" section, and a "results" section.[11]

Systematic review has generally been applied to treatment and drug trials, but has also become a powerful tool when adapted to the clinical and radiological presentations of important oral and maxillofacial lesions.[12–25] These are the ameloblastoma (Figure 1.38),[12–13] odontogenic myxoma

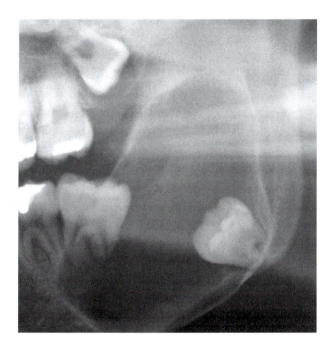

Figure 1.37. Panoramic radiograph displaying a *unicystic ameloblastoma* associated with the third molar (this attachment is apical to the cementoenamel junction). The second molar's roots appear to have been displaced anteriorly. The lowest border of the lesion has expanded down past the still undisplaced and largely intact lower border of the mandible. Although this phenomenon is generally a feature of the ameloblastoma, it has also been observed for orthokeratinized odontogenic cysts. **Note 1:** The secondary image of the contralateral mandible has conferred a ground-glasslike appearance on the upper two-thirds of the lesion. The lower third displays a truer degree of radiodensity. **Note 2:** The erupting maxillary third molar exhibits an enlarged follicular space.

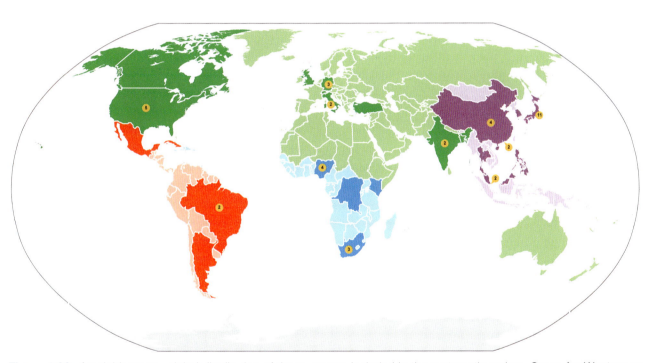

Figure 1.38. Ameloblastoma: global distribution of those reports included in the systematic review. Green for Western or predominantly Caucasian communities, blue for sub-Saharan African communities, violet for East Asian communities, and red for Latin American communities. The lighter shades denote each of the four global groups or regions, whereas the darker shade denotes a systematic review-included report for a particular state. If more than one such report exists, the number over one is inserted for that state. Acknowledgment: James Pagnotta: Media support analyst: Faculty of Dentistry; University of British Columbia.

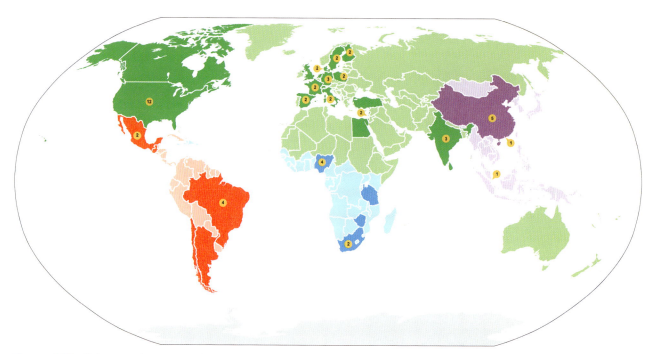

Figure 1.39. Odontogenic myxoma (updated March 2010): global distribution of those reports included in the systematic review. Note that both Swedish reports are binational: 1 with Denmark and 1 with Finland. Green for Western or predominantly Caucasian communities, blue for sub-Saharan African communities, violet for East Asian communities, and red for Latin American communities. The lighter shades denote each of the four global groups or regions, whereas the darker shade denotes a systematic review-included report for a particular state. If more than one such report exists, the number over one is inserted for that state. Acknowledgment: James Pagnotta: Media support analyst: Faculty of Dentistry; University of British Columbia.

(Figure 1.39),[14] keratocystic odontogenic tumor (Figure 1.40),[15] dentigerous cyst (Figure 1.41),[16] orthokeratinized odontogenic cyst (Figure 1.42),[17] glandular odontogenic cyst (Figure 1.43),[18] fibrous dysplasia (Figure 1. 44),[19] ossifying fibroma (Figure 1.45),[20] florid osseous dyplasia (Figure 1.46),[21] Focal osseous dysplasia (Figure 1.47),[22] idiopathic osteosclerosis (Figure 1.48),[23] central giant cell granuloma,[24] and cleidocranial dysostosis.[25]

Global Groups

In order to determine deeper patterns within the systematic review, its reports are divided into four Global groups based broadly on ethnicity; these are East Asian (predominantly represented in the SR by Chinese and Japanese), sub-Saharan African (predominantly Black Africa, including Jamaica), Western/Caucasian (North America and Europe, Middle East, North Africa, and India), and Latin American (including Cuba). Although the Western group is predominantly White (Caucasian; classically of European origin) it contains significant non-White minorities, particularly from sub-Saharan Africa. The population of the United States was at the last census 69.1% White.[26] Reports from the Indian subcontinent are included in the Western/Caucasian group, because 95% of Indians are Caucasian (Indo-Aryans and Dravidians). Although these four global groups are cartographically represented by four almost discrete regions, they are not primarily regional, because variable socioeconomic and other ethnocultural factors also play important roles that affect the availability and provision of diagnostic and therapeutic services. For example, the South Asian nations, including India, although largely Caucasian nations, are still developing their economies, along with many of those of sub-Saharan Africa. Although Africa itself is divided between a Caucasian North and a substantially Black sub-Saharan South, it is the latter that constitutes both the bulk of the population of the African continent and the African diaspora (Jamaica is 90%

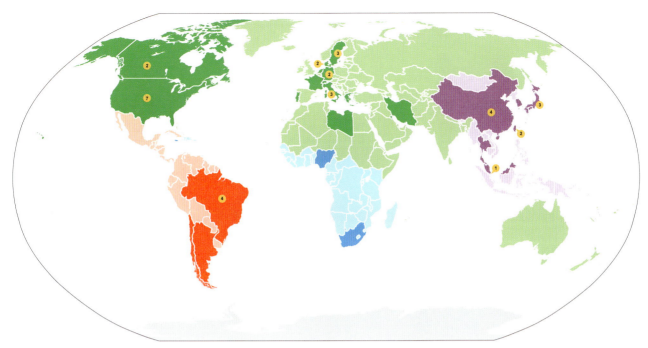

Figure 1.40. Keratocystic odontogenic tumor: global distribution of those reports included in the systematic review. Green for Western or predominantly Caucasian communities, blue for sub-Saharan African communities, violet for East Asian communities, and red for Latin American communities. The lighter shades denote each of the four global groups or regions, whereas the darker shade denotes a systematic review-included report for a particular state. If more than one such report exists, the number over one is inserted for that state. Acknowledgment: James Pagnotta: Media support analyst: Faculty of Dentistry; University of British Columbia.

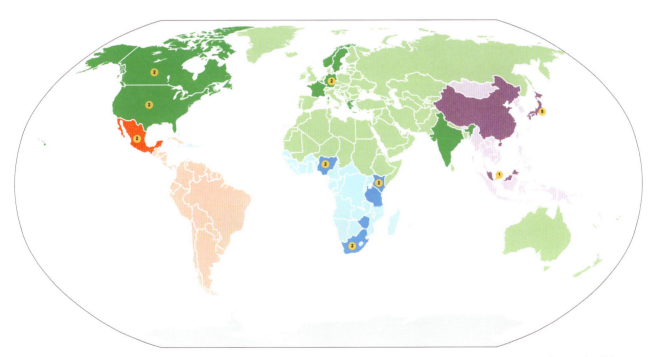

Figure 1.41. Dentigerous cyst: global distribution of those reports included in the Systematic review. Green for Western or predominantly Caucasian communities, blue for sub-Saharan African communities, violet for East Asian communities, and red for Latin American communities. The lighter shades denote each of the four global groups or regions, whereas the darker shade denotes a systematic review-included report for a particular state. If more than one such report exists, the number over one is inserted for that state. Acknowledgment: James Pagnotta: Media support analyst: Faculty of Dentistry; University of British Columbia.

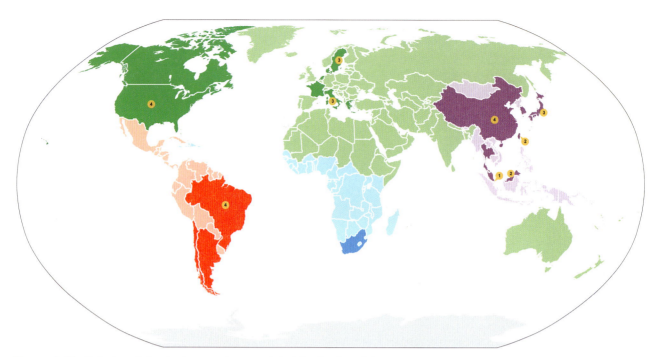

Figure 1.42. Orthokeratinized odontogenic cyst: global distribution of those reports included in the systematic review. Green for Western or predominantly Caucasian communities, blue for sub-Saharan African communities, violet for East Asian communities, and red for Latin American communities. The lighter shades denote each of the four global groups or regions, whereas the darker shade denotes a systematic review-included report for a particular state. If more than one such report exists, the number over one is inserted for that state. Acknowledgment: James Pagnotta: Media support analyst: Faculty of Dentistry; University of British Columbia.

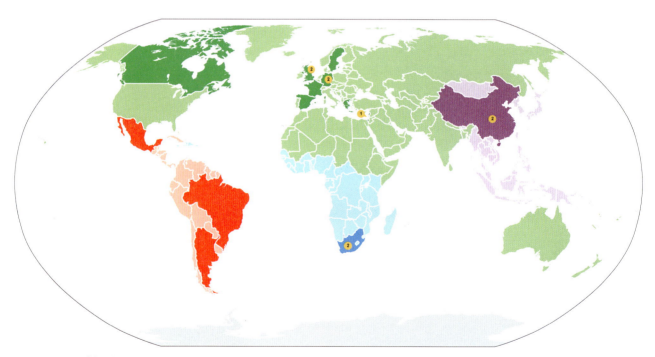

Figure 1.43. Glandular odontogenic cyst: global distribution of those reports included in the systematic review Green for Western or predominantly Caucasian communities, blue for sub-Saharan African communities, violet for East Asian communities, and red for Latin American communities. The lighter shades denote each of the four global groups or regions, whereas the darker shade denotes a systematic review-included report for a particular state. If more than one such report exists, the number over one is inserted for that state. Acknowledgment: James Pagnotta: Media support analyst: Faculty of Dentistry; University of British Columbia.

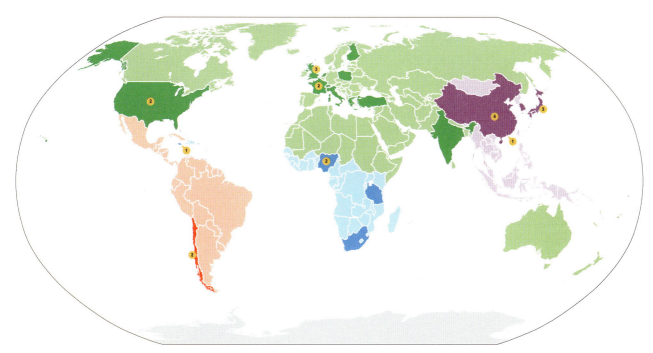

Figure 1.44. Fibrous dysplasia: global distribution of those reports included in the Systematic review. Green for Western or predominantly Caucasian communities, blue for sub-Saharan African communities, violet for East Asian communities, and red for Latin American communities. The lighter shades denote each of the four global groups or regions, whereas the darker shade denotes a systematic review-included report for a particular state. If more than one such report exists, the number over one is inserted for that state. Acknowledgment: James Pagnotta: Media support analyst: Faculty of Dentistry; University of British Columbia.

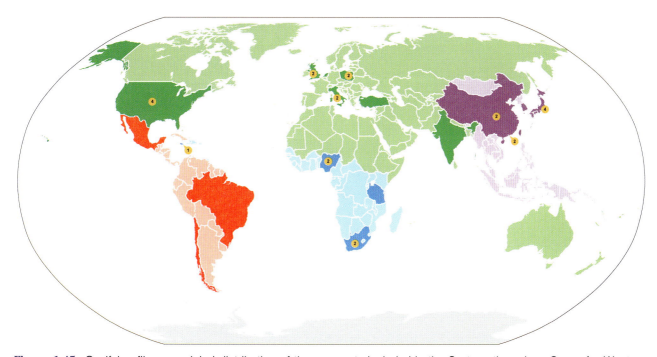

Figure 1.45. Ossifying fibroma: global distribution of those reports included in the Systematic review. Green for Western or predominantly Caucasian communities, blue for sub-Saharan African communities, violet for East Asian communities, and red for Latin American communities. The lighter shades denote each of the four global groups or regions, whereas the darker shade denotes a systematic review-included report for a particular state. If more than one such report exists, the number over one is inserted for that state. Acknowledgment: James Pagnotta: Media support analyst: Faculty of Dentistry; University of British Columbia.

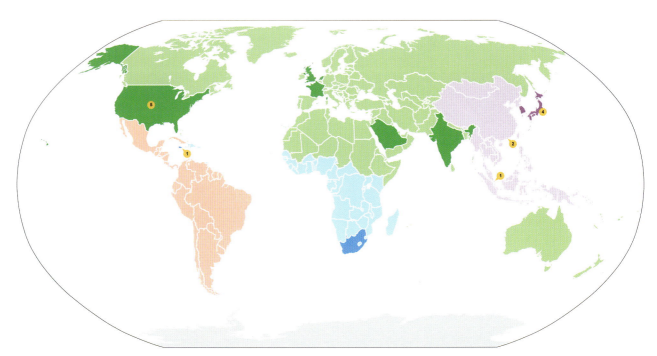

Figure 1.46. Florid osseous dysplasia: global distribution of those reports included in the systematic review. Green for Western or predominantly Caucasian communities, blue for sub-Saharan African communities, violet for East Asian communities, and red for Latin American communities. The lighter shades denote each of the four global groups or regions, whereas the darker shade denotes a systematic review-included report for a particular state. If more than one such report exists, the number over one is inserted for that state. Acknowledgment: James Pagnotta: Media support analyst: Faculty of Dentistry; University of British Columbia.

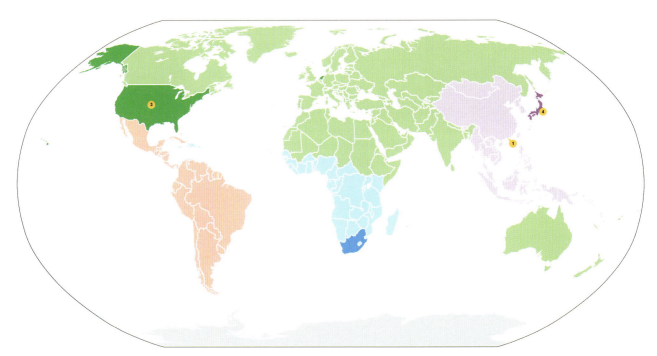

Figure 1.47. Focal osseous dysplasia: global distribution of those reports included in the systematic review. Green for Western or predominantly Caucasian communities, blue for sub-Saharan African communities, violet for East Asian communities, and red for Latin American communities. The lighter shades denote each of the four global groups or regions, whereas the darker shade denotes a systematic review-included report for a particular state. If more than one such report exists, the number over one is inserted for that state. Acknowledgment: James Pagnotta: Media support analyst: Faculty of Dentistry; University of British Columbia.

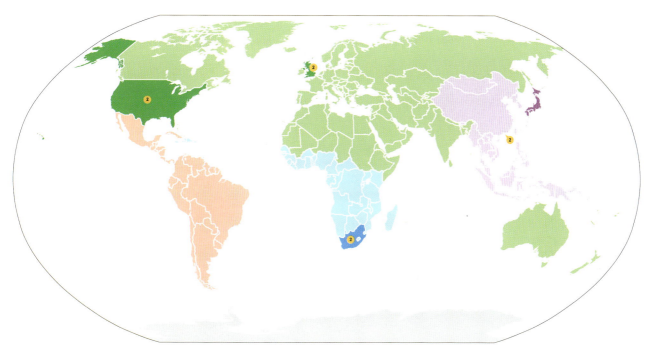

Figure 1.48. Idiopathic osteosclerosis: global distribution of those reports included in the systematic review. Green for Western or predominantly Caucasian communities, blue for sub-Saharan African communities, violet for East Asian communities, and red for Latin American communities. The lighter shades denote each of the four global groups or regions, whereas the darker shade denotes a systematic review-included report for a particular state. If more than one such report exists, the number over one is inserted for that state. Acknowledgment: James Pagnotta: Media support analyst: Faculty of Dentistry; University of British Columbia.

of sub-Saharan African origin). Although the global distribution for each lesion is largely determined by the number size and quality of the systematic review-included reports, the lesions for some communities are likely to be underreported. These are discussed in Chapters 9 and 10.

References

1. Holsten v Card, 1999 CanLII 1358 (BC S.C.) Available from URL: http://www.canlii.org/bc/cas/bcsc/1999/1999bcsc11805.html
2. Wenig BM, Mafee MF, Ghosh L. Fibro-osseous, osseous, and cartilaginous lesions of the orbit and paraorbital region. Correlative clinicopathologic and radiographic features, including the diagnostic role of CT and MR imaging. *Radiol Clin North Am* 1998;36:1241–1259, xii.
3. Pretty IA, Maupomé G. A closer look at diagnosis in clinical dental practice: part 2. Using predictive values and receiver operating characteristics in assessing diagnostic accuracy. *J Can Dent Assoc* 2004;70:313–316.
4. Lu Y, Xuan M, Takata T, Wang C, He Z, Zhou Z, Mock D, Nikai H. Odontogenic tumors. A demographic study of 759 cases in a Chinese population. *Oral Surg Oral Med Oral Pathol Oral Radiol Endod* 1998;86:707–714.
5. Pindborg JJ. Kramer IRH, Torloni H. Histological Typing of Odontogenic Tumours, Jaw Cysts, and Allied Lesions. *WHO Geneva*, 1971.
6. Barnes L, Eveson JW, Reichert P, Sidransky D. Pathology and Genetics of Head and Neck Tumours. *World Health Organization Classification of Tumours— International Agency for Research on Cancer. IARC Press*, Lyon 2005: pp 284–327.
7. Langland OE, Langlais RP, McDavid WD, DelBalso AM. Panoramic Radiography. 2nd ed. Lea and Febiger, Philadelphia 1989: pp 52–57.
8. MacDonald-Jankowski DS. Keratocystic odontogenic tumour in a Hong Kong community; the clinical and radiological presentations and the outcomes of treatment and follow-up. *Dentomaxillofacial Radiol* (in press).
9. Slootweg PJ, Müller H. Differential diagnosis of fibro-osseous jaw lesions. A histological investigation on 30 cases. *J Craniomaxillofac Surg* 1990;18:210–214.
10. Sackett DL, Strauss SE, Richardson WS, Rosenberg W, Haynes RB. Evidence-based medicine. *How to Practice and Teach EBM.* 2nd ed. Churchill-Livingstone, Edinburgh 2000: pp 133–136.

11. MacDonald-Jankowski DS, Dozier MF. Systematic review in diagnostic radiology. *Dentomaxillofac Radiol* 2001;30:78–83.
12. MacDonald-Jankowski DS, Yeung R, Lee KM, Li TK. Ameloblastoma in the Hong Kong Chinese. Part 1: systematic review and clinical presentation. *Dentomaxillofac Radiol* 2004;33:71–82.
13. MacDonald-Jankowski DS, Yeung R, Lee KM, Li TK. Ameloblastoma in the Hong Kong Chinese. Part 2: systematic review and radiological presentation. *Dentomaxillofac Radiol* 2004;1:141–151.
14. MacDonald-Jankowski DS, Yeung R, Lee KM, Li TK. Odontogenic myxomas in the Hong Kong Chinese: clinico-radiological presentation and systematic review. *Dentomaxillofac Radiol* 2002;31:71–83.
15. MacDonald-Jankowski DS. Keratocystic odontogenic tumour: a systematic review. *Dentomaxillofac Radiol* 2011;40:1–23.
16. MacDonald-Jankowski DS, Chan KC. Clinical presentation of dentigerous cysts: systematic review. *Asian J Oral Maxillofac Surg* 2005;15:109–120.
17. MacDonald-Jankowski DS. Orthokeratinized odontogenic cyst: a systematic review. *Dentomaxillofac Radiol* 2009;39:455–467.
18. MacDonald-Jankowski DS. Glandular odontogenic cyst: a systematic review. *Dentomaxillofac Radiol* 2010;39:127–139.
19. MacDonald-Jankowski DS. Fibrous dysplasia: a systematic review. *Dentomaxillofac Radiol* 2009;38:196–215.
20. MacDonald-Jankowski DS. Ossifying fibroma: a systematic review. *Dentomaxillofac Radiol* 2009;38:495–513.
21. MacDonald-Jankowski DS. Florid cemento-osseous dysplasia: a systematic review. *Dentomaxillofac Radiol* 2003;32:141–149.
22. MacDonald-Jankowski DS. Focal cemento-osseous dysplasia: a systematic review. *Dentomaxillofac Radiol* 2008;37:350–360.
23. MacDonald-Jankowski DS. Idiopathic osteosclerosis in the jaws of Britons and of the Hong Kong Chinese: radiology and systematic review. *Dentomaxillofac Radiol* 1999;28:357–363.
24. Stavropoulos F, Katz J. Central giant cell granulomas: a systematic review of the radiographic characteristics with the addition of 20 new cases. *Dentomaxillofac Radiol* 2002;31:213–217. Erratum in *Dentomaxillofac Radiol* 2002;31:394.
25. Golan I, Baumert U, Hrala BP, Müssig D. Dentomaxillofacial variability of cleidocranial dysplasia: clinico-radiological presentation and systematic review. *Dentomaxillofac Radiol* 2003;32:347–354. Erratum in *Dentomaxillofac Radiol* 2004;33:422.
26. United States 2000 census http://www.censusscope.org/us/chart_race.html

Chapter 2
Viewing conditions

Introduction

Conventional imaging is represented by both digital and analog technologies, whose main features are outlined and compared in Table 2.1. Conventional imaging almost invariably precedes advanced imaging, because it imparts less radiation dose, is generally cheaper, is more widely available and—particularly in regard to intraoral technology—affords the best spatial resolution. This last feature has so far not been remotely achieved by the enormous advances in *cone-beam computed tomography* (CBCT).

The optimal viewing conditions of the images produced by *conventional radiography* employed by specialist radiologists may differ markedly from those used by many other clinicians. The latter may generally review their radiographic imaging in brightly lit clinical areas. Although this bright lighting assists in the proper evaluation of the patient, it can catastrophically degrade the displayed radiographic image. Because most dentists and oral and maxillofacial specialists are their own radiologists, they should use the same viewing conditions as specialist radiologists for primary readings (radiological diagnosis) of their patients' images.

Image Display

Although Krupinski et al. found no difference between the performance of radiologists using monitors of differing luminance, the dwell-time (time spent reviewing the image prior to diagnosis) was significantly longer when it was read under suboptimum viewing conditions.[1] The two factors affecting display are the quality of the monitor and the ambient lighting at the time of the review of the image displayed on that monitor.

Oral and Maxillofacial Radiology: A Diagnostic Approach,
David MacDonald. © 2011 David MacDonald

MONITORS

In medicine conventional radiographs taken for diagnosis are read by radiologists on *medical-grade diagnostic grayscale* (MGDG) monitors under reduced ambient lighting. These monitors are monochromatic. The main advantage of MGDG monitors is their high luminence, which makes it easier to see the entire grayscale from black to white (Figure 2.1). The radiologist produces a report, which accompanies the images. The referring clinician, using a "point-of-care" monitor (usually a standard commercially available monitor of variable quality, which may have a color display) has the radiologist's report to guide him/her (Figure 2.1c). Other important features of a MGDG monitor are an optimal spatial resolution (image detail, measured in line-pairs per millimeter (lp/mm)) and contrast resolution (discerning the difference between two adjacent densities and commonly expressed in bit-depth or gray levels), high brightness, adjustment for the human eye's nonlinear perception, and self-calibration (Table 2.2.).

The displayed image should fully represent all the data captured by the detector. Ideally, the display of each pixel of the image captured by the detector should be represented by a corresponding pixel on the monitor display in order to optimize the detector's spatial resolution. This is a 1-to-1 or 1:1 display. Therefore, information contained within the captured image may not be displayed on the monitor if the display is not 1:1. Haak et al. reported that ratios of 1:1 and 2:1 were significantly better for detection of approximal caries than a ratio of 7:1.[2] In their comparison of a standard desktop with a dedicated medical monitor, Gutierrez et al. found that the standard desktop display was clearly inadequate for diagnostic radiology.[3]

These MGDG monitors are technologically complex. Only the main features will be overviewed here. The MGDG monitor's *grayscale standard-display function* (GSDF) is based on a phenomenon called *human-contrast sensitivity* (HCS), which

Table 2.1. A comparison between the imaging technologies available to dentists*

	Imaging Technologies-DETECTORS			
	Film	Solid State		Phosphor Plate
		Charge Couple Device (CCD)	Complementary Metal-Oxide-Semiconductor (CMOS)	
Brief description partly provided by Parks (2008)	Silver bromide developed to silver, the density of which provides the grayscale image.	X-rays cause emitted electrons to collect in electron wells converted to grayscale image.	Array of field effect transmitters with a polysilicon gate	Scanned by red light laser and emit blue light
Vulnerability to damage?	No—unless poorly stored—heat fogs it	Yes—by dropping and autoclaving	Yes—by dropping and autoclaving	Yes—frequently unusable after 50 uses
Basic costs of detectors, not including operating systems or software	Cheapest Note that the film is completely consumed in a single use.	On average 10,000–20,000 Euros	On average 10,000–20,000 Euros	Although 40–50 Euros each they last 50 uses and the scanners are expensive 10,000 Euros
	The above costs were derived and converted into euros from the following URL: http://www.cliniciansreport.org/page/additonal-studies-archive This is also discussed by Parks (2007).			
Immediate Image?	No—chemical development of latent image	Yes	Yes	No—needs to be scanned into the patient record
Likelihood of image degradation if delayed?	No—unless reexposed before developed	N/A	N/A	Yes—deteriorates with delay before scanned
Special room required?	Yes (dark room)	No	No	Yes (dim room)
Noxious chemicals?	Yes	No	No	No
Whole surface available for image capture?	Yes	No	No	Yes
Spatial resolution (detail) in line pairs per millimeter?	Kodak InSight 20lp/mm	Kodak RVG-ui 20lp/mm	Kodak RVG 6000 20lp/mm	Planmeca Dixi 16lp/mm
	All above are from the 2005 report by Farman AG and Farman TT, who compared 17 detectors.[#]			
Dynamic range?	Narrow	Narrow	As wide as the phosphor plate?	Wide

Shorter exposure time?	Yes—if E and F speed	Same as E and F speed	Same as E and F speed	Yes—potential to be shorter
More exposures required for full-mouth survey?	No—optimum	Yes—smaller area available for image capture	Yes—smaller area available for image capture	No—same area available for image capture as film
Retakes more likely?	No	Yes—cone-cuts more likely	Yes—cone-cuts more likely	No
Patient comfort?	Yes	No—bulky and inflexible	No—bulky and inflexible	Yes—same as film
Permit taking of vertical bitewings?	Yes	No—bulky and inflexible	No—bulky and inflexible	Yes—same as film
Occlusal size available?	Yes	No	No	Yes
Panoramic radiograph?	Yes	Yes	Not yet available	Yes
Lateral cephalogram?	Yes	Yes	Not yet available	Yes
Infection control challenges?	No—disposal after single use	Yes	Yes	Yes
Integration with an electronic patient record (EPR)?	No—also scanned image contains a fraction of the information	Yes	Yes	Yes
Image display	Bright-light viewing box	All 3 digital technologies under high brightness medical diagnostic-grade grayscale monitor		
Optimal viewing conditions?	All under reduced ambient lighting			
Ease of image enhancement?	No—brightness only	Yes	Yes	Yes
Integrity of original image (vulnerability to fraud)?	No	No—almost all modern systems preserved original image; any subsequent amendments are preserved as date-stamped editions		
Long-term storage?	Yes—if properly developed	Unknown	Unknown	Unknown
Vulnerability of data in the image to loss?	Not if properly developed and stored, but will be destroyed if surgery is destroyed	Can be vulnerable to computer viruses. If the data is backed up to a remote facility it can survive destruction of the surgery.		
Telemedicine?	No	Yes	Yes	Yes

*Adapted from MacDonald DS. Factors to consider in the transition to digital radiological imaging. *Journal of the Irish dental Association.* 2009;55:26–34.
#Farman AG, Farman TT. A comparison of 18 different x-ray detectors currently used in dentistry. *Oral Surg Oral Med Oral Pathol Oral Radiol Endod* 2005;99:485–489.

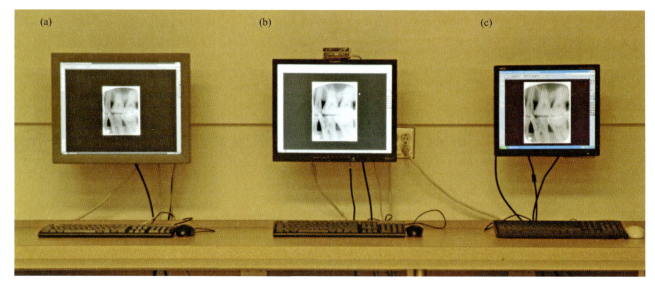

Figure 2.1. Monitors. Monitor A is a 3-megapixel medical-grade diagnostic grayscale monitor used for primary radiological diagnosis; monitor B is a 2-megapixel medical-grade diagnostic grayscale monitor used for primary radiological diagnosis; monitor C is the point-of-care color monitor used at the chairside for accessing the electronic patient record, which integrates those images already read on monitor A or B. Note that the lighting chosen was to optimize photography and is not the reduced ambient lighting optimal for primary reading.

Table 2.2. Comparison between medical diagnostic grayscale monitor and ordinary general purpose monitors

	Medical Diagnostic Grade Grayscale Monitor	General Purpose (Ordinary Commercially Available) Monitors
Color	No (monochromatic)*	Yes
Electronic patient record	Yes (but without color)*	Yes
Primary radiographic diagnosis (primary read)	Yes	No—used only as aide-memoire at point-of-care
Primary read in reduced ambient lighting	Yes—essential	No—advisable but not essential
DICOM	Yes 12-bit	No 8-bit
Brightness (luminance)	500 cd/mm	Maximum is 250 cd/mm; most are much lower
Grayscale with Windows operating system (OS)	8-bit	8-bit
Maximum grayscale possible	10.5-bit for mammography	8-bit
Spatial resolution	2 megapixel = 1600 × 1200 3 megapixel = 2560 × 2048	1080 × 1024 but can be higher
Grayscale standard display function	Yes	No
Luminance fades over time	Yes	Yes
Self-calibration	Yes (if has 12-bit needed for DICOM)	No (almost all have only 8-bit)

*There are some color medical diagnostic-grade monitors.

takes the human eye's nonlinear perception into account. The human eye easily sees relatively small changes in brighter areas than in darker areas. The GSDF adjusts the brightness so that all areas have the same level of perceptibility.[4]

Although the monitors employed for medical diagnosis use 12-bit–depth technology, if they operate within an *operating system* (OS) such as Windows, they will resolve to only 8-bit depth (or 256 gray-level used by ordinary monitors). Despite this, the medical monitors do require the 12-bit–depth technology for accurate self-calibration, which is performed to *digital imaging and communications in medicine* (DICOM) standards.

Seto et al. results "indicate that medical display systems must be carefully ... calibrated to ensure adequate image quality."[5] Self-calibration of the monitor's luminance (brightness) ensures that every time the dentist, in his/her essential role as radiologist, reviews an image it is of optimal quality. MGDG monitors are exceptionally bright, optimally about 500 candela (candles) per square meter (cd/m^2). As all monitors fade with time, this self-calibration ensures optimal and standardized brightness until the backlight brightness falls below the threshold and needs to be replaced.

AMBIENT LIGHTING (ILLUMINANCE)

The illuminance (reduced ambient lighting) essentially goes in tandem with monitor brightness. Recommendations for reduced ambient lighting in diagnostic reading stations for conventional analog (and digital) radiographs are 2–10 lx (illuminance is commonly expressed in *lux* or more simply *lx*), in comparison with 200–250 lx in clinical viewing stations in hospitals.[6] The evidence for the need for reduced ambient lighting for dentistry is provided by Haak et al.[7] They found that differences in monochromatic intensity were detected significantly earlier if the ambient lighting was reduced (70 lx versus the 1000 lx recommended for the dental operatory). More recently Hellen-Halme et al. demonstrated that when the reduced ambient lighting is less than 50 lx there is a significant increase in the accuracy of diagnosing approximal caries.[8] Although both monitors used by Haak et al. did not reach the *National Electrical Manufacturers Association's* (NEMA) standards for DICOM,[7] it was found that the flat screen monitor performed better than the *cathode ray tube* (CRT) in the dental operatory, probably because the flat screen monitor was brighter. Note that the type of monitor, whether CRT and *liquid crystal display* (LCD), functioned equally well provided they comply with DICOM standards.[4]

Image Enhancement

Image enhancement of the captured image is clearly an advantage that the digital technologies have over film. Parks recently displayed and discussed several enhancements: these are density (brightness), contrast, measurement, image inversion, magnification, flashlight and pseudocolor.[9] Although altering the brightness can lighten overexposed images, underexposed images should be retaken. Therefore, the need to optimally expose a solid-state detector is just as important as it is for film. As indicated earlier, images should be reviewed at a 1:1 ratio. This may not be always possible, particularly for detectors with very high spatial resolutions or large images such as panoramic radiographs. In such cases a 1:1 ratio will magnify (*magnification*) the image requiring the clinician to scroll or pan through the image. Haak et al. demonstrated that review of radiographic images at higher magnification improves accuracy.[2] Perhaps one of the most desired features of digital radiology is measurement; nevertheless, Kal et al. found that all processing algorithms provided significantly shorter measurements of the endodontic file lengths than their true length.[10]

Koob et al. compared different image processing modes or filters on the reproducibility and accuracy of the assessment of approximal caries viewed in CCDs.[11] Although they found there were no significant differences in reproducibility, the exposure time influences the overall accuracy of the central depth measurement of the approximal carious lesion. Haiter-Neto et al. found that the accuracy for the detection of noncavitated approximal caries among seven solid-state detectors was not significant.[12]

Storage and Compression of Images

The need to review (read) the primary images under optimal conditions also requires that these images must be faithfully preserved and stored so that they can be reaccessed and reread later. This may be necessary as part of the continued management of the patient not only by the primary clinician, but also for referral to another. These images may also be required by the clinician to defend him/herself against a legal suit like the one overviewed in Chapter 1.

Adopting digital technology does not alleviate the problem of long-term storage of all existing films (analog images). Fundamentally, the storage of electronic dental records must accurately preserve the original content of the record (e.g., text, image or chart).[13] The record must include complete information about the creation of any modification of the record (author, date, time, and exact source of the record, such as workstation). The format must be "read only" and protected from unauthorized alteration, loss, damage, or any other event that might make the patient information it contains inaccessible. Many jurisdictions require that digital clinical data be backed up to a remote server. The advantage of this is that this data is preserved if the surgery has been destroyed by fire or natural catastrophe. The advantage to both the dentist and his/her patients is that this data can be retrieved and treatment quickly recommenced at an alternative venue. This is particularly important as the value of a practice is still based in part on the goodwill represented by active patient records.

Many jurisdictions require retention of dental records for at least 10 years. The dentist considering adopting digital radiography needs to consider this as it is likely that during that period at least for some of his/her patients, s/he may need to convert to a different system at least once. It is a common experience that *information technology* (IT) changes rapidly with time with a risk that different generations may become incompatible. Therefore, in order to ensure that data survives transfer from one system to another, the dentist must ensure that not only should the systems be DICOM-compatible, but also that all digital images are transferred into the new record system without a loss of data. So far there does not appear to be a report to confirm that this can actually be achieved in dentistry.

Although not much of an issue for a single practitioner, the storage of images may present a much greater challenge for a large group practice that uses CBCT data for implants and orthodontic cephalometry. Intraoral images account for only hundreds of bytes of storage and panoramic radiographs for only a few thousand. The very large image files required for CBCT data quickly exhaust even a very generous storage capacity, measured in picabytes.

Compression of image files is an alternative to increasing storage. Two systems are used for compression, *lossless* and *lossy*. Lossless compression does not result in a loss of data. Lossy compression, however, involves an irrevocable loss of data. Although Eraso et al. reported that loss of image quality is not a factor unless the file size is reduced to 4% or less,[14] Fidler et al., who systematically reviewed the literature on lossy compression, reported that the amount of information lost is difficult to express and standardize.[15] Therefore, until lossy compression has been definitively tested, all data contained in a clinical image file must be preserved. Furthermore, the format of the image at the time of creation remains the original.[16] Therefore, scanning a film, even on a medical-grade scanner, only creates a copy, the film is the original image and must be preserved. Furthermore, those images created digitally, remain the original images, although they may have been printed onto the appropriate quality of paper or transparencies by medical-grade printers. These printouts are just copies. It also follows that any modification of the original image can only be an edition of the original, which must remain unaltered. The later edition should be automatically date-stamped with the date of its later creation.

The dentist must understand that the image s/he views on his/her monitor is not the original image captured by the detector, This captured image is in itself not the raw image captured by the detector, but instead it is the image, which has been automatically preprocessed so as to compensate for defects such as nonfunctioning pixels. The programs, which perform this preprocessing, cannot be accessed and modified by the dentist. It is this preprocessed or "presented" image that constitutes the "original" image from a legal perspective.

All referred digital images should ideally be delivered on CDs or DVDs rather than emailed if a *local area network* (LAN) is not available. Email is not only not secure, but often requires lossy compression of the original images. All image files should be saved and stored as TIFF rather than as JPEG images. The latter is prone to further lossy compression each time it is opened.

References

1. Krupinski E, Roehrig H, Furukawa T. Influence of film and m44onitor display luminance on observer performance and visual search. *Acad Radiol* 1999;6:411–418.
2. Haak R, Wicht MJ, Nowak G, Heilmich M. Influence of displayed image size on radiographic detection of approximal caries. *Dentomaxillofac Radiol* 2003;32:242–246.
3. Gutierrez D, Monnin P, Valley JF, Verdun FR. A strategy to qualify the performance of radiographic monitors. *Radiat Prot Dosimetry* 2005;114:192–197.
4. National Electrical Manufacturers Association (NEMA). Digital imaging and communications in medicine (DICOM). Part 145: Grayscale Standard Function. 2006. Available from URL: www.nema.org/stds/complimentary-docs/upload/PS3.14.pdf
5. Seto E, Ursani A, Cafazzo JA, Rossos PG, Easty AC. Image quality assurance of soft copy display systems. *J Digit Imaging* 2005;18:280–286.
6. Samei E, Badano A, Chakraborty D, Compton K, Cornelius C, Corrigan K et al. Assessment of display performance for medical imaging systems: executive summary of AAPM TG18 report. *Med Phys* 2005;32:1205–1225.
7. Haak R, Wicht MJ, Heilmich M, Nowak G, Noack MJ. Influence of room lighting on grey-scale perception with a CRT and a TFT monitor display. *Dentomaxillofac Radiol* 2002;31:193–197.
8. Hellén-Halme K, Petersson A, Warfvinge G, Nilsson M. Effect of ambient light and monitor brightness and contrast settings on the detection of approximal caries in digital radiographs: an *in vitro* study. *Dentomaxillofac Radiol* 2008;37:380–384.
9. Parks ET. Digital radiographic imaging: is the dental practice ready? *J Am Dent Assoc* 2008;139:477–481.
10. Kal BI, Baksi BG, Dündar N, Sen BH. Effect of various digital processing algorithms on the measurement accuracy of endodontic file length. *Oral Surg Oral Med Oral Pathol Oral Radiol Endod* 2007;103:280–284.
11. Koob A, Sanden E, Hassfeld S, Staehle HJ, Eickholz P. Effect of digital filtering on the measurement of the depth of proximal caries under different exposure conditions. *Am J Dent* 2004;17:388–393.
12. Haiter-Neto F, dos Anjos Pontual A, Frydenberg M, Wenzel A. Detection of non-cavitated approximal caries lesions in digital images from seven solid-state receptors with particular focus on task-specific enhancement filters. An ex vivo study in human teeth. *Clin Oral Investig* 2008;12:217–223.
13. Fefergrad I. Recordkeeping in dentistry. In: Downie J, McEwen K, MacInnes W, eds. *Dental Law in Canada*. Butterworths LexisNexis Canada Inc., Markham (ON) 2004: pp 271–278.
14. Eraso FE, Anaioui M, Watson AB, Rebeschini R. Impact of lossy compression on diagnostic accuracy of radiographs for periapical lesions. *Oral Surg Oral Med Oral Pathol Oral Radiol Endod* 2002;93:621–625.
15. Fidler A, Likar B, Skaleric U. Lossy JPEG compression: easy to compress, hard to compare. *Dentomaxillofac Radiol* 2006;35:67–73.
16. Goga R, Chandler NP, Love RM. Clarity and diagnostic quality of digitized conventional intraoral radiographs. *Dentomaxllofac Radiol* 2004;33:103–107.

Chapter 3
Physiological phenomena and radiological interpretation

Once the optimally made image is viewed under optimal conditions by the trained and experienced clinician, that image should be optimally interpreted. Unfortunately, physiological phenomena within each clinician may exert their influence.[1] As an educator and consultant in oral and maxillofacial radiology I have observed two such phenomena that are particularly important, certainly with regard to student and general clinicians. These are the *reversible (or ambiguous) figures* and the *Mach band effect*.

Perhaps the best known of such reversible (or ambiguous) figures is the "two faces or vase" figure. When such a figure is viewed. only one orientation can be perceived. This phenomenon is believed to arise from the transmission of sensory data to the visual cortex by way of at least two alternate pathways. The clinical importance of this phenomenon is that the clinician's perception of a particular image may change due to fatigue that occurs in one pathway compelling the image data to be transmitted by the other pathway. Figure 3.1 shows this phenomenon with some of my undergraduate students in formal examinations. Although it is expected that the clinician's experience of this phenomenon will decline with further training and practice, fatigue will always be a problem when the workload increases. Examination-induced fatigue may have contributed to these students' response to this type of image.

The Mach band effect is ubiquitous in radiology and occurs at the junction of two regions of differing radiodensity (Figure 3.2), particularly where there are superimposed structures. Although this phenomenon is most frequently perceived on the panoramic radiograph, it also is frequently experienced upon reviewing the *enamel-dentinal junction* (EDJ); the enamel margin immediately adjacent to the EDJ is white, whereas the dentine trends to black. The Mach band effect arises within the retina and results from the physiological process of lateral inhibition. With stimulation of a receptor (either a rod or a cone) in addition to initiating the transmission of an impulse toward the visual cortex, inhibitory impulses are being transmitted to neighboring receptors preventing them from also discharging an impulse back to the visual cortex.

The Mach band effect can be observed in any panoramic radiograph, particularly in relation to the secondary image of the contralateral lower border of the mandible (Figure 3.3), but its effect can be seen clearest in Figures 1.24 and 1.34. The Mach band effect is a two-edged sword. Although it is invaluable in assisting the clinician to perceive otherwise small differences between adjacent densities, it can create artifacts, particularly "fracture lines" (Figure 3.3).

Oral and Maxillofacial Radiology: A Diagnostic Approach, David MacDonald. © 2011 David MacDonald

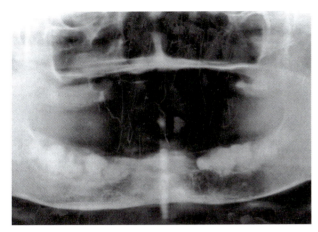

Figure 3.1. Panoramic radiograph of florid osseous dysplasia (FOD). Some students and general dental clinicians report the subjacent radiolucency rather than FOD. Reprinted with permission from MacDonald-Jankowski DS. Gigantiform cementoma occurring in two populations; London and Hong Kong. *Clinical Radiology* 1992;45: 316–318.

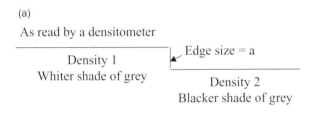
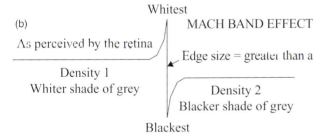

Figure 3.2. Mach band effect: Schematic diagram of (a) two adjacent radiodensities as read by a densitometer and (b) two adjacent radiodensities as perceived by the human retina. The size of the step (a) is now greater, accentuating the difference between the two densities and making the step now easier to perceive. Reprinted with permission from MacDonald-Jankowski DS. Fibrous dysplasia in the jaws of a Hong Kong population: radiographic presentation and systematic review. *Dentomaxillofacial Radiology* 1999; 28:195–202.

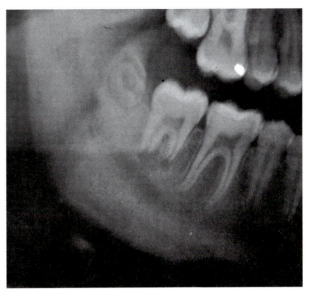

Figure 3.3. Panoramic radiograph displaying the Mach band effect. It accentuates the soft-tissue shadows created by the soft palate, dorsum of the tongue, and palatine tonsil to mimic a fracture through the angle of the mandible. **Note 1:** The secondary image of the contralateral lower border of the mandible is delineated by the Mach band effect. **Note 2:** The white fingerprint marks are most likely to arise from the fixer-contaminated fingers of the person taking it out of the cassette and placing it into the processor.

Reference

1. Daffner RH. Visual illusions in the interpretation of the radiographic images. *Curr Probl Diagn Radiol* 1989;18: 62–87.

Part 2
Advanced imaging modalities

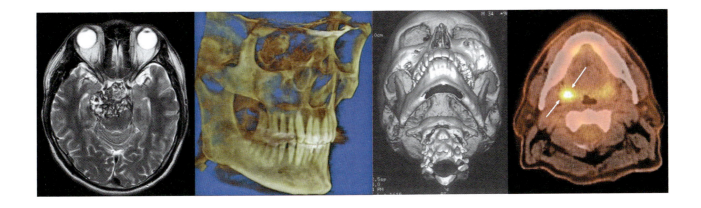

Chapter 4
Helical computed tomography

Introduction

Computed tomography (CT) can be divided broadly into fan-beam CT (including *helical computed tomography* [HCT] and its subsets), and *cone-beam computed tomography* (CBCT) (Figure 4.1); the latter is addressed in Chapter 5.

Computed tomography, particularly HCT, is increasingly available for the investigation of face and jaw lesions. This chapter introduces the various types of computed tomography, while concentrating on HCT, and covers window level and width, pitch, multidetector computed tomography, *3-dimensional* (3-D) reformatting, and the limitations of HCT. The indications for an increased need for HCT are discussed.

Why Do We Need Computed Tomography?

Formerly, clinicians relied on a clinical examination and "conventional radiography" (traditional 2-dimensional imaging) to assess and diagnose lesions affecting the jawbones. Unfortunately, conventional radiography generally reveals images that lack the sensitivity to display small changes in the bone. Conventional radiography also presents only as a *2-dimensional* (2-D) image the superimposition of all structures within the 3-D volume of the region examined.

What Are the Basic Construction and Principles of Computed Tomography?

The CT unit has 3 main components, as shown in Figure 4.2. The CT unit itself consists of the gantry (some of which may be angled up to 30°) and the patient table (or bed or couch) that moves the patient through the aperture in the gantry (Figure 4.2a), and the control console (Figure 4.2b). There are currently 2 types of CT unit available: the third- and fourth-generation units (Figure 4.3). The former constitutes the vast majority of CT equipment. For the third-generation CT, the X-ray tube and the detectors, which occupy an arc, are fixed in opposing positions within the gantry and rotate as a unit around the patient when in operation (Figure 4.3a), whereas for the fourth-generation unit, the X-ray tube alone rotates within a complete stationary ring of detectors (Figure 4.3b). The advantage of the fourth-generation unit is that the detectors have time to recover before being irradiated again.

How is the Computed Tomographic Image Displayed?

The display is a digital image reconstructed by the computer as pixels (picture elements), which represent a 3-D block of tissue. The voxel is the pixel size multiplied by the slice thickness (the voxel's length is from as low as 1 mm in some units to 20 mm). Each pixel is assigned a CT number (see later) representing tissue density. This density is proportional to the degree to which the material within the voxel has attenuated the X-ray beam. The resultant *attenuation coefficient* of a particular voxel reflects the mean of all tissues within it, the proportion of hard to soft tissues, and the voxel length (slice thickness).

There are 3 planes: X, Y, and Z. X and Y together represent the axial plane, a transverse plane through the patient. The Z plane represents the head-to-toe long axis of the patient. Upon inspection of an axial section, the pixels are represented by 2-D squares in the axial (XY) plane of a 3-D section of tissue of a thickness in the Z plane.

Oral and Maxillofacial Radiology: A Diagnostic Approach, David MacDonald. © 2011 David MacDonald

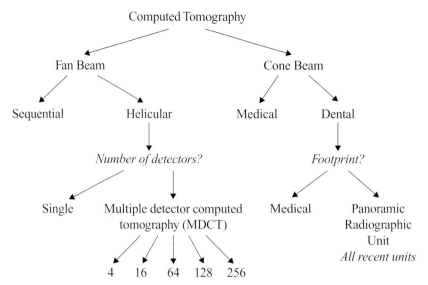

Figure 4.1. The classification and nomenclature of computed tomography.

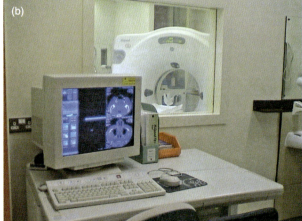

Figure 4.2. The 3 components of the computed tomography unit. (a) The gantry, containing the X-ray tube and detectors, and the table upon which the patient lies and is progressively advanced through the gantry; and (b) the control console with monitor. This is separated from the CT unit by a lead wall, including a lead glass window. It is crucial to observe the patient and gantry throughout the entire exposure, in case the exposure needs to be terminated. Reprinted with permission from MacDonald-Jankowski DS, Li TK. Computed tomography for oral and maxillofacial surgeons. Part 1: Spiral computed tomography. *Asian Journal of Oral Maxillofacial Surgery* 2006;18:68–77.

What is Helical Computed Tomography?

HCT is also known as *volume acquisition CT*. As Hounsfield's genius introduced and developed the concept of CT in 1968, that of Kalender introduced HCT. HCT violates a previous firm tenet of radiology: the patient should not move during the exposure. Instead, HCT requires that the patient, who remains motionless, be moved through the aperture of the gantry during the generation of X-rays by the rotating X-ray head (Figure 4.4a), creating a helix or spiral of data. This is in contrast to the separate incremental slices, which are stacked like coins, of the conventional sequential slice CT (the original technology of the 1970s and 1980s), now frequently called "sequential CT" (SeqCT) (Figure 4.4b).

Chapter 4: Helical computed tomography 51

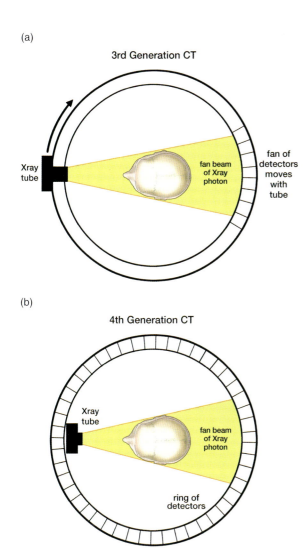

Figure 4.3. The 2 types of computed tomography units currently available. (a) The third-generation computed tomography unit permits a fan array of detectors to rotate around the patient in tandem with the rotating X-ray tube—the X-ray beam is fan-shaped; and (b) the fourth-generation computed tomography unit only permits rotation of the X-ray tube within the continuous, but stationary, array of detectors—the X-ray beam is fan-shaped. Reprinted with permission from MacDonald-Jankowski DS, Li TK. Computed tomography for oral and maxillofacial surgeons. Part 1: Spiral computed tomography. *Asian Journal of Oral Maxillofacial Surgery* 2006;18:68–77.

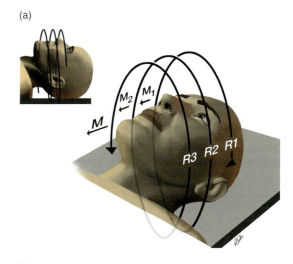

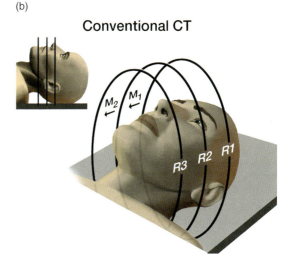

Figure 4.4. Helical and sequential computed tomography. (a) In helical computed tomography, the rotating X-ray tube describes a helix or spiral as it exposes the patient on the bed continuously moving through the gantry; and (b) in sequential computed tomography, the rotating X-ray tube can only describe complete loops between incremental movements of the patient's table through the gantry. Reprinted with permission from MacDonald-Jankowski DS, Li TK. Computed tomography for oral and maxillofacial surgeons. Part 1: Spiral computed tomography. *Asian Journal of Oral Maxillofacial Surgery* 2006;18:68–77.

How is Helical Computed Tomography Better Than Sequential Computed Tomography?

Because there is a continuous string of data encompassing a volume of the patient with HCT, this data can be readily reconstructed to give 3-D images. To achieve the same for SeqCT, the patient would need to undergo a second exposure overlapping with the first exposure, thus doubling the radiation dose. Because the data produced by HCT represents a continuous volume of the patient, it can be readily reconfigured to produce slices in any plane, including the coronal plane. However, the generation of coronal sections by SeqCT would require a reexposure of the patient through a coronal head position.

What Does the Data Found on the Image Represent?

Reviewers of HCT images should understand these terms: bone and soft-tissue windows, window width (WW) and level (WE), and pitch.

BONE AND SOFT-TISSUE WINDOWS

Bone and soft-tissue windows and their widths and levels are expressed in *Hounsfield units* (HU), which are also called "CT numbers." These range from a minimum of −1000 HU representing air (fixed point), through 0 HU representing water (fixed point), up to 3000 HU representing dense metal or bone. Bone and soft-tissue windows (Figure 4.5) are 2 of the 3 standard protocols for viewing the data captured by CT; the air window is the third protocol and is used mainly by respiratory physicians. Each of these protocols optimizes viewing of tissue types by appropriately adjusting the WL and WW. The soft-tissue window for face and jaw lesions is sited close to that of water (0 HU), WL at 40 to 60 HU, and WW at 250 HU, whereas the bone windows for such lesions are WL at 250 to 500 HU and WW 1000 to 2000 HU or greater. The *level* may be defined as equivalent to tuning a radio into the desired frequency, whereas the *width* is

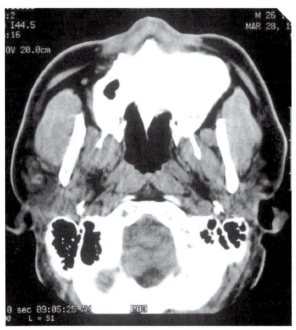
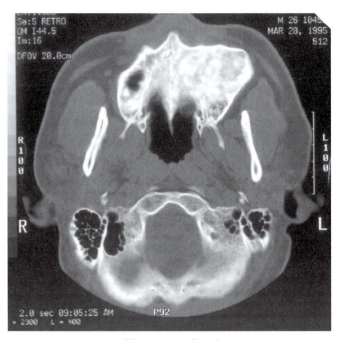

S/T window   **Bone window**

Figure 4.5. Soft-tissue and bone windows. The soft-tissue window (left image) displays cell-rich structures such as the muscles, skin, salivary glands, brainstem, and blood vessels as "gray" structures, whereas the fatty subcutaneous tissues and fascia appear almost as black as the air filled pharynx and mastoid air-cells—the bony structures appear as homogeneous white areas; and (right image) the bone window displays the bony structures in such detail that trabeculae could be discerned. Note that bony structures appear slightly smaller in area than they do in the soft-tissue window. The bony window displays soft tissue, but fat is appreciated as a darker gray shape in comparison to the nonfatty structures. **Note:** It is noticed that, with the same slice thickness, the area covered by the bone image on the soft-tissue window appears larger than that on the bone window. This is due to differences in windowing and levelling as well as the sharpness of the edges as defined by the spatial filter. Reprinted with permission from MacDonald-Jankowski DS, Li TK. Computed tomography for oral and maxillofacial surgeons. Part 1: Spiral computed tomography. *Asian Journal of Oral Maxillofacial Surgery* 2006;18:68–77.

equivalent to a filter. Formerly, the latter varied greatly depending upon individual radiologists who would then have the images formatted as a hard copy for onward transmission to the referring clinician along with the radiologist's report. These contained about 12 images per sheet and were reviewed on a lightbox. It is now almost routine for the complete dataset to be downloaded onto a CD/DVD and forwarded to the clinician along with the radiologist's report. The referring clinician, with the appropriate software, can handle the data to suit his or her own requirements. In either scenario, the accompanying radiologist's reports should be referred to while reviewing the images.

PITCH

Pitch is the tightness of the helix and affects the spatial resolution (detail) that would be visible on reconstruction. Because the string of data will be longer for a pitch of 2 : 1 in comparison for one of 3 : 1 for a given volume of patient, it follows that the radiation to the patient will be higher, although the detail will be better when viewed on thin reconstructed slices. For severe facial trauma a 1 : 1 pitch is best.

What is *Multidetector Computed Tomography?*

Multidetector computed tomography (MDCT) is a subset of HCT with up to 256 sets of X-ray tubes and corresponding sensor arrays. This means that the time needed to acquire data from a given volume of the patient is correspondingly reduced.

What is 3-Dimensional Reformatting and Why is it Required?

Each original voxel is divided into cubes, called *cuberilles*, by interpolation; each cuberille has the same mean attenuation coefficient of the original voxel (Figure 4.6). The need for this interpolation arises because the original voxel's resolution is best in the axial plane, where the density of pixels is greatest. Only those cuberilles that represent the surface of the *object of interest* (OI) are projected onto the monitor. The 3-D reconstructions are

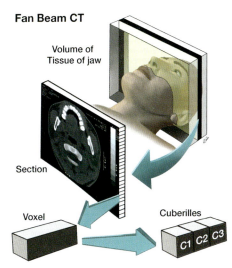

Figure 4.6. Fan-beam computed tomography achieves 3-dimensional reconstruction by dividing the voxel into cuberilles, each with the same attenuation coefficient as the original voxel. Reprinted with permission from MacDonald-Jankowski DS, Li TK. Computed tomography for oral and maxillofacial surgeons. Part 1: Spiral computed tomography. *Asian Journal of Oral Maxillofacial Surgery* 2006;18:68–77.

capable of being rotated to display the reconstruction from any point of view (Figure 4.7).

The OI is broadly defined by, and selected according to, its CT number. By fine adjustment of the former, in addition to supplementary functions such as "edit" with its "scalpel" (a digital freehand tool perfectly analogous to the physical scalpel), exquisite images are possible, particularly if they are assigned different colors. Furthermore, the images can be rotated about any axis to display any surface of the OI for both further edition or definitive viewing. The OI can be copied back into a second 3-D reconstruction of the affected jaw and can be used for treatment planning (Figure 4.8).

Already, it is becoming commonplace for neuroradiologists and neurosurgeons to collaborate to identify cerebral aneurysms, define their extent and associated tissue supplied by their end-arteries, and determine the optimal site for intervention. Furthermore, the 3-D reconstruction can facilitate *computer-assisted design/computer-assisted manufacturing* (CAD/CAM) reconstruction of a face following extensive ablative surgery or severe

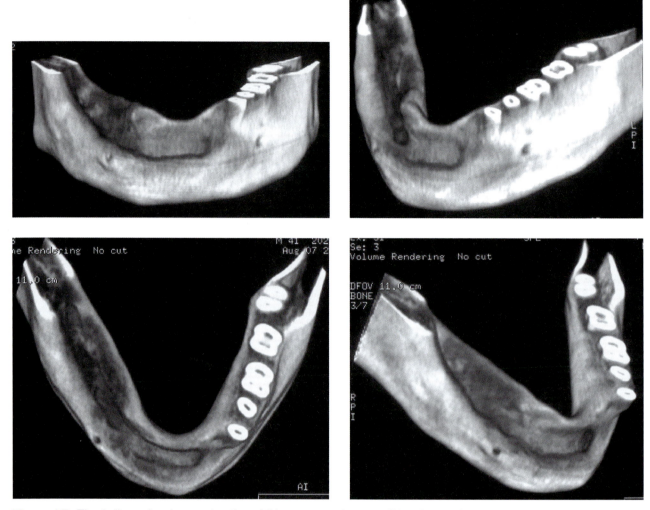

Figure 4.7. The 3-dimensional reconstruction of this postoperative mandible allows a fuller evaluation for definitive reconstruction. Reprinted with permission from MacDonald-Jankowski DS, Li TK. Computed tomography for oral and maxillofacial surgeons. Part 1: Spiral computed tomography. *Asian Journal of Oral Maxillofacial Surgery* 2006;18:68–77.

trauma. Blank and Kalender have précised the principles and issues of virtual images.[1]

What Are the Limitations of Helical Computed Tomography?

REDUCED RESOLUTION IN ALL PLANES EXCEPT THE AXIAL PLANE

Before we develop this point, we must advise readers that this limitation no longer applies to the most-modern scanners using up to 256 detectors.

Scarfe reported that *multiplanar reformatting* (MPR), especially in the coronal plane, was inadequate for the assessment of severe facial trauma primarily oriented in the axial plane[2] because the spatial resolution is greatest in the axial (XY) plane.[3] Hoeffner et al. suggested that specific protocols are required for obtaining coronal MPR of data acquired axially.[4] Nevertheless, although the spatial resolution of HCT is poorer than that of *conventional radiography*, the problem is caused by having anisotropic cuberilles, which is intrinsic to most fan-beam types of CT to which HCT belongs. Only CBCT and the most-modern 4 to 256

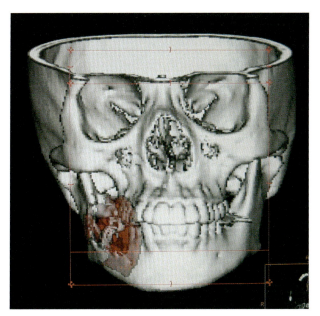

Figure 4.8. Color-coded 3-dimensional reformatting displays the extent of the ameloblastoma (represented in red) within the mandible. It has perforated the alveolar bone in 2 places (red). This reconstruction was produced by dissecting out the neoplasm and then replacing it within a second reconstruction of the bony mandible. Reprinted with permission from MacDonald-Jankowski DS, Li TK. Computed tomography for oral and maxillofacial surgeons. Part 1: Spiral computed tomography. *Asian Journal of Oral Maxillofacial Surgery* 2006;18:68–77.

Figure 4.9. An axial computed tomographic section displaying substantial metallic spraying arising from extensive metal restorations. Figure courtesy of Dr. Montgomery Martin, British Columbia Cancer Agency.

detectors MDCT units avoid this. They have isotropic cuberilles (explained in Chapter 5.)

METAL STREAK OR SPRAY ARTIFACTS

Although streak or spray artifacts can degrade the HCT image (Figure 4.9), as they do most other imaging modalities, this can be reduced by *metal artifact-reduction* (MAR) software.[5] Furthermore, Baum et al. suggested that a short additional HCT parallel to the body of the mandible reduces artifacts behind the dental arch and improves the overall diagnostic quality.[6]

INTRAOPERATIVE IMAGING IN THE OPERATING THEATER

During an operation, it may be desirable to obtain more images. So far it has not been possible to make a mobile HCT for use in the operating theater.

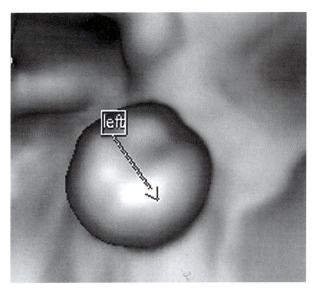

Figure 4.10. This example of virtual antroscopy by use of the "navigator" program displays a 3-dimensional evaluation of a lesion arising from the roof of the maxillary antrum. Reprinted with permission from MacDonald-Jankowski DS, Li TK. Computed tomography for oral and maxillofacial surgeons. Part 1: Spiral computed tomography. *Asian Journal of Oral Maxillofacial Surgery* 2006;18:68–77.

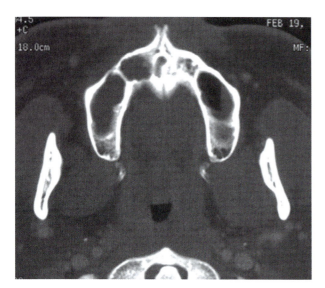

Figure 4.11. An axial computed tomographic section (bone window) displaying enhancement of the normal blood vessels of the face.

In Chapter 5, we discover that CBCT has successfully addressed this issue.

LOW SENSITIVITY FOR IDENTIFICATION OF SMALL TUMORS

Although HCT has a high specificity for metastatic lesions, which, according to van den Brekel's review

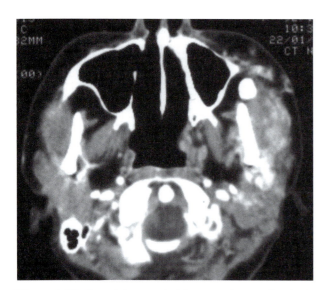

Figure 4.12. An axial computed tomographic section (soft-tissue) displaying enhancement of a large hemangioma affecting the face. The large blood vessels are also enhanced.

is higher than MRI[7] it has a lower sensitivity. This is largely due to the fact that the necrosis, which is pathognomic for metastasis, is rarely visible in small lesions. Therefore, the sensitivity of HCT is optimal only for larger lesions, which are thereby more likely to be associated with a poorer prognosis.

What Are the Indications for an Increased Need for Helical Computed Tomography?

In addition to osseointegrated implants, which have transformed prosthodontics, there is a need to evaluate complex fractures and to accurately stage carcinoma. The last two are addressed in Chapters 14 and 18, respectively.

What Are the Other Functions of Helical Computed Tomography?

DENTASCAN

In addition to its more usual role of preimplant planning, Au-Yeung et al.'s pictorial review displayed this program's capacity also to evaluate lesions affecting the jaw, ranging from squamous cell carcinoma to infection.[8]

VOLUME RENDERING

Volume rendering is a technique that uses the concept of opacity, which quickly reconstructs a 3-D volume acquired on CT or MRI. The end result is similar to a virtual anatomical dissection and can assist surgical planning for a particular patient. Cavalcanti and Antunes compared *volume rendering* with surface rendering for 20 patients and found that the former improved visualization in comparison to the latter.[9] It was also more sensitive for the diagnosis of maxillofacial lesions—in particular, those that were primarily intraosseous.

COLOR-CODED 3-DIMENSIONAL REFORMATTING

Color-coded 3-D reformatting may be done for extensive lesions[10] by ascribing a separate color to the lesion, the bone, and adjacent soft tissues. This has been applied to an ameloblastoma in Figure 4.8.

NAVIGATOR

This function of perspective volume rendering permits virtual antroscopy to evaluate the surface contours of antral lesions (Figure 4.10), virtual arteroscopy for defining vascular lesions, and virtual pharyngoscopy and laryngoscopy. This function has been applied to the maxillofacial region by Tao et al.[11]

COMPUTER-ASSISTED DESIGN/COMPUTER-ASSISTED MANUFACTURING

CAD/CAM technology is such that it can be adopted in any hospital for daily use. A Hong Kong group used a 4-stage process to produce a "quantitative osteotomy simulation bone model" that could predict the postoperative appearance with photo-realistic quality.[12] CAD/CAM can generate 3-D models by laser or by milling.

COMPUTED TOMOGRAPHY ANGIOGRAPHY

Although Tipper et al. reported that the specificity of *computed tomographic angiography* (CTA) for the internal carotid artery approximates to that of digital subtraction anglography,[13] Teksam et al. commented that the presence of small aneurysms may be easier to detect if they are aligned according to the patient's long axis rather than axially.[14] A review by Tomandl et al. of the postprocessing of intracranial CTA is relevant for the maxillofacial region.[15]

INTRAVENOUS CONTRAST

Intravenous contrast enhances blood vessels (Figure 4.11). For optimal vascular and tissue contrast, Baum et al. recommend that 150 mL of *contrast medium* (CM) be delivered at 2.5 mL/second flow rate with a start delay of 80 seconds.[6] It should be appreciated that mild hypersensitivity reactions occur in up to 12.7% with ionic CM and 3.1% with the lower osmolar nonionic CM, but that the death rates for both are equal at 1 per 100,000 investigations.[16] Bettmann has addressed the frequently asked questions on CM-induced allergies, nephropathy, and other risks.[17] Figure 4.12 displays the use of intravenous contrast rendering obvious a large hemangioma.

Whenever contrast is used, regardless of modality, precontrast images should be made first.

Conclusion

Although much objective work is required to fully evaluate the quality of the predictive aspects of HCT images, it cannot be denied that HCT has completely transformed medical imaging. The clinician is provided with a detailed preview of the patient and his or her disease, thus minimizing the risk of hidden features complicating both the procedure and its successful outcome. In this way, HCT has the potential to enhance treatment and procedure planning. The images, when appropriately prepared, should be able to facilitate collaboration between head and neck specialists, who may be called together to treat a lesion in the most effective manner.

References

1. Blank M, Kalender WA. Medical volume exploration: gaining insights virtually. *Eur J Radiol* 2000;33:161–169.
2. Scarfe WC. Imaging of maxillofacial trauma: evolutions and emerging revolutions. *Oral Surg Oral Med Oral Pathol Oral Radiol Endod* 2005;100:75–96.
3. Rosenthal E, Quint DI, Johns M, Peterson B, Hoeffner E. Diagnostic maxillofacial coronal images reformatted from helically acquired thin-section axial CT data. *AJR Am J Roentgenol* 2000;175:1177–1181.
4. Hoeffner EG, Quint DJ, Peterson B, Rosenthal E, Goodsitt M. Development of a protocol for coronal reconstruction of the maxillofacial region from axial helical CT data. *Br J Radiol* 2001;74:323–327.
5. Lemmens C, Faul D, Nuyts J. Suppression of metal artifacts in CT using a reconstruction procedure that combines MAP and projection completion. *IEEE Trans Med Imaging* 2009;28:250–260.
6. Baum U, Greess H, Lell M, Nomayr A, Lenz M. Imaging of head and neck tumors—methods: CT, spiral-CT, multislice-spiral-CT. *Eur J Radiol* 2000;33:153–160.
7. van den Brekel MW. Lymph node metastases: CT and MRI. *Eur J Radiol* 2000;33:230–238.
8. Au-Yeung KM, Ahuja AT, Ching AS, Metreweli C. Dentascan in oral imaging. *Clin Radiol* 2001;56:700–713.
9. Cavalcanti MG, Antunes JL. 3D-CT imaging processing for qualitative and quantitative analysis of maxillofacial cysts and tumors. *Pesqui Odontol Bras* 2002;16:189–194.
10. Greess Fl, Nomayr A, Tomandl B, Blank M, Lell M, Lenz M, Bautz WA. 2D and 3D visualisation of head and neck tumours from spiral-CT data. *Eur J Radiol* 2000;33:170–177.

11. Tao X, Zhu F, Chen W, Thu S. The application of virtual endoscopy with computed tomography in maxillofacial surgery. *Chin Med J* (Engl) 2003;116:679–681.
12. Xia J, Ip HH, Samman N, Wong HT, Gateno J, Wang D, Yeung RW, Kot CS, Tideman H. Three-dimensional virtual- reality surgical planning and soft-tissue prediction for orthognathic surgery. *IEEE Trans Inf Technol Biomed* 2001;5:97–107.
13. Tipper G, U-King-Im JM, Price SJ, Trivedi RA, Cross JJ, Higgins NJ, Farmer R, Wat J, Kirollos R, Kirkpatrick PJ, Antoun NM, Gillard SF1. Detection and evaluation of intracranial aneurysms with 16-row multislice CT angiography. *Clin Radiol* 2005;60:565–572.
14. Teksam M, McKinney A, Cakir B, Truwit CL. Multi-slice CT angiography of small cerebral aneurysms: is the direction of aneurysm important in diagnosis? *Eur J Radiol* 2005;53:454–462.
15. Tomandl BF, Kostner NC, Schempershofe M, Huk WJ, Strauss C, Anker L, Hastreiter P. CT angiography of intracranial aneurysms: a focus on postprocessing. *Radiographics* 2004;24:637–655.
16. Brockow K, Christiansen C, Kanny G, Clement O, Barbaud A, Bircher A, Dewachter P, Gueant JL, Rodriguez Gueant RM, Mouton-Faivre C, Ring J, Romano A, Sainte-Laudy J, Demoly P, Pichler WJ. Management of hypersensitivity reactions to iodinated contrast media. *Allergy* 2005;60:150–158.
17. Bettmann MA. Frequently asked questions: iodinated contrast agents. *Radiographics* 2004;24:3–10.

Chapter 5
Cone-beam computed tomography

What Is Cone-Beam Computed Tomography?

Cone-beam computed tomography (CBCT) uses a cone-shaped beam of X-ray photons rather than the fan-shaped beam used by *helical CT (HCT)* (Figure 5.1). The cone's shape can be either round or rectangular. CBCT can scan the region of interest in up to a single 360° rotation in contrast to the multiple rotations required by HCT using a fan beam. CBCT interrogates a much smaller volume of tissue than HCT. CBCT is also frequently called *dental computed tomography (DCT)* or *cone-beam volumetric tomography (CBVT)* or *volumetric computed tomography (VCT)*. Farman prefers CBCT because this describes the principles of operation rather than its application in dentistry or the resulting dataset."[1] CBCT has been applied to respiratory,[2] breast,[3] and cardiac imaging.[4] CBCT has also been used during craniospinal surgery.[5]

Why Is Cone-Beam Computed Tomography Desirable?

The advantage of CBCT is its superior spatial resolution (the ability to identify separately 2 minute points very close together), relative to HCT, of structures with high-contrast mineralized tissue such as teeth and bone. CBCT, in addition to producing 3-D images, can produce 2-D images similar to the panoramic and cross-sectional reconstructions produced for preimplant assessment. Figure 5.2 displays both the 2-D and 3-D reconstructions of a torus palatinus and exostoses.

Although HCT is an improvement on *conventional radiography* with regard to elimination

---

Oral and Maxillofacial Radiology: A Diagnostic Approach, David MacDonald. © 2011 David MacDonald

superimposition, 3-D reconstruction, and its excellent contrast resolution, its spatial resolution is still inadequate, particularly in comparison to intraoral radiography. The current best spatial resolution for CBCT is 6.5 line pairs per mm (0.076 mm minimum voxel size), whereas for intraoral conventional radiography it is at least 2 times that (see Table 2.1). Therefore, intraoral images will continue to display fine features such as the ground-glass and peau d'orange (orange peel) appearance of bone, the poorer-defined margins associated with fibrous dysplasia, and the fine cortices of other lesions. On the other hand, the CBCT's panoramic reconstruction is superior to that of a conventional panoramic X-ray because it displays no superimposed secondary images and is geometrically more accurate.

How Does Cone-Beam Computed Tomography Differ from Helical Computed Tomography?

HCT uses a planar geometry and 2-D reconstruction, whereas CBCT performs nonplanar geometry and a 3-D reconstruction.[6] CBCT reconstructs the 3-D images by generating cuberilles directly, each with its own attenuation coefficient (Figure 5.3; compare to Figure 4.6). This allows 3-D reconstructions with better spatial resolution in the Z plane (patient's long axis), in addition to the axial (XY) plane. Except perhaps for the most modern *Multidetector computed tomographic* (MDCT) units, HCT can only produce cuberilles secondarily from voxels (see Figure 4.6). CBCT's cuberilles are called *isotropic voxels*, which result in highly detailed 3-D images (Figure 5.4).

Although many reconstruction algorithms for CBCT have not achieved the performance observed for the best MDCT, this is an area of active research that will continue to improve the quality of CBCT technology, with particular regard to its software.[7]

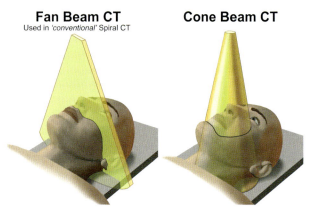

Figure 5.1. The fan beam upon which *helical computed tomography* is based interrogates only a slice of tissue, whereas the cone beam of cone-beam computed tomography interrogates a 3-dimensional region within a 360° rotation. Reprinted with permission from MacDonald-Jankowski DS, Orpe E. Computed tomography for oral and maxillofacial surgeons. Part 2: Cone-beam computed tomography. *Asian Journal of Oral Maxillofacial Surgery* 2006;18:85–92.

Vannier described "craniofacial" CBCT as a "disruptive technology" that changed the entire course of an industry.[8] Because CBCT is relatively cheap and compact, it can be readily accommodated in the practitioner's office where space and operating costs are a small fraction of those incurred by HCT.[8] Unfortunately, this ease of access facilitates other significant problems, particularly inappropriate criteria for its use. This is discussed later in this chapter.

Most CBCT makes now use *flat panel detectors* (FPDs) constructed of amorphous silicon, rather than the image intensifiers frequently used in HCT. Some CBCTs also use solid state chips.[1] FPDs have the advantage of not producing the geometric distortion that occurs with image intensifiers. In addition, FPDs have a wider dynamic range, better *signal to noise ratio* (SNR), and better spatial resolution.[9]

Patient Posture During a Cone-Beam Computed Tomography Examination

With the exception of the NewTom 3G, which exposes the patient in the supine position, almost all CBCT units expose the patient seated or standing upright (Figure 5.5). The gantry used can range from the one designed solely for the CBCT unit, such as the iCAT in Figure 5.5a, to that used by the Promax 3D which is the same as that used for panoramic tomography and cephalometry (Figure 5.5b.).

Effects of Imaging Advances in Clinical Practice

These advances have enhanced communications between practitioners[10] and between practitioners and their patients. This has been accelerated by the introduction of PACS and DICOM standards, eliminating the physical transport of hardcopy as printed transparencies, or softcopy as DVDs or CDs. The DVD/CD is preferred because it contains all the original data generated during the investigation, rather than selected and manipulated images printed on transparencies. The data on the DVD/CD can be downloaded by clinicians, providing the treatment and reconstruction according to individual requirements. However, this does need the appropriate software, for example Simplant (Materialise, Leuven, Belgium) and NobelGuide (Nobel Biocare. Zurich, Switzerland).

How Does Cone-Beam Computed Tomography Compare with Helical Computed Tomography?

The advantages of CBCT over HCT are better spatial resolution, lower radiation dose, smaller footprint, and lower cost.

BETTER SPATIAL RESOLUTION

The best spatial resolution of CBCT is 0.076 mm[11] minimum voxel size, whereas the best size for the most modern HCT, is 0.35 mm.

BETTER 3-DIMENSIONAL RECONSTRUCTION

CBCT, by virtue of its isotropic voxels (see Figure 5.4), also allows better 3-D reconstruction in the Z

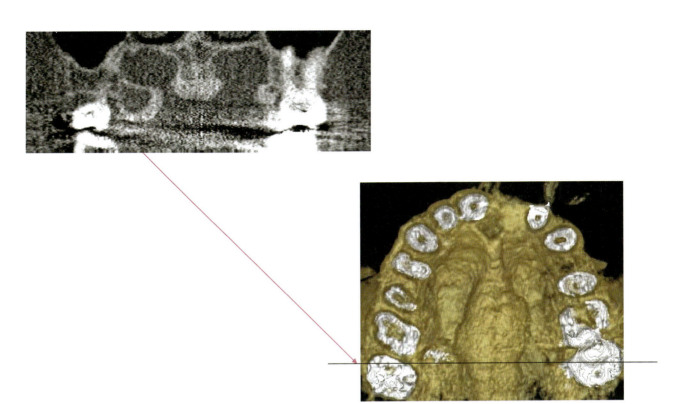

Figure 5.2. Display of the coronal (top); and 3-dimensional (bottom) reconstruction of a torus palatinus and bilateral exotoses on the palatal aspect of the maxillary tuberosities. The metallic artifact observed on the coronal reconstruction at the level of the black line on the bottom-right figure was minimized in the 3-dimensional reconstruction by excluding most of the crowns, hence the pulp chambers rather then occlusal surfaces. There is still some residual metallic artifact observed as white streaks on both exostoses. The unit used was an iCAT with a 0.2 mm voxel size. Reprinted with permission from MacDonald-Jankowski DS, Orpe E. Computed tomography for oral and maxillofacial surgeons. Part 2: Cone-beam computed tomography. *Asian Journal of Oral Maxillofacial Surgery* 2006;18:85–92.

plane in addition to the XY, thus allowing equally high spatial resolution regardless of the plane of reconstruction.

BETTER BONE IMAGING; POORER CONTRAST RESOLUTION OR SOFT-TISSUE IMAGING

CBCT is very good for intrinsically high-contrast structures such as bone but, although the soft-tissue outlines can be silhouetted by the air-filled space outside and within them, differences within the soft tissue cannot be resolved.[12] The main reason for this is that CBCT generally uses 12–14-bit technology rather than the 16–24-bit technology used in HCT. Because the main indication for CBCT is currently preimplant evaluation, a lack of a soft-tissue image will have little impact. The main features pertinent to implant assessment are bone thickness, depth, and quality.

CONE-BEAM COMPUTED TOMOGRAPHY ARTIFACTS

Unlike HCT, CBCT does not use *Hounsfield units* (HU). Therefore, it is difficult to analyze bone density. In addition, structures outside the area of interest may cause beam-hardening.[13] This is particularly a problem for small *field of view* (FOV), such as the Accuitomo used in that study.[13] CBCTs designed to scan a large FOV with a large detector and a primary beam with a higher kVp not only

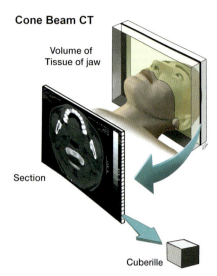

Figure 5.3. Cone-beam computed tomography reconstructs the 3-dimensional images by generating cuberilles directly, each with its own attenuation coefficient. This allows 3-dimensional reconstructions with better resolution in the Z axis, in addition to the axial (XY) plane. Helical computed tomography, except the most modern MDCT, can only produce cuberilles secondarily from voxels (see Figure 4.6). Reprinted with permission from MacDonald-Jankowski DS, Orpe E. Computed tomography for oral and maxillofacial surgeons. Part 2: Cone-beam computed tomography. *Asian Journal of Oral Maxillofacial Surgery* 2006;18:85–92.

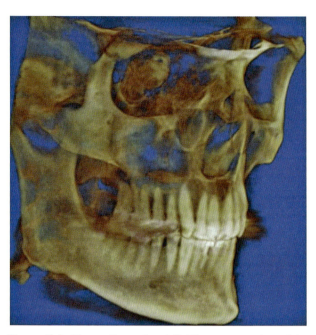

Figure 5.4. Three-dimensional (3-D) reconstruction of the data captured by a cone-beam computed tomographic (CBCT) unit (iCAT). Because this image is constructed of isotropic cuberilles, the spatial resolution is very good in all three dimensions. Figure courtesy of Dr. Babak Chehroudi, Faculty of Dentistry, University of British Columbia.

appear to obviate these artifacts but may make possible the application of HU.[13] Katsumata et al. (Asahi) suggest FPD produced fewer artifacts than intensifying screen.[14]

LOWER RADIATION DOSE

Ludlow et al. compared the radiation imparted by 8 different CBCT units to the tissues of the head and neck according to the FDA's recent dose limits. They vary widely in terms of the radiation doses of a panoramic radiograph from 4 to 54 panoramic radiographs.[15]

The lower radiation dose of CBCT is in part related to the short exposure time of its single 360° rotation (or 180° for the mobile CBCT units for use in the operating or emergency room or theatre[12]). This short range of scanning time minimizes risk of movement artifacts. Tsiklakis et al. reported that the already reduced effective dose inherent to CBCT (DVT 9000, QR) can be further reduced to the thyroid and cervical spine if lead shielding is used.[16]

ACCURACY OF MEASUREMENTS

An *in vitro* study (iCAT) by Moshiri et al. reported that CBCT 2-D lateral cephalograms are more accurate than conventional lateral cephalograms for most linear measurements in the sagittal plane.[17] An *in vivo* study (iCAT) by Grauer et al. found no systematic differences between the average landmark coordinates of both modalities.[18] Instead, Grauer et al. reported that, when both modalities are used in longitudinal studies for the same patient, method error could produce clinically significant differences.[18] A systematic review (including 8 reports of which only 1 was *in vivo*) by Lou et al. concluded that each orthodontic landmark contributed a characteristic error.[19] This error contributed to inaccuracy of the measurement and was reduced by repeated practice.[19]

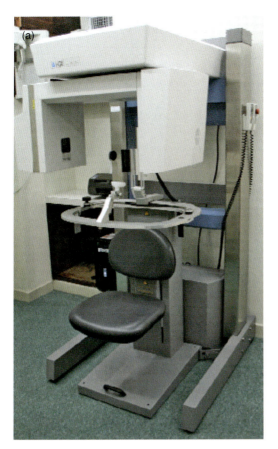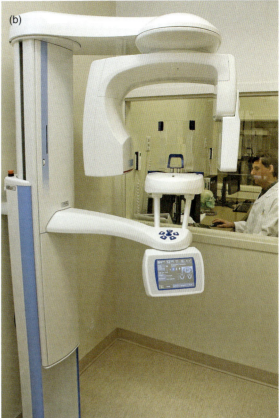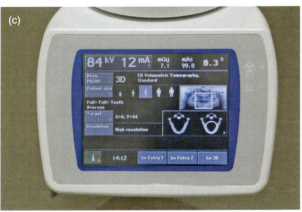

Figure 5.5. There are many types of cone-beam computed tomographic units currently available. Two examples are displayed. The robust iCAT (a) has a seat, whereas the Promax 3D (b) uses a panoramic radiographic gantry in which the patient stands. The control for positioning the patient within the Promax 3D is shown in (c). The exposure of all units should be directed from outside the room in which the CBCT unit is sited. Such a control console is apparent through the window behind the Promax 3D. The operator seated at the monitor also can observe the unit and patient during the entire exposure.

An *in vitro* report (iCAT) by Brown et al. on the reliability and accuracy of linear measurements between cephalometric landmarks on CBCT 3-D images with direct measurements on dried skulls reported that a reduction of the number of projections for the 3-D reconstruction did not reduce accuracy; instead it may reduce radiation dose.[20]

Cone-Beam Computed Tomography for Use in the Operating or the Emergency Room During a Procedure

The development of mobile CBCT units with a C-arm allows any part of the body to be investigated and, therefore, is usable in the emergency or operating room, where the surgeon can perform 3-D real-time reconstruction. The radiation dose for a jaw scan is equivalent to a panoramic X-ray for a 180° scan, rather than the 360° scan common to other CBCT units. These units deliver an even lower dose to the parotid glands than a panoramic X-ray.[21] Pohlenz et al. have successfully used such a unit on 179 oral and maxillofacial cases in the operating room.[22]

What Are the Potential Uses for Cone-Beam Computed Tomography?

Clearly, preoperative evaluation for implant placement does and will continue to constitute a majority of the referrals for CBCT for the foreseeable future. Such referrals should be accompanied by stents (see Figures 15.11 and 15.12) to ensure the proper correlation of the site imaged to that in the patient, thus ensuring optimum results. Hamada et al. suggested that CBCT (Asahi) is suitable for the assessment of alveolar bone grafting before and after placement of dental implants or orthodontic treatment of teeth adjacent to clefts.[23] (Hitachi). Hamada et al. also reported that 3-D CBCT displayed the integrity of the bone adjacent to the subsequent implant.[23] Oberoi et al. (Hitachi) followed the eruption of canines into the cleft site treated by bone grafts by 3-D CBCT.[24]

Neugebauer et al. (Galileos) reported the importance of 3-D imaging with a high spatial resolution of those third molars whose roots are in contact with the mandibular canal.[25]

An HCT report by Sharan et al. revealed that a substantial proportion of teeth associated with the floor of the maxillary antrum are intimately associated with it.[26] The superior spatial resolution of CBCT would better determine the frequency of this phenomenon.

Tsiklakis et al. reported that CBCT (DVT 9000 QR) produced high-quality diagnostic images of the TMJ for a lower radiation dose than HCT.[27] Maki et al. outlined the application of computational simulations using CBCT (MercuRay) to simulate condylar growth, bone formation, and orthognathic surgery.[28] Cevidanes et al. suggested that their 3-D colored displacement mapping (DVT 9000, QR) could be applied to determine bone remodeling following surgery.[29]

What Are the Restrictions of Cone-Beam Computed Tomography?

THE USE OF CONTRAST MEDIA

Due to CBCT's inherently poorer soft-tissue imaging, the definitive assessment of malignant and locally invasive lesions such as *ameloblastomas* and *myxomas* requires the use of an intravenous contrast medium. This must be delivered only to patients in the supine position (see Figure 4.2a). Therefore, apart from the NewTom 3G, all other dental CBCTs, which investigate patients in the seated or standing positions, are unsuitable for the intravenous delivery of contrast to the patient. Regardless of the patient's position, such lesions are best investigated by HCT, not only because these are sited in medical facilities that are better able to deal with adverse reactions, but HCT itself also provides soft-tissue windows that may become essential to determine the extension of these lesions into the adjacent soft tissue if they have breached the cortex.

ANATOMY AND PATHOLOGY BEYOND THE ORAL AND MAXILLOFACIAL REGION

Recently, the American Academy of Oral and Maxillofacial Radiology[30] and the European Academy of Dental and Maxillofacial Radiology[31] have separately issued their guidelines for the use of CBCT. Clinicians using this technology should evaluate the entire dataset of each CBCT examination. How and who should do this are detailed in each publication. An essential principle is that those datasets derived from medium and large FOVs, which include extragnathic structures such as the base of the skull and the neck should be reported by radiologists. The important lesions that can arise in these extragnathic structures are detailed in Chapters 16 to 18.

Conclusion

In the short time since CBCT first appeared on the market, tremendous advances have been made: a reduction in the footprint, an overall reduction in radiation dose imparted by this technology, improved spatial resolution, and increased versatility of the software. We can expect these to continue in the near future. CBCT has already transformed imaging of the face and jaws within the last decade since the technology first emerged.

References

1. Farman AG. Commentary to Comparison of cone-beam volumetric imaging and combined plain radiographs for localization of the mandibular canal prior to removal of impacted lower third molars by Neugebauer et al. *Oral Surg Oral Med Oral Pathol Oral Radiol Endod* 2008;105:643.
2. Sonke 11, Zijp L, Remeijer P, van Herk M. Respiratory correlated cone beam CT. *Med Phys* 2005;32:1176–1186.
3. Zhong J, Ning R, Conover D. Image denoising based on multiscale singularity detection for cone beam CT breast imaging. *IEEE Trans Med Imaging* 2004;23:696–703.
4. Manzke R, Grass M, Hawkes D. Artefact analysis and reconstruction improvement in helical cardiac cone beam CT. *IEEE Trans Med Imaging* 2004;23:1150–164.
5. Hott JS, Deshmukh VR, Klopfenstein JD, Sonntag VK, Dickman CA, Spetzler RF, Papadopoulos SM. Intra-operative Iso-C C-arm navigation in craniospinal surgery: the first 60 cases. *Neurosurgery* 2004;54:1131–1136; discussion 1136–1137.
6. Kalender WA. *Computed Tomography: Fundamentals, System Technology, Image Quality, Applications*. Publicis MCD Verlag, Munich 2000: p 63.
7. Vannier MW. Craniofacial imaging informatics and technology development. *Orthod Craniofac Res* 2003;6:73–81; discussion 179–182.
8. Vannier MW. Craniofacial computed tomography scanning: technology, applications and future trends. *Orthod Craniofac Res* 2003;6:23–30; discussion 179–182.
9. Baba R, Ueda K, Okabe M. Using a flat-panel detector in high resolution cone beam CT for dental imaging. *Dentomaxillofac Radiol* 2004;33:285–290.
10. Scarfe WC. Imaging of maxillofacial trauma: evolutions and emerging revolutions. *Oral Surg Oral Med Oral Pathol Oral Radiol Endod* 2005;100:575–596.
11. Michetti J, Maret D, Mallet JD, Diemer F. Validation of cone beam computed tomography as a tool to explore root canal anatomy. *J Endod* 2010;36:1187–1190.
12. Heiland M, Schmelzle R, Hebecker A, Schulze D. Intraoperative 3D imaging of the facial skeleton using the SIREMOBIL Iso-C3D. *Dentomaxillofac Radiol* 2004;33:130–132.
13. Katsumata A, Hirukawa A, Okumura S, Naitoh M, Fujishita M, Ariji E, Langlais RP. Effects of image artifacts on gray-value density in limited-volume cone-beam computerized tomography. *Oral Surg Oral Med Oral Pathol Oral Radiol Endod* 2007;104:829–836.
14. Katsumata A, Hirukawa A, Okumura S, Naitoh M, Fujishita M, Ariji E, Langlais RP. Relationship between density variability and imaging volume size in cone-beam computerized tomographic scanning of the maxillofacial region: an *in vitro* study. *Oral Surg Oral Med Oral Pathol Oral Radiol Endod* 2009;107:420–425.
15. Ludlow JB, Ivanovic M. Comparative dosimetry of dental CBCT devices and 64-slice CT for oral and maxillofacial radiology. *Oral Surg Oral Med Oral Pathol Oral Radiol Endod* 2008;106:106–114.
16. Tsiklakis K, Donta C, Gavala 5, Karayianni K, Kamenopoulou V, Hourdakis CJ. Dose reduction in maxillofacial imaging using low dose cone beam CT. *Eur J Radiol* 2005;56:413–417.
17. Moshiri M, Scarfe WC, Hilgers ML, Scheetz JP, Silveira AM, Farman AG. Accuracy of linear measurements from imaging plate and lateral cephalometric images derived from cone-beam computed tomography. *Am J Orthod Dentofacial Orthop* 2007;132:550–560.
18. Grauer D, Cevidanes LS, Styner MA, Heulfe I, Harmon ET, Zhu H, Proffit WR. Accuracy and landmark error calculation using cone-beam computed tomography–generated cephalograms. *Angle Orthod* 2010;80:286–294.
19. Lou L, Lagravere MO, Compton S, Major PW, Flores-Mir C. Accuracy of measurements and reliability of landmark identification with computed tomography (CT) techniques in the maxillofacial area: a systematic review. *Oral Surg Oral Med Oral Pathol Oral Radiol Endod* 2007;104:402–411.
20. Brown AA, Scarfe WC, Scheetz JP, Silveira AM, Farman AG. Linear accuracy of cone beam CT derived 3D images. *Angle Orthod* 2009;79:150–157.
21. Schulze D, Heiland M, Thurmann H, Adam G. Radiation exposure during midfacial imaging using 4- and 16-slice computed tomography, cone beam computed tomography systems and conventional radiography. *Dentomaxillofac Radiol* 2004;33:83–86.

22. Pohlenz P, Blessmann M, Blake F, Heinrich S, Schmelzle R, Heiland M. Clinical indications and perspectives for intraoperative cone-beam computed tomography in oral and maxillofacial surgery. *Oral Surg Oral Med Oral Pathol Oral Radiol Endod* 2007;103:412–417.
23. Hamada Y, Kondoh T, Noguchi K, Iino M, Isono H, Ishii H, Mishima A, Kobayashi K, Seto K. Application of limited cone beam computed tomography to clinical assessment of alveolar bone grafting: a preliminary report. *Cleft Palate Craniofac J* 2005;42:128–137.
24. Oberoi S, Gill P, Nguyen A, Hatcher D, Vargervik K. Three-dimensional assessment of the eruption path of the canine in individuals with bone grafted alveolar clefts using cone beam computed tomography. *Cleft Palate Craniofac J* 2010 (Feb 24) [Epub ahead of print].
25. Neugebauer J, Shirani R, Mischkowski RA, Ritter L, Scheer M, Keeve E, Zöller JE. Comparison of cone-beam volumetric imaging and combined plain radiographs for localization of the mandibular canal before removal of impacted lower third molars. *Oral Surg Oral Med Oral Pathol Oral Radiol Endod* 2008;105:633–642; discussion 643.
26. Sharan A, Madjar D. Correlation between maxillary sinus floor topography and related root position of posterior teeth using panoramic and cross-sectional computed tomography imaging. *Oral Surg Oral Med Oral Pathol Oral Radiol Endod* 2006;102:375–381.
27. Tsiklakis K, Syriopoulos K, Stamatakis HC. Radiographic examination of the temporomandibular joint using cone beam computed tomography. *Dentomaxillofac Radiol* 2004;33:196–201.
28. Maki K, lnou N, Takanishi A, Miller AJ. Computer-assisted simulations in orthodontic diagnosis and the application of a new cone beam X-ray computed tomography. *Orthod CraniofacRes* 2003;6:95–101; discussion 179–182.
29. Cevidanes LH, Bailey U, Tucker GR Jr, Styner MA, Mol A, Phillips CL, Proffit WR, Turvey T. Superimposition of 3D cone-beam CT models of orthognathic surgery patients. *Dentomaxillofac Radiol* 2005;34:369–375.
30. Carter L, Farman AG, Geist J, Scarfe WC, Angelopoulos C, Nair MK, Hildebolt CF, Tyndall D, Shrout M. American Academy of Oral and Maxillofacial Radiology. American Academy of Oral and Maxillofacial Radiology executive opinion statement on performing and interpreting diagnostic cone beam computed tomography. *Oral Surg Oral Med Oral Pathol Oral Radiol Endod* 2008;106:561–562.
31. European Academy of Dental and Maxillofacial Radiology. *Basic Principles for Use of Dental Cone Beam CT*. January 2009.

Chapter 6
Magnetic resonance imaging

Part 1: Basic Principles

INTRODUCTION

During the 32 years since *magnetic resonance imaging* (MRI) was first used to investigate a patient in Aberdeen, Scotland,[1] it has become the first-choice investigation for a number of lesions affecting the face and jaws because of its ability to distinguish clearly soft-tissue lesions from adjacent healthy tissue. However, *computed tomography (CT)* remains the primary cross-sectional modality for the majority of cases because, among others (better for evaluating cortical bone, less susceptible to motion artifacts and so on) of its shorter scanning time.[2] MRI already plays an important role in the evaluation and diagnosis of temporomandibular joint disorders and neoplasms of the face and jaws. Part 1 of this series introduces the principles of MRI and the terms most likely to be used in a surgeon-radiologist dialogue. Part 2 will explain the jargon and other MRI features that are most likely to appear in a radiologist report. Part 3 will focus on the pathology of the face and jaws displayed by MRI and other clinical applications (such as for preimplant evaluation/planning).

The MRI signal intensity depends on many factors, including the sequence used. The most common sequences used to image the jaws and their lesions are T1-weighted and T2-weighted, which will be explained later. New users must address a steep learning curve, which not only includes an understanding of this technology, but also of the images it produces; for example, most oral and maxillofacial clinicians are used to seeing bone as white, not black. Oral and maxillofacial clinicians already familiar with CT images will need to adjust further to the grayscale for soft tissues because fat shows up as "very bright" in contrast to black in the bone and soft-tissue windows of CT.

WHAT IS MAGNETIC RESONANCE IMAGING?

Although the MRI scanner and its images bear a superficial resemblance to CT, the two imaging modalities are completely different. MRI does not use ionizing radiation, but rather a radiowave "dialogue" with the patient's tissues and lesion within a magnetic field. Its efficacy as a clinical imaging modality is based primarily upon the soft tissues being proton-rich; they are composed of between 70% and 90% water, which is concentrated hydrogen nuclei or protons. "MRI images are obtained by measuring how rapidly hydrogen nuclei of different tissues return to their resting energy states after being excited by a strong magnetic field."[3]

The properties and amount of water within a tissue can alter drastically with disease or injury; MRI is very sensitive to the former and, therefore, is a very sensitive diagnostic modality.[3] MRI images display a better definition between the lesion and the adjacent normal tissue than other imaging modalities[2] generally available to the oral and maxillofacial clinician.

WHY IS MAGNETIC RESONANCE IMAGING SO SENSITIVE?

MRI detects subtle changes in the magnetism of the atomic nucleus and thus probes much deeper than X-rays, which interact only with the electron shells. In addition to merely imaging anatomy and pathology, MRI can, at its most advanced, investigate organ function, probing *in vivo* chemistry and visualizing brain activity.[3]

Oral and Maxillofacial Radiology: A Diagnostic Approach,
David MacDonald. © 2011 David MacDonald

COMPONENTS OF THE MAGNETIC RESONANCE IMAGING SUITE

The MRI suite has 3 components: the magnet room, the technical room, and the console room. The "magnet room" contains the magnet and *radiofrequency* (RF) coils. It is enclosed within a faraday cage, which limits the influence of radiofrequencies arising outside the magnet preventing their interference with those generated by the magnet. This cage is an RF-shielded enclosure made of copper, aluminum, or steel sheets, and includes the door and windows, which are made of special wire-embedded glass. The door should be closed at all times, save for entry or exit. All equipment present or entering the magnet room must be MR compatible. Figure 6.1 is a schematic diagram displaying the principle components of the MRI scanner.

The magnet

The magnet is the main component of the MRI scanner. There are 4 types of magnets: air-cored resistive magnets, iron-cored electromagnets, permanent magnets, and superconducting magnets. The superconducting magnets produce excellent field uniformity and stability. They are the most frequently encountered type. They are liquid nitrogen or helium-cooled to 0° kelvin. This allows the Niobium-Aluminum alloy magnet to lose all its resistance to electrical current, enhancing its field strength. Most magnets in clinical use are within the midfield range of 0.5 to 1.5 tesla (T; a measure of magnetic field strength), which is suggested by Langlais et al. to be the optimum range.[4] Nevertheless, Tutton and Goddard's pictorial review[5] of 29 figures of "MRI of teeth" displayed very good images created on a 0.2 T scanner (low-field strength). The patient, or part to be imaged, is placed within a "tunnel." Magnets of less than 0.5 T generally do not have to have a tunnel and are "open."[1] The *signal-to-noise ratio* (SNR) is proportional to the field strength; the higher the field strength the greater the SNR.[6] Runge et al. advocated routine brain imaging at 3 T because this offers superior SNR and *contrast-to-noise ratio* (CNR; see the later section on coils and image quality) compared with those for 1.5 T.[7] However, low field strength is preferred for implant planning to reduce artifacts.

DISADVANTAGES OF MAGNETIC RESONANCE IMAGING

Even when produced by the same manufacturer, each magnet has its own inhomogeneities, which uniquely affect the images produced. Therefore, if the patient were to be reinvestigated on a different MRI scanner, the exact images are unlikely to be repeated; this is unlike CT. Studies comparing the results derived from different scanners are very few; Westwood et al. reported the difference between three European institutions assessing myocardial iron in patients with thalassemia.[8]

Air and bone appear to be black, because they have fewer protons and give virtually no signal. This is a problem for the maxillary antrum in which the air-filled antrum is normally separated from the thin bony walls by a thin antral mucosa, which may not always be obvious depending both on its thinness and spatial resolution of the MRI scanner.

Like any other type of medical imaging, movement has a deleterious effect on the resultant images. This is particularly important in view of the relatively long scanning times; a routine MRI scan can take 30 to 40 minutes compared with only 10 to 15 minutes for CT.[2] Unlike CT, in which the entire volume data is available to manipulate, for example, to produce bone and soft-tissue windows and their

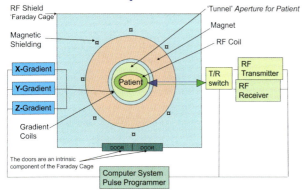

Figure 6.1. Schematic displaying the main elements of the magnetic resonance imaging scanner properly juxtaposed. The faraday cage is formed within the walls of the magnet room and includes the doors. The radiofrequency coils are applied directly to the part of the patient to be investigated, whereas the gradient coils remain part of the scanner. Reprinted with permission from MacDonald-Jankowski DS, Magnetic resonance imaging for oral and maxillofacial surgeons. Part 1: Basic principles. *Asian Journal of Oral Maxillofacial Surgery* 2006;18:165–171.

variable widths, MRI has to scan each sequence separately. Therefore, sections reformatted from the T1-weighted (anatomy) and T2-weighted (pathology) scans may not necessarily be from exactly the corresponding anatomical sections, due to some change in patient position between the sequences. Nevertheless, the MRI technicians take very great care to ensure that the patient's head is effectively, but comfortably, immobilized throughout the entire MRI investigation.

PULSE SEQUENCE

All MR images are produced using a pulse sequence, which is stored on the scanner computer. The sequence contains RF pulses and gradient pulses that have carefully controlled durations and timings. There are many different types of sequence, but they all have timing values that can be modified to obtain the required image contrast.[5] All sequences are dependent on the two *timing values*, *echo time* (TE) and *repetition time* (TR).[3]

PROTON DENSITY

MRI is based on the natural magnetism that is induced in the human body when it is placed in the scanner's magnetic field. The strength of this magnetism depends on the *proton density* (PD) of the tissues. PD is dependent on the number of hydrogen atoms in the tissues; fluids and fats both have high PDs so contrast between them is poor, unless "fat suppressed," discussed later in Part 2 (Figure 6.2). Nevertheless, PD is used to display the roots of the cervical spinal cord.[3] The PD sequence uses a short TE (less than 40 ms) and a long TR (longer than 1500 ms).[3]

WHAT HAPPENS DURING A MAGNETIC RESONANCE IMAGING SEQUENCE?

When a scanning sequence begins, the magnetism within the tissues is knocked out of alignment with the main field of the scanner (Z or longitudinal axis) by the RF excitational pulses into the XY or transverse plane. The XY realigned magnetism has the potential to create its own signal. Characteristics of this signal are determined by the gradient, timing, and RF of the sequence.[3] Once the RF signal is turned off, the protons relax by giving off energy (this energy loss is called *free induction decay*, FID) to get back to the equilibrium position along the Z axis. This energy loss is transferred to

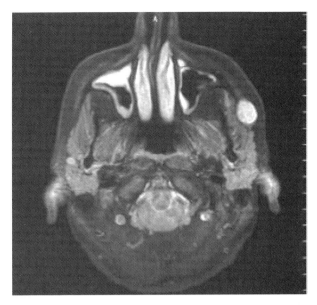

Figure 6.2. Proton density scan through the maxilla. Because fat and soft tissues (containing water) are rich in protons (hydrogen nuclei), they have a similar intensity; both display gray shades. Contrast between fat and soft tissues has been enhanced by "fat suppression," which is discussed in Part 3. Note the bright round "lump" on the left face. This painless, gradually enlarging swelling was a pleomorphic salivary adenoma. Compare its appearance in Figures 6.3 and 6.4. Compare also the hyperintense inflamed and nasal and antral mucosae with their presentations in Figures 6.3 and 6.4. Reprinted with permission from MacDonald-Jankowski DS. Magnetic resonance imaging for oral and maxillofacial surgeons. Part 1: Basic principles. *Asian Journal of Oral Maxillofacial Surgery* 2006;18:165–171.

surrounding molecules and results in a T1-weighted signal. When the magnetic moments interfere with adjacent protons, this causes dephasing and loss of transverse magnetism resulting in a T2-weighted signal.[5] Relaxation, T1-weighted, and T2-weighted will be discussed in more detail later in this part.

RADIOFREQUENCY COIL

The RF coil is made up of two separate coils; one to transmit the RF pulse to the patient's tissues and lesion and the other to receive the MRI signal from the patient's tissues and lesion. The RF coil's shape and size is appropriate to the part of the body to be investigated. There is a whole range of coils for the head, parts of the spine, and major joints, and even one for the temporomandibular joint.[3]

HOW CAN THE RADIOFREQUENCY COIL AFFECT IMAGE QUALITY?

Selection of the appropriate coil is of paramount importance to achieve a high SNR and, consequently, high spatial resolution. A coil that is able to completely cover the region of interest so as to produce an image with the highest SNR and homogeneity should be chosen. The coil must be correctly positioned because the signal decreases as the distance from the coil increases. The decrease in signal also produces a decrease in SNR; therefore all coils must be precisely tuned.[3,5]

The *bandwidth* (BW) of RF frequencies needed to encode the spatial positions of signals, is also important for the SNR—the wider the BW, the noisier the image and the lower the SNR.[7] If the SNR is too low (the image appears fuzzy when very low), the contrast will be poor and subtle contrast changes may be obscured. Therefore, according to McRobbie et al., the CNR is the most important aspect of the MRI's image quality.[3] SNR and CNR are not equal, in that a thick slice can give a good SNR but poor CNR. However, if the CNR or SNR are too low, the noise will obscure the spatial resolution.

The magnetic field strength is also an important factor in image quality. In dentistry, low field strength reduces artifacts from metal,[9] which is often more important. However, this may now be less of an issue because of new strategies and programs. Kakimoto et al. reported that metal artifacts rendered the CT images for 2 of their 9 patients useless, whereas those of the MRIs were unaffected.[10] Artifacts are briefly discussed later.

DISPLAYING AND REVIEWING THE IMAGES

Although many clinicians will still receive the images printed out in a 12-image format to be viewed on a standard viewing box, an increasing number will receive them downloaded onto a CD/DVD, which can be uploaded onto their computer monitors. The images of both formats should be read in conjunction with the radiologist's report.

RELAXATION

Relaxation is the process the magnetization experiences as it fades (like a tuning fork) or precesses (like a spinning top) from the XY plane to the Z axis. There are two relaxation processes: "spin-lattice" and "spin-spin." Spin-lattice controls the growth/recovery of the magnetization along the Z axis, and is also known as the "longitudinal proton relaxation time" or the T1-weighted sequence. Spin-spin controls the decay of the signal in the XY plane, and is also known as "the transverse proton relaxation time" or the T2-weighted sequence. Spin-lattice and spin-spin will now be referred to as T1-weighted and T2-weighted, respectively.[5]

T1-WEIGHTING

T1-weighted scans are often known as "anatomy scans" because their images display excellent contrast and most clearly show the boundaries between different tissues (Figure 6.3). These boundaries are made obvious by the fat-filled fascial planes; the T1-weighted scan is also termed "fat scans."[3] The central role of T1-weighted to display anatomy has been illustrated by Li et al. in their report on the presurgical evaluation of the parapharyngeal space. The parameters of this space are defined by important structures such as blood vessels, which can be difficult to evaluate by clinical and other radiological modalities.[11] Further discussion of this and other spaces is found in Chapter 16. T1-weighted on its own has been applied to the TMJ to determine the position of the articular disc (see Figure 12.4). For a T1-weighted image a short TE (less than 40 ms) and a short TR (less than 750 ms) are required.

T2-WEIGHTING

T2-weighted images take longer to acquire than T1-weighted images. T2-weighted images are often termed "pathology scans" because collections of abnormal fluid are bright against the darker normal tissue (Figure 6.4). T2-weighted images are also called "water images."[3] T2-weighted images are useful for detecting infection, hemorrhage, and neoplasms.[1] For a T2-weighted image a long TE (greater than 75 ms) and a long TR (less than 1500 ms) are required. Further discussion of T2-weighted is deferred to Part 3. T1-weighted and T2-weighted processes are independent of each other.

The reader should be aware that not all lesions are hyperintense on the T2-weighted sequence, nor are they always distinguishable. The more important of the strategies developed to

Chapter 6: Magnetic resonance imaging 71

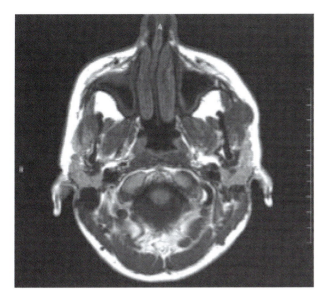

Figure 6.3. T1-weighted scan of the same section displayed in Figure 6.2. The subcutaneous fat and pharyngeal fat appear white ("bright" or "hyperintense"), whereas the other soft tissues display a range of grays, from very light ("hypointense") to those represented by skeletal muscle and the pleomorphic salivary adenoma ("isointense"). The black areas represent air-filled spaces, bone, or blood vessels containing fast-flowing blood. Reprinted with permission from MacDonald-Jankowski DS. Magnetic resonance imaging for oral and maxillofacial surgeons. Part 1: Basic principles. *Asian Journal of Oral Maxillofacial Surgery* 2006;18:165–171.

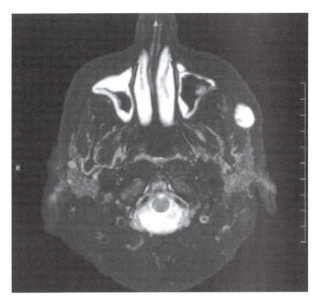

Figure 6.4. T2-weighted scan of the same section displayed in Figures 6.2 and 6.3. This scan accentuates water-rich tissues, which includes most lesions such as the pleomorphic adenoma, and renders them hyperintense. In some cases, it may be difficult to differentiate these areas from fat, which is also hyperintense on T2-weighted scans. Frequently, the fat signal is suppressed on T2-weighted scans to further accentuate the water-containing features. This image displays "fat suppression," which is discussed in Part 3. Reprinted with permission from MacDonald-Jankowski DS Magnetic resonance imaging for oral and maxillofacial surgeons. Part 1: Basic principles. *Asian Journal of Oral Maxillofacial Surgery* 2006;18:165–171.

address these feature later in this chapter. They are contrast media, fat suppression (particularly "fat saturation") and other sequences.

WHAT ARE SPIN ECHO AND GRADIENT ECHO SEQUENCES?

The signal that comes back from the patient is collected as an echo. Echoes are produced in two ways by the pulse sequence. The most common pulse sequence is the *spin echo* (SE) because it makes the best quality images. Because the SE takes a long time—minutes rather than seconds—*fast (or turbo) spin echo* (FSE), a variation of SE, is often used to speed up the process.[3]

The gradient echo is represented by a whole family of different pulse sequences. Examples of these are *fast low angle shot* (FLASH), *fast imaging with steady state precession* (FISP). These have been developed to provide rapid acquisition, in seconds rather than minutes. This minimizes movement and physiological artifacts, increases patient throughput, and allows dynamic studies.[6]

ARTIFACTS

Gray et al. identified two types of artifacts that may be produced in the MRI of dental structures,[1] those due to patient motion (Figure 6.5) and those due to inhomogeneities in the magnetic field caused by magnet susceptibility effects. Patient motion is seldom the cause of artifacts when the patient has been well instructed and is comfortably positioned.

The susceptibility artifacts presenting as geometric distortions are caused by air/tissue or bone/tissue artifacts and by the effect of ferromagnetic metals. The former is more likely to be a problem in high-field scanners (those operating above 1.5 T), but Gray et al. list strategies to minimize this effect.[1]

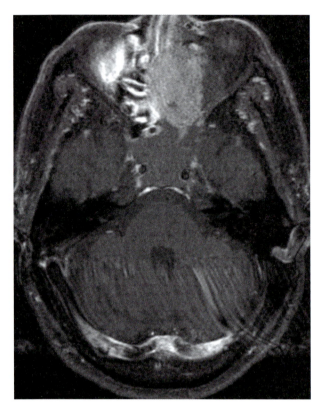

Figure 6.5. Axial MRI displaying movement artifact. Figure courtesy of Dr. Montgomery Martin, British Columbia Cancer Agency.

The latter, occasioned by metal, are usually localized areas of signal blackout adjacent to the metal structure (Figure 6.6) except for orthodontic bands, which may cause severe geometric distortion. Concerns that the MRI image may have been distorted, can be checked by using SPAMM (spatial modulation of magnetization). This has been done by Bridcut et al., using a 0.95T (Siemens) scanner, "to provide a qualitative estimate of the accuracy of the MRI when planning dental implants."[12]

Eggers et al. (1.5 T)[13] reported that all restorative materials particularly amalgam and titanium implants were magnetic resonance–compatible, Starcuková et al. (1.5 T)[14] reported that titanium was less so. Nevertheless its artifact was limited only to the immediate area. This could be minimized by shimming (see later). Lee et al.[15] reported the orientation of metallic orthopedic implants was minimized if its long axis was parallel to the direction of the main magnetic field. As dental implants are, or are close to being, parallel to the main magnetic field, this criterion should be achievable in most cases.

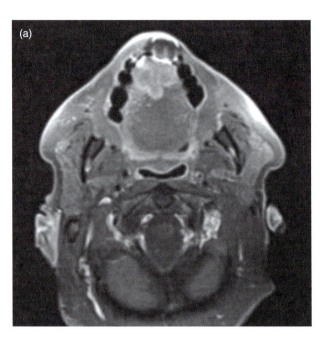

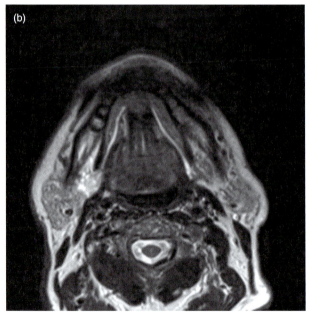

Figure 6.6. MRI images of a squamous cell carcinoma of the floor of the mouth. Both T1-weighted (a) and T2-weighted (b) images of the lesion have been degraded by an artifact arising from restorations in the dentition immediately adjacent to it. Figure courtesy of Dr. Montgomery Martin, British Columbia Cancer Agency.

The presence of implants within patients requiring an MRI investigation for an unrelated matter raises serious safety concerns among medical radiologists and their staffs. Sawyer-Glover and Shellock state that dental implants "are typically held in place with sufficient counterforces to prevent them from being moved or dislodged" up to 1.5 T (the usual maximum magnetic field strength at the time their review was published).[16]

SAFETY FIRST!

Although Clarke et al. showed that the provision of MRI can significantly reduce the CT collective dose,[17] we have to consider MRI safety. Although this imaging technique is not "invasive" there are potential dangers. To date, there have been at least 13 reported deaths due to MRI accidents[18]; 10 people with pacemakers died during a sequence and 3 patients died following entry into the core by ferrous equipment, including an oxygen cylinder, while the patient was present.[3] To minimize such events, Kanal et al. recently set out the *American College of Radiology's Guidance Document for Safe MR Practices*. Appliances are labeled *MR-Safe* (square label), *MR-Conditional* (triangular label; safe up to 1.5 T) and *MR-Unsafe* (round label).[19]

THE DANGERS OF MAGNETIC RESONANCE IMAGING

The dangers, in decreasing order of importance, are the magnet, RF exposure, and acoustic noise.

The magnet

Semiconducting magnets are on all the time. In addition to attracting ferrous metal equipment into the core and causing malfunction of pacemakers, the magnet can displace arterial clips and metallic foreign bodies (such as shrapnel) near the eye or major vessels.[18] Although tattoos can cause skin reactions, they are not a contraindication for MRI.[14] The magnet is associated with claustrophobia, which can be experienced by as many as a quarter of all patients.[1] This often occurs for patients undergoing maxillofacial imaging, because the patient's head is enclosed in the tunnel-like aperture (indeed, the term "tunnel" is frequently used for this aperture). Although, most modern units have a two-way intercom to facilitate communications between patients and MRI staff, the patient should be discouraged from talking because movement would invalidate the images for that sequence. Occasionally, it may be necessary to sedate or even anesthetize a patient. To date, there have been no adverse effects following investigation of pregnant women although, as a precaution, such investigations are to be avoided unless absolutely necessary.[3] Extremely obese patients cannot fit in the aperture of many machines.[3]

Radiofrequency exposure

The main danger associated with RF is tissue heating due to RF exposure (akin to a microwave). The patient's temperature should not increase by more than 1°C. This is achieved automatically by monitoring and limiting the specific absorption rate based on the patient's weight; therefore, the patient's weight is ascertained prior to the procedure.[3]

Acoustic noise

The MR scanner is very noisy. It makes a "knocking" sound, which can provoke headaches. This noise is caused by the movement of the gradient coils against their mountings, during the scan. The noise can be more than 100 dB for some sequences. Hearing protection is recommended for patients during scanning. Noise reduction is an active area of development for manufacturers.[3]

Part 2: MRI Terminology Most Frequent in the Radiologist's Report

In addition to T1-weighted and T2-weighted, the clinician is likely to see reference to intensity, fat suppression (usually fat saturation) and gadolinium.

MAGNETIC RESONANCE IMAGING JARGON

Hyperintensity can be used to describe a bright structure, due to fat in case of T1-weighted (Figure 6.7) and T2-weighted images, and fluid-filled lesions in case of T2-weighted. *Hypointense* structures, both in T1-weighted and T2-weighted, are dense calcified tissues (including calcified lesions) and fast-flowing blood. *Isointensity* is exhibited by the skeletal muscle in T1-weighted images (Figure 6.7).

FAT SUPPRESSION

Fat normally appears hyperintense on both T1-weighted and T2-weighted scans. Unfortunately,

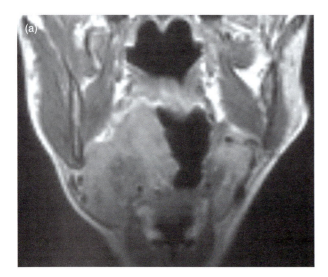

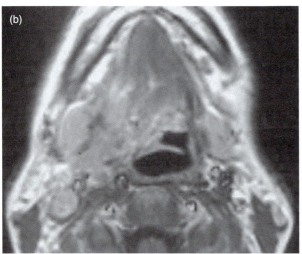

Figure 6.7. T1-weighted gadolinium-enhanced images of an advanced squamous cell carcinoma of the tongue. The hyperintensity of fat in both images has not been fat-suppressed. (a) Coronal T1-weighted showing the bilateral lesion partially obturating the oropharynx and invading the submandibular space. **Note:** The center of this lesion is not very hyperintense because it is relatively not enhanced and is as isointense as the skeletal muscle; and (b) axial T1-weighted showing the lesion to have extensively invaded the lateral pharyngeal space and obturated the ipsilateral vallecula. Reprinted with permission from MacDonald-Jankowski DS, Li TK, Matthew I. Magnetic resonance imaging for oral and maxillofacial surgeons. Part 2: Clinical applications. *Asian Journal of Oral Maxillofacial Surgery* 2006;18:236–247.

particularly in T2-weighted scans, most lesions (neoplastic, cystic, and inflammatory) are also hyperintense. Therefore, in order to identify them, the hyperintense fat signal has to be suppressed. When contrast media is used in T1-weighted, fat suppression is required to detect subtle lesions that normally occur within fat-filled structures such as bone marrow. Figure 6.8 displays T1-weighted and T2-weighted images that have been fat-suppressed; the hyperintense signal of the subcutaneous fat has been suppressed in both scans. Another reason for fat suppression is that a very hyperintense fat signal can create an artifact a few pixels distant to its real position; this is called the *chemical shift*.[3,20] The clinical implications of this phenomenon will be revealed later in this chapter.

There are four fat suppression methods, *fat saturation* (FatSat or simply FS), *short T1 inversion recovery* (or STIR), *water excitation*, and *subtraction*.[21] Although *fat saturation* is frequently used, it operates best in strong fields, which better distinguish between fat and water. *Short T1 inversion recovery* (STIR), using a completely different mechanism, takes advantage of fat's shorter T1 and completely eliminates the fat signal. The advantage of the latter is that the magnet need not be perfectly *shimmed*. *Shimming* is the reduction of inhomogeneities within the magnet or ferromagnetic metals outside it. Shimming can be *fixed* or *dynamic*. *Fixed* can be *active (using shimming coils)* or *passive*.[3]

CONTRAST MEDIA

Contrast media have already been encountered with regards to *helical CT*, where they are used to enhance blood vessels and vascularized tissues.

Why is contrast media required in magnetic resonance imaging?

In MRI a contrast medium is used to obtain an investigation within a reasonable time frame.[3] Furthermore, although MRI is a very sensitive imaging modality for pathology, it is not always specific for a particular type. Therefore, contrast media can improve specificity with only a slight increase in the patient's scan time. The available contrast media include gadolinium and iron oxide (*superparamagnetic iron oxide*, SPIO).[2] Gadolinium is often simply referred to by radiologists as "gad."

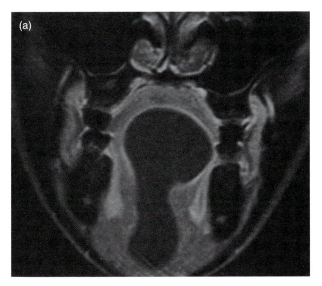

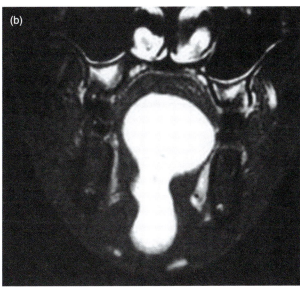

Figure 6.8. Fat-suppressed T1-weighted and T2-weighted images of a 7-year-old with painless progressive sublingual swelling. A dermoid cyst is situated between the genioglossus. Other differential diagnoses were thyroglossal cysts and cystic hygroma. (a) Gadolinium and fat-suppressed coronal T1-weighted showing gadolinium-enhanced blood vessels in the capsule around the superior aspect of the cyst. **Note:** The normally hyperintense signal of the subcutaneous fat has been suppressed (compare with that of Figure 6.5); and (b) fat-suppressed coronal T2-weighted showing the hyperintense signal completely occupying the hypointense structure observed on the T1-weighted scan. The bilateral teardrop-shaped structures are the saliva-filled sulci of the floor of the mouth. Reprinted with permission from MacDonald-Jankowski DS, Li TK, Matthew I. Magnetic resonance imaging for oral and maxillofacial surgeons. Part 2: Clinical applications. *Asian Journal of Oral Maxillofacial Surgery* 2006;18:236–247.

How does contrast media work?

All tissues have some degree of magnetic susceptibility, i.e., the degree to which they can get magnetized when placed in a strong magnetic field. Magnetic susceptibility ranges from diamagnetic (from zero for bone and air to some for most other tissues, including fully oxygenated hemoglobulin), paramagnetic (deoxyhemoglobulin and gadolinium), superparamagnetic (ferritin, hemosiderin, and SPIO) to ferromagnetic (iron, nickel, and cobalt).[3]

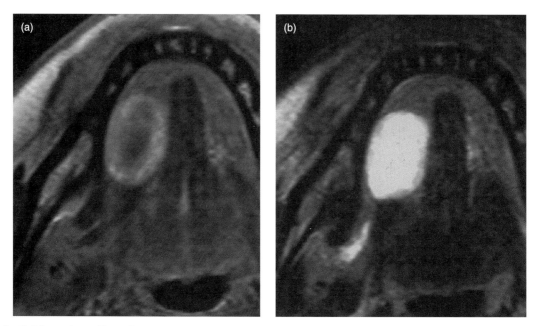

Figure 6.9. Sublingual swelling of 4–5 years duration, with occasional mild pain. Neuroma. (a) Gadolinium and fat- suppressed axial T1-weighted image showing "ring" enhancement of peripheral blood vessels; and (b) fat-suppressed axial T2-weighted image shows that the entire lesion is hyperintense. Most lesions are water-rich and will be hyperintense on T2-weighted. Reprinted with permission from MacDonald-Jankowski DS, Li TK, Matthew I. Magnetic resonance imaging for oral and maxillofacial surgeons. Part 2: Clinical applications. *Asian Journal of Oral Maxillofacial Surgery* 2006;18: 236–247.

After gadolinium is injected into the body, it is rapidly distributed in the blood and is gradually excreted by the kidneys over 24 hours. Gadolinium enhances T1-weighted signals in the tissues as it accumulates in richly vascular lesions (such as most neoplasms) that become hyperintense (Figures 6.9a, 6.10a, 6.11a; see also Whyte and Chapeikin's Figure 18[22]). Because gadolinium and fat are hyperintense, the definition of gadolinium-enhanced tissue within a fat-rich region is further enhanced by fat suppression.[2]

Gadolinium has recently shown to induce *nephrogenic systemic fibrosis* (NSF).[23] NSF may rapidly progress to wheelchair dependence in many, intractable pain syndrome in some, and occasionally death.[19] Although it is particularly the *gadodiamides* that are most likely to provoke this, caution on their use has been extended to other forms, particularly if the patient has moderate severe to severe renal insufficiency. The reason why the gadodiamides are considerably more associated with NSF is that they are thermodynamically stable causing release of the unbound (or free) gadolinium, which is toxic.[24] American[23] and European[24] authorities have suggested that the at-risk patient should undergo dialysis within 3 hours or 36 hours, respectively, of the investigation.

SPIO is available in several preparations; it reduces the intensity in T2-weighted of those tissues it accumulates in, most commonly the liver and spleen (readily taken up by the Kupffer cells). Therefore, suppression of the normal signal of the adjacent normal tissue allows the lesion to become conspicuous.[3] Anzai et al. have applied SPIO to oral cancer (see later).[25] Only 6% of cases investigated with SPIO contrast showed side effects, but these complications were not serious and were readily managed. Contrast media for MRI may also induce adverse reactions similar to those already discussed earlier for CT.

To determine whether enhancement has occurred, it is standard to compare the contrast images (usually T1-weighted with gadolinium) with the previously taken noncontrast images.

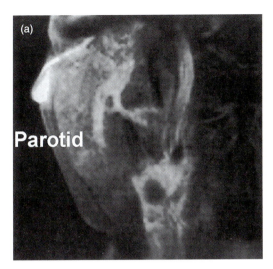 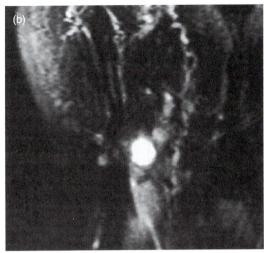

Figure 6.10. New, enlarged jugulodigastric node after partial glossectomy for squamous cell carcinoma. (a) Gadolinium and fat-suppressed coronal T1-weighted image shows the gadolinium (hyperintense) "ring" enhancement of jugulodigastric node, suggesting central necrosis indicative of metastasis; and (b) fat-suppressed coronal T2-weighted image confirms hyperintense signal from the area of central necrosis, which has been defined by fat suppression. Reprinted with permission from MacDonald-Jankowski DS, Li TK, Matthew I. Magnetic resonance imaging for oral and maxillofacial surgeons. Part 2: Clinical applications. *Asian Journal of Oral Maxillofacial Surgery* 2006;18:236–247.

Part 3: The Pathology of the Face and Jaws That May Be Apparent on Magnetic Resonance Imaging

This part focuses on the pathology of the face and jaws displayed by MRI.

FACIAL PAIN

Goh et al. (1.0T) reported that 14% of their 42 *trigeminal neuralgia* (TN) cases were associated with structural lesions; of these, 2 were vestibular neuromas (formerly acoustic neuromas; see Chapter 17).[26] Therefore, they suggest that a routine MRI should be considered for all TN patients. This suggestion is supported by Tanaka et al. (1.5T; *3D-Fast Asymmetry Spin Echo*, FASE), who reported that the majority of their 150 patients with TN had neurovascular compression of the nerve root in the cerebellopontine angle cistern.[27] Six patients in their series had brain tumors.

Schmidt et al. suggested that a functional MRI technique called *blood oxygen level dependent* (BOLD) may improve our understanding of central nervous system sites involved in pain transmission and processing.[28] BOLD takes advantage of the differing magnetic susceptibilities of fully oxygenated hemoglobulin (diamagnetic) and deoxyhemoglobulin (paramagnetic). The resultant changes in blood oxygenation alter T2-weighted decay allowing localization of distinct areas of the brain activated during delivery of painful stimuli.

FACIAL SWELLINGS

Browne et al. (1.5T) reported that MRI was sufficiently accurate to diagnose or exclude some neoplasms.[29] MRI also appeared to be effective in the diagnosis of inflammatory disease, including infections (Figure 6.12).

MALIGNANCIES

Diagnosis and preoperative assessment

Despite many advances in our understanding of head and neck cancer, the survival rates are still

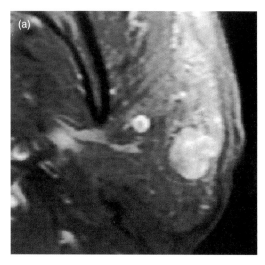

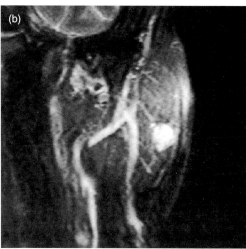

Figure 6.11. Painless gradually enlarging swelling of the left parotid. Needle biopsy showed pleomorphic salivary adenoma treated by superficial parotidectomy. (a) Gadolinium and fat-suppressed axial T1-weighted image showing gadolinium-enhanced blood vessels within this neoplasm; and (b) fat-suppressed coronal T2-weighted image showing the blood vessel–filled component of this lesion that contributes to its hyperintensity. Reprinted with permission from MacDonald-Jankowski DS, Li TK, Matthew I. Magnetic resonance imaging for oral and maxillofacial surgeons. Part 2: Clinical applications. Asian Journal of Oral Maxillofacial Surgery 2006;18:236–247.

poor. This is largely a result of advanced disease upon first presentation. Figure 6.7 presents one such case. Most head and neck malignancies are associated with *squamous cell carcinomas* (SCC), predominantly of the mouth and pharynx. Although imaging is infrequently the primary mode of diagnosis, it is invaluable in assisting the surgeon to stage the lesion and determine the optimum course of treatment (including palliation).

Nevertheless, spatial resolution could be a limiting factor. Daisne et al. determined that neither CT, MRI (1.5 T) nor *positron emission tomography* (PET) scanning adequately depicted superficial tumor extension.[30] Furthermore, false positives were seen for cartilage, extralaryngeal, and pre-epiglottic extensions.

SCC is displayed on a T1-weighted scan as isointense, which (particularly in case of the tongue) can make it difficult to distinguish from adjacent skeletal muscle. Figure 6.7 compares the central relatively unenhanced area with skeletal muscle. Hsu et al. (1.5 T) maintain that direct laryngoscopy is the most accurate in evaluating the mucosal surface of the aerodigestive tract.[31] Gadolinium-enhanced T1-weighted images can display the lingual lesion (Figures 6.7 and 6.13a). On a T2-weighted scan, SCC is hyperintense and can be readily distinguished without fat suppression because the tongue, which is a muscular structure, contains little fat (Figure 6.13b).

Although MRI displays contrast enhancement better than CT, this may be nullified by movement artifacts produced during the longer scanning time required to produce higher resolution and better quality images.[32]

Involvement of adjacent bone

Bolzoni et al. (1.5 T) reported that in a series of MRIs of 43 patients with SCC before mandibulectomy, 16 (of which 2 were false positives) displayed mandibular involvement.[33] Although the MRIs of the remaining 37 patients displayed cortical integrity, one mandible was found microscopically to have become involved, and therefore a false negative.

Imaizumi et al. (1,5T) reported that the specificity of MRI for cortical invasion was significantly lower than for CT (p <0.001).[20] They postulated that most of these false positives were due to the chemical shift artifact of the bone marrow fat. Rumboldt et al. recommend STIR (see fat suppression) to display bone marrow invasion as a hyper-

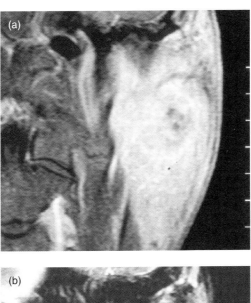

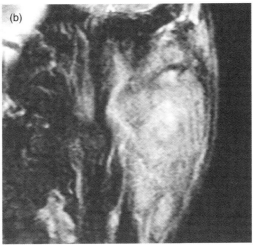

Figure 6.12. Left buccal swelling and acute pain with no pus or discharge. It was considered to be a pleomorphic salivary adenoma with malignant transformation, adenocystic carcinoma, and mucoepidermoid carcinoma. Aspirate showed actinomyces-like organisms. (a) Gadolinium and fat-suppressed coronal T1-weighted image shows that gadolinium enhances the blood supply to the parotid gland, which is very hyperintense. Few hypointense areas are visible at the center; and (b) fat-suppressed coronal T2-weighted image: the hypointensities on T1-weighted now appear hyperintense. Reprinted with permission from MacDonald-Jankowski DS, Li TK, Matthew I. Magnetic resonance imaging for oral and maxillofacial surgeons. Part 2: Clinical applications. *Asian Journal of Oral Maxillofacial Surgery* 2006;18:236–247.

intense signal (see Figures 5 and 6b in Rumboldt et al.).[32]

ASSESSMENT OF REGIONAL LYMPH NODES

Metastasis to the regional lymph nodes is frequent. Management of almost all cases must therefore include these nodes even in those cases where lymph node involvement is not apparent clinically,[34] because the risk of occult metastasis is high for oral, pharyngeal, and nasopharyngeal cancers.[35]

Central necrosis within a lymph node appears within a prominent "ring" due to the enhancement by gadolinium on a T1-weighted scan. This is a key indicator of metastasis (see Figure 6.10). Furthermore, a neoplastic node is more likely to be round, with the shortest diameter of it being greater than 1 cm. Reactive or inflammatory nodes are more likely to be ovoid, with the shortest diameter being less than 1 cm (Figure 6.14). This is discussed more fully in Chapter 18.

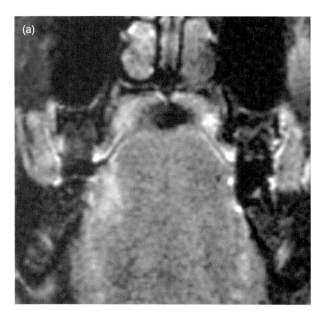

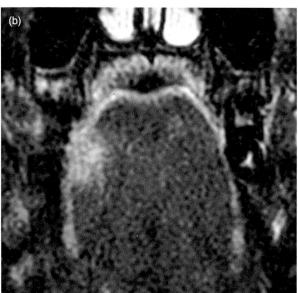

Figure 6.13. Fat-suppressed T2-weighted scan displaying squamous cell carcinoma of the tongue more prominently than that displayed in a T1-weighted scan enhanced by gadolinium and fat suppression. (a) Gadolinium fat-suppressed coronal T1-weighted image showing gadolinium-enhanced blood vessels at the periphery of the lesion; and (b) fat-suppressed coronal T2-weighted image shows fat suppression accentuates the water-rich neoplasm. **Note:** This enhancement is less within the tongue because the tongue is a muscular structure that contains very little fat. Reprinted with permission from MacDonald-Jankowski DS, Li TK, Matthew I. Magnetic resonance imaging for oral and maxillofacial surgeons. Part 2: Clinical applications. *Asian Journal of Oral Maxillofacial Surgery* 2006;18:236–247.

ASSESSMENT OF DEEPLY PLACED OR INVADING CANCERS

The deep extension of laryngeal and pharyngeal SCC, including submucosal extension beyond the obvious primary mucosal lesion, cannot be accurately assessed by endoscopy. Hsu et al. (1.5 T) demonstrated (in a retrospective and blinded study of 75 patients with advanced cancers) that unenhanced T1-weighted scans are valuable to predict the absence of fixation of head and neck cancers to the prevertebral fascia.[31]

POSTTHERAPEUTIC IMAGING

Lell et al. (1.5 T) reported that, in 39 cases of advanced SCC, helical CT is better at displaying posttherapeutic changes and tumor recurrences, whereas MRI's vaunted ability to differentiate tumors is compromised by edema following radiotherapy.[36]

VASCULAR LESIONS

Bentz et al. reported that 86.7% of the 324 consecutive cases of noninflammatory masses in the salivary gland region of American children were vascular (two-thirds were *hemangiomas* and one-third were *lymphangiomas*).[37] High-flow vascular malformations give a hypointense signal both on T1-weighted and T2-weighted sequences.[4] CT and *digital subtraction angiography* (DSA) are more suited for investigation of arterial malformations than *arteriovenous malformations* (AVMs). CT can also readily display whether the lesion is uni- or multilocular. Of the 9 patients reported by Kakimoto et al. (1.5 T), 8 were investigated using fat-suppressed gadolinium-enhanced images, which afforded very good detection of the vascular lesions.[10] Additionally, they reported that although *phleboliths* were best detected on CT, in 2 cases the CT images were degraded by the metal artifacts caused by dental restorations. The fat suppression allowed the tumor to show greater contrast than the adjacent tissues (Figure 6.15a).

Tanaka et al. (1.5 T) combined 3-D phase-contrast MRI and T2-weighted with 3-D-FASE; they were able to display the 3-D structure of the hemangiomas and the feeding arteries without using contrast media.[38] Figure 6.15b displays MR angiography.

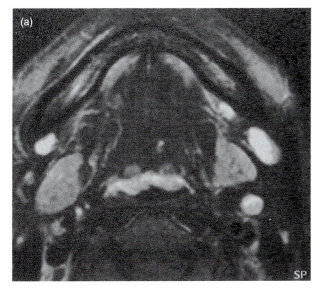

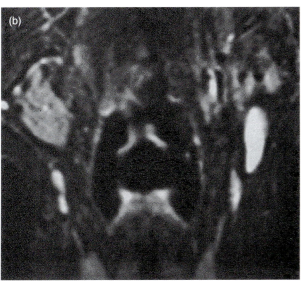

Figure 6.14. An indurated ulcer is observed on the floor of the mouth. It is 0.5 cm in diameter. The jugulodigastric lymph node and those nodes adjacent to submandibular gland are displayed. **Note:** They are oval-shaped and may be reactive rather than neoplastic; this is further indicated by the shortest axis being less than 1 cm. (a) Fat-suppressed axial T2-weighted and (b) fat-suppressed coronal T2-weighted images. Reprinted with permission from MacDonald-Jankowski DS, Li TK, Matthew I. Magnetic resonance imaging for oral and maxillofacial surgeons. Part 2: Clinical applications. *Asian Journal of Oral Maxillofacial Surgery* 2006;18:236–247.

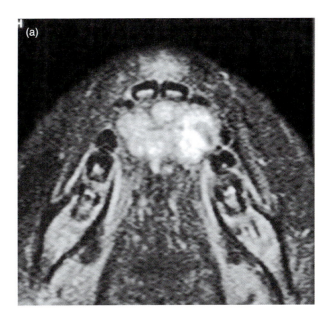

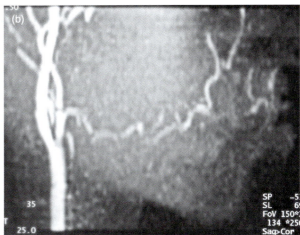

Figure 6.15. Hemangioma of the tongue. The labial lesion had already been treated by laser. (a) Fat-suppressed axial T2-weighted shows the extent of the hemangioma is a clearly defined hyperintense signal; and (b) magnetic resonance angiography image of the facial artery shows a dilated facial artery and superior labial branch **Note:** The inferior labial branch is not visible because of previous laser treatment. Reprinted with permission from MacDonald-Jankowski DS, Li TK, Matthew I. Magnetic resonance imaging for oral and maxillofacial surgeons. Part 2: Clinical applications. *Asian Journal of Oral Maxillofacial Surgery* 2006;18:236–247.

INFECTIONS

Infections of the face was touched on briefly earlier (Figure 6.12). Kito et al. (1.5 T) applied the FASE variant of diffusion-weighed MRI (normally used for strokes[3]) for locating head and neck abscesses prior to drainage.[39] Miller et al. (1.5 T; GE Medical Systems) did not consider that the routine use of gadolinium for investigating musculoskeletal infection was necessary.[40] They found that fat-suppressed T2-weighted scans gave comparable results. Instead, they suggested that gadolinium-enhanced T1-weighted scans be reserved for clinically suspected infection in or around the joint and in cases not responding to treatment due to possible abscess formation.

References

1. Gray CF, Redpath TW, Smith F, Staff RT. Advanced imaging: magnetic resonance imaging in implant dentistry. *Clin Oral Implants Res* 2003;14:18–27.
2. Silvers A. Imaging of the neck. In Van de Water TR, Straecker H, eds. *Otolaryngology: Basic Science and Clinical Review*. Thieme, New York 2006: p 667–681.
3. McRobbie DW, Moore EA, Graves MJ, Prince MR. *MRI: From Picture to Proton*. 2nd ed. Cambridge University Press, Cambridge 2007.
4. Langlais RP, van Rensburg U, Guidry J, Moore WS, Miles DA, Nortje CJ. Magnetic resonance imaging in dentistry. *Dent Clin North Am* 2000;44:411–426.
5. Tutton LM, Goddard PR. MRI of the teeth. *Br J Radiol* 2002;75:552–562.
6. English PT, Moore C. *MPJ for Radiographers*. Springer, Berlin 1995.
7. Runge VM, Patel MC, Baumann SS, Simonetta AB, Ponzo JA, Lesley WS, Calderwood GW, Naul LG. T1-weighted imaging of the brain at 3 tesla using a 2-dimensional spoiled gradient echo technique. *Invest Radiol* 2006;41:68–75.
8. Westwood MA, Firmin DN, Gildo M, Renzo G, Stathis G, Markissia K, Vasii B, Pennell DJ. Intercentre reproducibility of magnetic resonance T2*-weighted measurements of myocardial iron in thalassaemia. *Int J Cardiovasc Imaging* 2005;21:531–538.
9. Gray CE, Redpath TW, Smith FW. Low-field magnetic resonance imaging for implant dentistry. *Dentomaxillofac Radiol* 1998;27:225–229.
10. Kakimoto N, Tanimoto K, Nishiyama H, Murakami S, Furukawa S, Kreiborg S. CT and MR imaging features of oral and maxillofacial hemangioma and vascular malformation. *Eur J Radiol* 2005;55:108–112.
11. Li QY, Zhang SX, Liu ZJ, Tan LW, Qiu MG, Li K, Cui GY, Guo YL, Yang XP, Zhang WG, Chen XH, Chen JH, Ding SY, Chen W, You J, Wang YS, Deng JH, Tang ZS. The pre-styloid compartment of the parapharyngeal space: a three-dimensional digitized model based on the Chinese Visible Human. *Surg Radiol Anat* 2004;26:411–416.
12. Bridcut RR, Redpath TW, Gray CE, Staff RT. The use of SPAMM to assess spatial distortion due to static field inhomogeneity in dental MRI. *Phys Med Biol* 2001;46:1357–1367.
13. Eggers G, Rieker M, Kress B, Fiebach J, Dickhaus H, Hassfeld S. Artefacts in magnetic resonance imaging caused by dental material. *MAGMA* 2005;18:103–111.
14. Starcuková J, Starcuk Z Jr, Hubálková H, Linetskiy I. Magnetic susceptibility and electrical conductivity of metallic dental materials and their impact on MR imaging artifacts. *Dent Mater* 2008;24:715–723.
15. Lee MJ, Kim S, Lee SA, Song HT, Huh YM, Kim DH, Han SH, Suh JS. Overcoming artifacts from metallic orthopedic implants at high-field-strength MR imaging and multi-detector CT. *Radiographics* 2007;27:791–803.
16. Sawyer-Clover AM, Shellock FG. Pre-MRI procedure screening: recommendations and safety considerations for biomedical implants and devices. *J Magn Reson Imaging* 2000;12:92–106.
17. Clarke JC, Cranley K, Kelly BE, Bell K, Smith PH. Provision of MRI can significantly reduce CT collective dose. *Br J Radiol* 2001;74:926–931.
18. Shellock FG, Crues JV. MR procedures: biologic effects, safety, and patient care. *Radiology* 2004;232:635–652.
19. Kanal E, Barkovich AJ, Bell C, Borgstede JP, Bradley WG Jr, Froelich JW, Gilk T, Gimbel JR, Gosbee J, Kuhni-Kaminski E, Lester JW Jr, Nyenhuis J, Parag Y, Schaefer DJ, Sebek-Scoumis EA, Weinreb J, Zaremba LA, Wilcox P, Lucey L, Sass N. ACR Blue Ribbon Panel on MR Safety. ACR guidance document for safe MR practices: 2007. *AJR Am J Roentgenol* 2007;188:1447–1474.
20. Imaizumi A, Yoshino N, Yamada I, Nagumo K, Amagasa T, Omura K, Okada N, Kurabayashi T. A potential pitfall of MR imaging for assessing mandibular invasion of squamous cell carcinoma in the oral cavity. *AJNR Am J Neuroradiol* 2006;27:114–122.
21. Purdy D. The skinny on FatSat. http://www.medical.siemens.com/siemens/en_US/gg_mr_FBAs/files/MRI_Hot_Topics/MRI_HotTopics_Skinny_on_FatSat_engl.pdf (last accessed 24th March 2008).
22. Whyte A, Chapeikin G. Opaque maxillary antrum: a pictorial review. *Australas Radiol* 2005;49:203–213.
23. Prasad SR, Jagirdar J. Nephrogenic systemic fibrosis/nephrogenic fibrosing dermopathy: a primer for radiologists. *J Comput Assist Tomogr* 2008;32:1–3.
24. Martin DR. Nephrogenic system fibrosis: A radiologist's practical perspective. *Eur J Radiol* 2008;66:220–224.
25. Anzai Y, Piccoli CW, Outwater EK, Stanford W, Bluemke DA, Nurenberg P, Saini S, Maravilla KR, Feldman DE, Schmiedl UP, Brunberg JA, Francis IR, Harms SE, Som PM, Tempany CM Group. Evaluation of neck and body metastases to nodes with ferumoxtran 10-enhanced MR

imaging: phase III safety and efficacy study. *Radiology* 2003;228:777–788.
26. Goh BT, Poon CY, Peck RH. The importance of routine magnetic resonance imaging in trigeminal neuralgia diagnosis. *Oral Surg Oral Med Oral Pathol Oral Radiol Endod* 2001;92:424–429.
27. Tanaka T, Morimoto Y, Shiiba S, Sakamoto E, Kito S, Matsufuji Y, Nakanishi O, Ohba T. Utility of magnetic resonance cisternography using three-dimensional fast asymmetric spin-echo sequences with multiplanar reconstruction: the evaluation of sites of neurovascular compression of the trigeminal nerve. *Oral Surg Oral Med Oral Pathol Oral Radiol Endod* 2005;100:215–225.
28. Schmidt BL, Milam SB, Caloss R. Future directions for pain research in oral and maxillofacial surgery: findings of the 2005 AAOMS Research Summit. *J Oral Maxillofac Surg* 2005;63:1410–1417.
29. Browne RF, Golding SJ, Watt-Smith SR. The role of MRI in facial swelling due to presumed salivary gland disease. *Br J Radiol* 2001;74:127–133.
30. Daisne JF, Duprez T, Weynand B, Lonneux M, Hamoir M, Reychler H, Gregoire V. Tumor volume in pharyngolaryngeal squamous cell carcinoma: comparison at CT, MR imaging, and FDG PET and validation with surgical specimen. *Radiology* 2004;233:93–100. Erratum in *Radiology* 2005;235:1086.
31. Hsu WC, Loevner LA, Karpati R, Ahmed T, Mong A, Battineni ML, Yousem DM, Montone KT, Weinstein GS, Weber RS, Chalian AA. Accuracy of magnetic resonance imaging in predicting absence of fixation of head and neck cancer to the prevertebral space. *Head Neck* 2005;27:95–100.
32. Rumboldt Z, Day TA, Michel M. Imaging of oral cavity cancer. *Oral Oncol* 2006;42:854–865.
33. Bolzoni A, Cappiello J, Piazza C, Peretti G, Maroldi R, Farina D, Nicolai P. Diagnostic accuracy of magnetic resonance imaging in the assessment of mandibular involvement in oral-oropharyngeal squamous cell carcinoma: a prospective study. *Arch Otolaryngol Head Neck Surg* 2004;130:837–843.
34. Castelijns JA, van den Brekel MW. Imaging of lymphadenopathy in the neck. *Eur Radiol* 2002;12:727–738.
35. Poon I, Fischbein N, Lee N, Akazawa P, Xia P, Quivey J, Phillips T. A population-based atlas and clinical target volume for the head-and-neck lymph nodes. *Int J Radiat Oncol Biol Phys* 2004;59:1301–1311.
36. Lell M, Baum U, Greess H, Nomayr A, Nkenke E, Koester M, Lenz M, Bautz W. Head and neck tumors: imaging recurrent tumor and post-therapeutic changes with CT and MRI. *Eur J Radiol* 2000;33:239–247.
37. Bentz BG, Hughes CA, Ludemann JP, Maddalozzo J. Masses of the salivary gland region in children. *Arch Otolaryngol Head Neck Surg* 2000;126:1435–1439.
38. Tanaka T, Morimoto Y, Takano H, Tominaga K, Kito S, Okabe S, Takahashi T, Fukuda J, Ohba T. Three-dimensional identification of hemangiomas and feeding arteries in the head and neck region using combined phase-contrast MR angiography and fast asymmetric spin-echo sequences. *Oral Surg Oral Med Oral Pathol Oral Radiol Endod* 2005;100:609–613.
39. Kito S, Morimoto Y, Tanaka T, Tominaga K, Habu M, Kurokawa H, Yamashita Y, Matsumoto S, Shinohara Y, Okabe S, Matsufuji Y Takahashi T, Fukuda J, Ohba T. Utility of diffusion-weighted images using fast asymmetric spin-echo sequences for detection of abscess formation in the head and neck region. *Oral Surg Oral Med Oral Pathol Oral Radiol Endod* 2006;101:231–238.
40. Miller TI, Randolph DA Jr, Staron RB, Feldman F, Cushin S. Fat-suppressed MRJ of musculoskeletal infection: fast T1-weighted techniques versus gadolinium-enhanced T1-weighted images. *Skeletal Radiol* 1997;26:654–658.

Chapter 7
Positron emission tomography

Introduction

The basics of *positron emission tomography* (PET) are introduced in this chapter; the clinical applications are further discussed in Chapter 18. PET is a nuclear imaging technique based on metabolic mechanisms. These mechanisms can be enzymatic, hormonal, or pharmacological. PET uses the unique decay characteristics of nucleotides that decay by positron emission. The labeled compound is introduced into the body, usually intravenously, and is distributed in the tissues in a manner determined by its biological properties

When the radioactive atom of a particular labeled molecule decays, a positron is emitted from the nucleus. The positron is an antimatter analogue of an electron. It has the same mass as an electron, but it carries a positive charge. It is produced when 18 isotope of fluorine decays to 18 isotope of oxygen. When a positron meets an electron they mutually annihilate each other, each completely converted into 2 photons each of energy 511 KeV (*annihilation radiation*) traveling in opposite directions.

Such an event will be detected when each of the two photons interact simultaneously with two detectors on opposite sides of the detector ring surrounding that part of the patient (Figure 7.1). These detectors must lie within a column joining the two, which passes through the site of annihilation. This is called *coincidence*.[1]

PET Scanner

At a first glance the PET scanner looks like a CT unit. Unlike CT, where the source of radiation arising from the unit itself is outside the patient, in PET the source of radiation arises from within the patient. The PET scanner merely detects this radiation. It achieves this by a stationary ring of detectors that surround the patient or part of the patient to be imaged. It converts and reconstructs the signals from the detectors into a three-dimensional image. The signal intensity of any particular image voxel is proportional to the amount of the radionucleotide (and hence the amount of molecule to which it is attached) in that voxel.

FDG Tracer

There are a number of potential tracers, but *18 fluorine fluoro-deoxyglucose* (FDG) is preferred because it detects the intense accumulation of FDG in malignant neoplasms. This reflects the increased glycolytic rate of the malignant neoplasm. Furthermore, its longer half-life (110 minutes) allows sufficient time to transport it from the cyclotron where it is produced to the hospital where it will be delivered to the patient.

Once FDG has passed into the cell and has been phosphorylated, unlike glucose-6-phosphate, it can no longer be metabolized and remains in the cell until it decays, thus marking the cell. During the examination the kidneys are removing FDG, which has not passed into the cell. This results in areas of increased avidity in the kidneys and bladder (Figure 7.2).

Standard Uptake Value

The *standard uptake value* (SUV) is the ratio between measured uptake in a *region of interest* (ROI) and the expected uptake if the FDG had been evenly distributed throughout the patient.[2] It determines the sites of high activity, particularly of metastatic lesions. Infections and rheumatoid arthritis can cause false-positives, whereas slow-growing adenocarcinomas or malignant neoplasms

Oral and Maxillofacial Radiology: A Diagnostic Approach,
David MacDonald. © 2011 David MacDonald

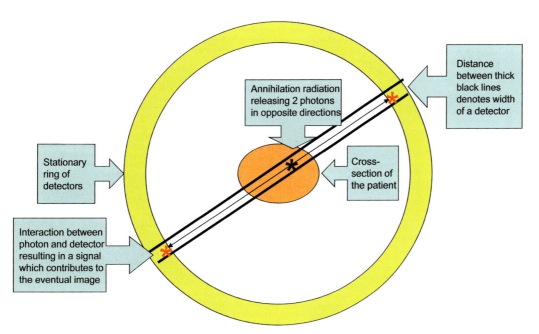

Figure 7.1. Coincidence. At the site of annihilation 2 photons of energy 511KeV are released and travel in opposite directions to interact with a detector at diametrically opposite sites on the stationary ring of detectors. These interactions result in signals that then contribute to the subsequent image.

that are smaller than 1 cm can produce false negatives.[2] Although FDG can also be used to distinguish between high-grade and low-grade lymphomas, the avidity of some low-grade lymphomas are so mild that a false-negative can ensue.[1]

The use of SUV by head and neck specialties at the *British Columbia Cancer Agency* (BCCA) has been largely restricted to *squamous cell carcinoma* (SCC). An SUV over 2.5 indicates that a lesion is more likely to be malignant. Nevertheless, there is a grey zone between frank malignant and frank benign lesions. Computed tomography plays a substantial role if further determination is needed as to whether the lesion is malignant. Lymph nodes that are greater than 1 cm wide, round in shape and enhancing with an SUV of 2.5 are considered to be malignant, whereas those that are no more than 0.8 cm, kidney-bean or oval shaped, and have a fatty hilum are considered to be benign.

Clinical Application

The main clinical application of PET in head and neck oncology is the diagnosis of SCC (Figure 7.3; see also Figures 18.32 and 18.33). A systematic review by Facey et al. revealed that FDG-PET improved the detection of occult head and neck tumors—a task all other modalities failed. They also observed that it improved the staging of clinically positive regional cervical lymph nodes. FDG-PET also improved the staging of lymphoma.[3] A systematic review by Isle et al. reported that PET was highly accurate in the detection of recurrent or persistent SCC of the head and neck. Nevertheless, it was less sensitive for recurrences within 10 weeks postoperatively.[4]

Although PET may also assist the clinician to differentiate between a residual neoplasm and any changes provoked by therapy, such as inflammation, periodontal and periapical pathology increase the uptake of FDG.[5] Therefore, care is required for interpretation of such hypermetabolic areas near suspected cancer.

Multimodality PET Imaging

This combines PET and *helical computed tomography* (HCT) technology. The advantage of this combined modality imaging is that the tissue density information provided by the HCT can now be used to calculate an attenuation correction for the PET

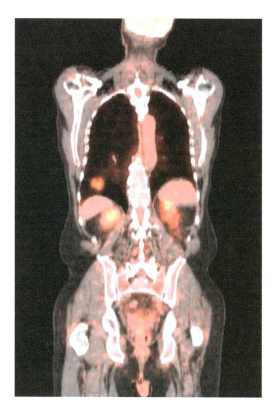

Figure 7.2. This image is derived from a combination of the modalities of positron emission tomography (PET) and helical computed tomography (HCT). This almost whole-body PET/CT image can display areas of increased metabolism. Such an area can be observed in one lung. Other areas are observed in both kidneys and the bladder suggesting that in addition to radioactive decay, which is necessary to produce the PET image, the radioisotope is being removed from the body in the usual way. This is particularly important after the examination has been completed so as to minimize radiation dose to the patient. Figure courtesy of Dr. Thomas Li, Hong Kong.

image.[6] Isles et al. reported that currently there were insufficient studies upon which to perform a systematic review.[4]

An outstanding issue is radiation dose, especially from the HCT.[6] Nevertheless, this combination of metabolic imaging with anatomical information has become an important modality for detecting hitherto unknown primary tumors and identifying distant metastasis.[7]

Currently, there is no single modality that can safely predict carcinoma invasion of the mandible. Nevertheless, this is likely to improve with development of better spatial resolution[8]; The discussion on PET and PET/CT is continued in Chapter 18.

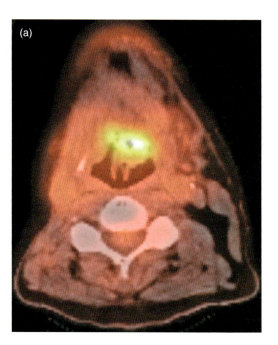

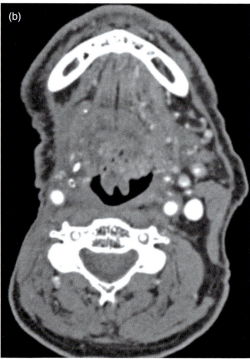

Figure 7.3. Axial fused modalities positron emission tomograph (PET) and helical computed tomography (HCT) (a) and a separate HCT examination (b) of a squamous cell carcinoma (SCC) affecting the base or dorsum of the tongue. The avidity of the SCC is clearly seen on the PET/CT image (a), whereas on the HCT with intravenous contrast (b), it is less obviously enhanced. **Note:** The sternocleidomastoid and the platysma muscles have been removed on one side. Figure courtesy of Dr. Montgomery Martin, British Columbia Cancer Agency.

References

1. Rohren EM, Turkington TG, Coleman RE. Clinical applications of PET in oncology. *Radiology* 2004;231:305–332.
2. Kuhlman JE, Perlman SB, Weigel T, Collins J, Yandow D 2nd, Broderick LS. PET scan-CT correlation: what the chest radiologist needs to know. *Curr Probl Diagn Radiol* 2004;33:171–188.
3. Facey K, Bradbury I, Laking G, Payne E. Overview of the clinical effectiveness of positron emission tomography imaging in selected cancers. *Health Technol Assess* 2007; 11:iii–iv, xi–267.
4. Isles MG, McConkey C, Mehanna HM. A systematic review and meta-analysis of the role of positron emission tomography in the follow up of head and neck squamous cell carcinoma following radiotherapy or chemoradiotherapy. *Clin Otolaryngol* 2008;33:210–222.
5. Shimamoto H, Tatsumi M, Kakimoto N, Hamada S, Shimosegawa E, Murakami S, Furukawa S, Hatazawa J. (18)F-FDG accumulation in the oral cavity is associated with periodontal disease and apical periodontitis: an initial demonstration on PET/CT. *Ann Nucl Med* 2008; 22:587–593.
6. Beyer T, Townsend DW. Putting "clear" into nuclear medicine: a decade of PET/CT development. *Eur J Nucl Med Mol Imaging* 2006;33:857–861.
7. Donta TS, Smoker WR. Head and neck cancer: carcinoma of unknown primary. *Top Magn Reson Imaging* 2007;18: 281–292.
8. Babin E, Desmonts C, Hamon M, Bénateau H, Hitier M. PET/CT for assessing mandibular invasion by intraoral squamous cell carcinomas. *Clin Otolaryngol* 2008;33: 47–51.

Chapter 8
Basics of ultrasound

Ultrasound or ultrasonography (US) has become an increasingly common applied modality in oral and maxillofacial radiology.[1] It is not used just in diagnosis but also in therapy. Therapeutically, diagnostic US can be used to guide microsurgery for calculi or strictures; therapeutic US can be used to break up larger calculi by lithotripsy.

Overview of US Technology and Terminology

US is a medical imaging technique that uses high frequency sound waves and their echoes. The range of diagnostic US lies between 1–20 MHz. A US scanner transducer converts electrical energy into sonic energy. The piezoelectric crystal, the most important component of the transducer, undergoes rapid changes in thickness in response to an electric current. Such changes induce the sound waves (ultrasound), which then enter the patient.

Each tissue has a different acoustic impedance, determining what proportion of the ultrasound energy is absorbed and how much is reflected back. It is this reflected energy which, upon reaching the transducer, carries the clinically important information. Upon reaching the transducer, it causes a change in the crystal's thickness. This information, after amplification and processing, is displayed. Modern equipment processes the reflected echoes with such rapidity that a perception of motion or real-time imaging can be appreciated. The acoustic impedance of a tissue changes with disease.

Oral and Maxillofacial Radiology: A Diagnostic Approach, David MacDonald. © 2011 David MacDonald

Clinical Applications of Ultrasound

US has been used to diagnose calculi and other pathology affecting the submandibular and parotid glands.[1-3] Figure 8.1 displays a dermoid cyst within the tongue and Figures 13.3c and 17.18c display mumps and Warthin's tumor respectively.

Calculi, if small enough, can be retrieved by a wire basket under US guidance. This can all be achieved under local anesthesia in the dental chair in an outpatient facility, The only surgery required is of the ostium to facilitate delivery of the basket containing the stone.[4]

Lithotripsy breaks the larger calculi by therapeutic ultrasound. *Extracorporeal shock wave lithotripsy* (ESWL) is a procedure used to shatter simple stones.[5] Pulses of ultrasonic waves are passed through the body until they strike the dense stones, which they pulverize so that they are more easily passed out of the gland and its duct by the salivary flow. It is generally now only used on parotid calculi, except in those cases of calculi within the submandibular gland. The lithotripter is placed on the overlying skin, with cotton-wool in the buccal sulcus to protect the teeth. Each session lasts 1 hour, with a week between sessions.

Other Applications Largely Used in Medicine

US-guided biopsy of the salivary glands is discussed in Chapter 13.

Doppler US has been used mostly to measure the rate of blood flow through the heart and major arteries. Figure 13.3d is an example of the use of Doppler US in the oral and maxillofacial region. The direction of blood flow is shown in different colors on the screen. Doppler US occurs when the object reflecting the US waves is moving, and it

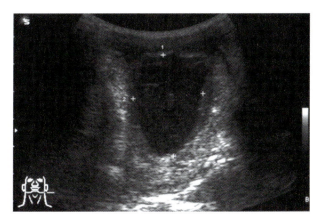

Figure 8.1. An ultrasound of a dermoid cyst. The magnetic resonance images of the case are displayed by Figure 6.8.

creates a higher frequency if it is moving toward the probe and a lower frequency if it is moving away from the probe. How much the frequency is changed depends upon how fast the object is moving.

References

1. Poul JH, Brown JE, Davies J. Retrospective study of the effectiveness of high-resolution ultrasound compared with sialography in the diagnosis of Sjogren's syndrome. *Dentomaxillofac Radiol* 2008;37:392–397.
2. Ching AS, Ahuja AT. High-resolution sonography of the submandibular space: anatomy and abnormalities. *AJR Am J Roentgenol* 2002;179:703–708.
3. Alyas F, Lewis K, Williams M, Moody AB, Wong KT, Ahuja AT, Howlett DC. Diseases of the submandibular gland as demonstrated using high resolution ultrasound. *Br J Radiol* 2005;78:362–369.
4. Iro H, Zenk J, Escudier MP, Nahlieli O, Capaccio P, Katz P. Brown J, McGurk M. Outcome of minimally invasive management of salivary calculi in 4,691 patients. *Laryngoscope* 2009;119:263–268.
5. Escudier MP, Brown JE, Drage NA, McGurk M. Extracorporeal shockwave lithotripsy in the management of salivary calculi. *Br J Surg* 2003;90:482–485.

Part 3
Radiological pathology of the jaws

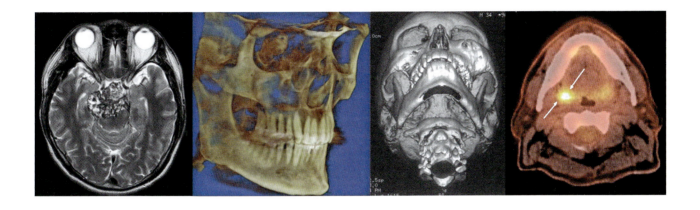

Chapter 9
Radiolucencies

Introduction

A radiolucency is the black or darker area on a conventional radiograph. It suggests an osteolytic process, particularly when it presents in bone. Most lesions associated with this process remain radiolucent, whereas some subsequently acquire a central opacity or opacities or eventually become completely radiopaque. The latter are more fully addressed in Chapter 10.

The flowcharts are based on the best evidence, either from recent systematic reviews or, where they are lacking, traditional narrative reviews or recently reported large case series. As seen, particularly in Figures 9.1–9.4, it generally flows from the most important clinical and radiological findings, addressing systemic lesions and malignancies first and then cysts and benign neoplasms. These flowcharts focus on the most common and important lesions and are not exhaustive with regard to the rarer lesions, particularly if they respond well to the initial treatment—i.e., do not recur.

Multiple radiolucencies, particularly if they are distributed throughout the jaws suggest a systemic cause, whereas the single radiolucency suggests a local cause.

The degree of marginal definition is crucially important to determine potentially serious disease. If it is well defined, the radiolucency is more likely to be benign; it is likely to be a benign neoplasm or a cyst. A poorly defined radiolucency on the other hand could represent a malignancy or infection.

Locularity is essentially a feature of radiolucencies; most are either unilocular or multilocular. It is important to differentiate between multilocular and scalloped radiolucencies because the latter is a variant of a unilocular radiolucency. Although unilocular radiolucencies are more likely to be odontogenic or simple bone cysts, multilocular radiolucencies almost always are odontogenic neoplasms. Nevertheless, the early, and therefore the dimensionally small, stage of some odontogenic neoplasms may present as a unilocular radiolucency. The multilocular radiolucency could be characterized according to three patterns. Although the soap-bubble and honeycomb would appear to be nonspecific, the tennis racket is virtually pathognomonic of the odontogenic myxoma.

The radiolucency's relationship to mandibular canal or the image of the hard palate (on panoramic or cephalometric radiographs) indicates whether it is likely to be of odontogenic origin. If the radiolucency is above the mandibular canal or below the image of the hard palate, it is within the dental alveolus and therefore likely to be of odontogenic origin. If the radiolucency is sited within the alveolus, its relationship to teeth is important to further refine the differential diagnosis. An association with the root of an erupted tooth, particularly if it has a large carious lesion or a large restoration, suggests the possibility of a necrotic pulp, and then the radiolucency is likely to have an inflammatory cause. If the radiolucency is associated with the crown of an unerupted tooth, a dentigerous cyst or an odontogenic neoplasm (assuming a secondary relationship to it) should be considered.

Lesions that commonly present as well-defined radiolucencies are cysts and neoplasms. Cysts are common and the majority are inflammatory. Almost all true cysts and most benign neoplasms expand by hydrostatic pressure and are therefore frequently spherical or nearly spherical in shape. This shape is achieved in larger cysts and neoplasms by displacing the buccal and lingual cortices and presenting as buccolingual expansion. Odontogenic lesions, arising within the alveolus, if sufficiently large will displace the mandibular canal downward. Additionally, all lesions can reduce the diameter of the canal or can completely erode its cortex (so that it is no longer visible) if they are enveloping it.

Oral and Maxillofacial Radiology: A Diagnostic Approach, David MacDonald. © 2011 David MacDonald

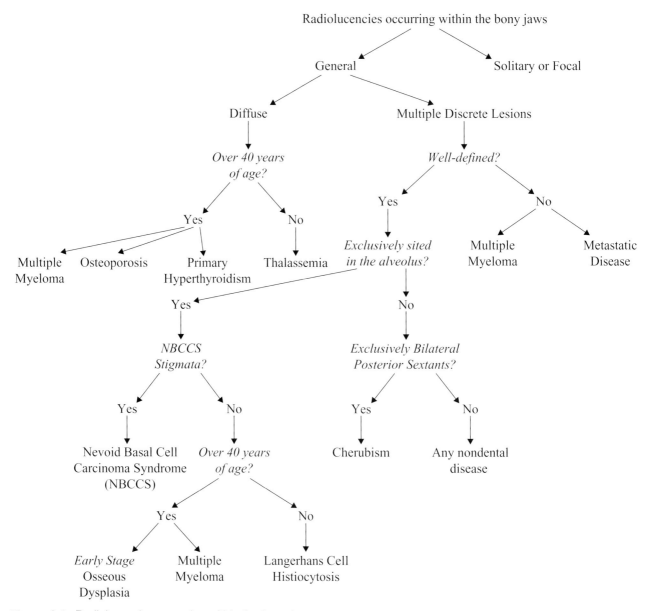

Figure 9.1. Radiolucencies occurring within the bony jaws.

The effect of the radiolucency on the adjacent teeth or anatomical structures is important. This effect is manifested by either displacement or erosion. The latter when applied to teeth, particularly their roots, is termed *root resorption*. Although all lesions presenting as radiolucencies may in due course cause root resorption, this would appear to be a particular feature of certain odontogenic neoplasms. Displacement of teeth and buccolingual cortices are universal to all expansile lesions.

Giant cell lesions and hemangiomas each have an extensive range of presentations. As a result they appear in the differential diagnosis of several lesions.

Some lesions that are generally understood to be radiopaque, may appear initially as radiolucencies in their earliest stage. This apparent inconsistency is analogous to the clearing of a building site and first excavating to establish the foundations of the new building to be erected.

In addition to *conventional radiography*, advanced imaging modalities such as *computed tomography* (CT) and *magnetic resonance* (MRI) are frequently used to investigate jaw lesions.

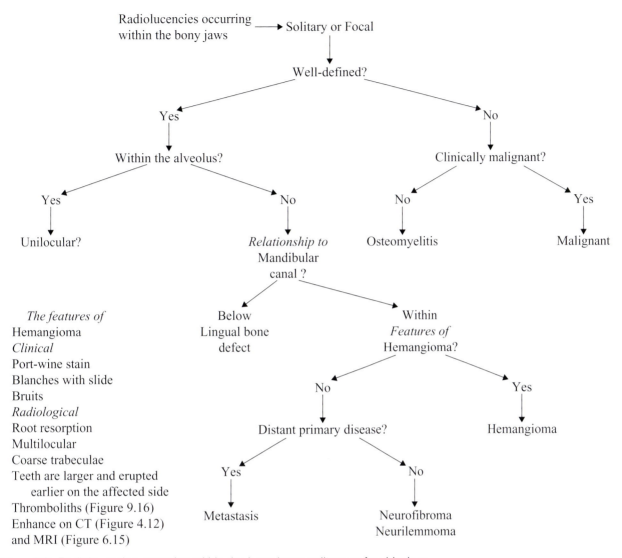

Figure 9.2. Radiolucencies occurring within the bony jaws; solitary or focal lesions.

Although osteolytic lesions and structures are still "black" on bone-window images made by either *helical computed tomography* (HCT, see Chapter 4) or *cone-beam computed tomography* (CBCT, see Chapter 5), they appear "white" on soft-tissue window HCT (see chapter 4) and MRI (see Chapter 6). Furthermore they may enhance with iodine-based or gadolinium intravenous contrast for HCT and MRI, respectively.

The term *significant* will be used only when the feature it is qualifying is $P < 0.05$.

Artifacts causing radiolucencies arise from three main sources, image development, normal anatomy and variants, and earlier treatment. Because these have already been addressed in other texts, they will not be considered further.

Multiple Radiolucencies

An early consideration in the review of the clinical and radiological findings is whether the patient is suffering from a generalized or systemic condition or whether it is of a local nature. Generalized disease can manifest itself in the context of radiolucencies by a generalized osteopenia—a reduction

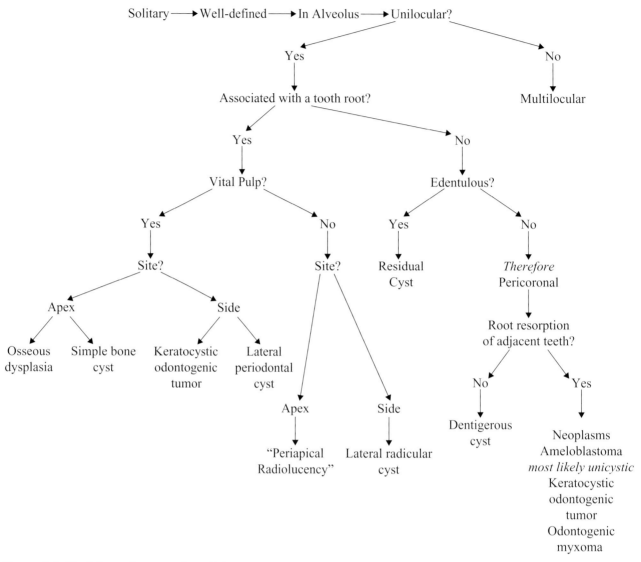

Figure 9.3. Radiolucencies occurring within the bony jaws; unilocular lesions

in trabeculae, both in their number and thickness, and thin cortices.

THALASSEMIA

In younger patients, generally of Meditereanean, Middle Eastern, and South Asian extraction, such a presentation may be indicative of thalassemia, the most common genetic disease. The classical features of thalassemia are hair-on-end appearance of the vault of the skull, obliteration of the air-sinuses and replacement of the normal trabecular pattern by fewer coarse straight trabeculae (Figure 9.5; see also Figures 11.6 and 17.22 for other images of the same patient).

The conventional radiological features of thalassemia affecting the jaws were reported in a Middle Eastern community by Hazza et al.[1,2] The maxillary antrum was obliterated in every case (see Figure 11.6). The teeth displayed "spiky" roots, an increased crown/root ratio, a reduction in the lamina dura and taurodontism. In addition the teeth display some delay in their development.

Thalassemia has not only a specific presentation upon diagnosis but also subsequently according to the mode of treatment. Hypertransfusions

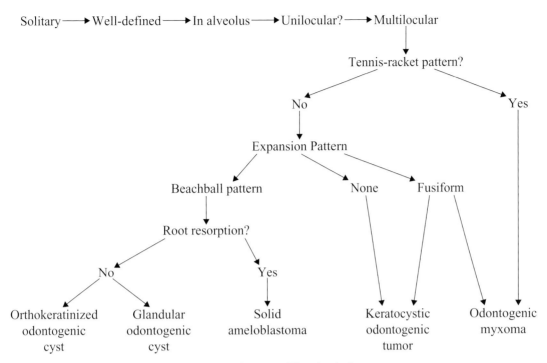

Figure 9.4. Radiolucencies occurring within the bony jaws; multilocular lesions.

and chelating agents also affect changes in both the skeleton and the extraskeletal organs,[3] which may be observed by radiology.

SICKLE CELL DISEASE

Although sickle cell disease is most frequently found among West Africans and their Afro-Caribbean and African-American descendants it is also found among Mediterranean and Middle Eastern communities. White et al.[4] reported that its detection by fourier analysis is more effective than counting struts. Fourier analysis revealed increased trabecular spacing in sickle cell disease.[5]

MULTIPLE MYELOMA

Multiple myeloma is a plasma cell malignancy. Its most frequent manifestation is that of widespread disseminated disease. The multiple radiolucencies affecting the vault of the skull classically confers a pepper-pot appearance (see Figure 18.1). Ninety percent of patients develop bone lesions, making it the most common cancer to affect bone.[6]

It accounted for 43% of all bone malignancies in one report's patient database.[7] Although the jaws are affected in 30% of cases, 16% of multiple myeloma first manifest in the jaws.[7] Generally whole-body scanning is achieved by *PET-CT* (see Chapter 7) using *18 fluorodeoxyglucose* (FDG) rather than technetium 99 m.[6] Sixty percent of patients develop pathological fractures.[6]

Observation of *solitary osseous plasmacytoma* (SOP) in a radiograph of the jaws may enhance the patient's prognosis by earlier local treatment prior to it becoming widespread disease. Seventy percent of cases, if untreated, progress to multiple myeloma. Thirty percent progress to multiple osteolytic (radiolucent) lesions without marginal sclerosis; in other words they present the classical "punched-out" radiolucent presentation of multiple myeloma going onto diffuse osteoporosis and finally diffuse osteosclerosis.[8] Witt et al. declared that neither diffuse osteoporosis nor diffuse osteosclerosis was observed in their case series.[8] Half of the SOPs affecting the jaws were observed on radiographs, they were radiolucent.[8] Pisaro et al. reported plasmacytoma of the oral cavity frequently appeared superimposed upon the roots of adjacent teeth and were accompanied by pain and a raised red mucosal lesion on the alveolar ridge. They need to be distinguished from a periapical radiolucency of inflammatory origin.[9]

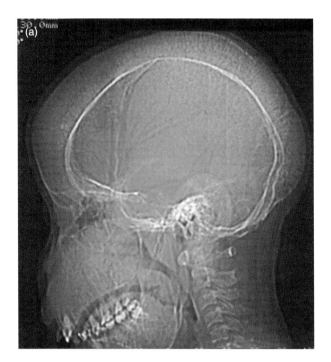

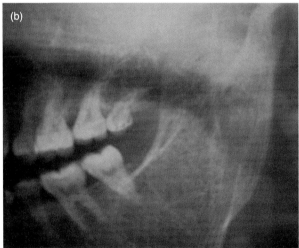

Figure 9.5. A lateral projection of the skull and a panoramic radiograph of thalassemia. (a) The lateral projection of the skull displaying typical radiological features of thalassemia; hair-on-end appearance of vault of skull, completely obturated air-sinuses (the frontal, ethmoidal sphenoidal, and maxillary). The mandible is also affected. (b) Panoramic radiograph exhibiting few, but coarse and linear, trabeculae within the enlarged mandible.

The literature emphasizes multiple myeloma's widespread manifestation of the skeleton. Nevertheless, it should be appreciated that this multisystem disease also affects soft tissues as made obvious by the MRI images of Figure 9.6a–c.

LANGERHANS CELL HISTIOCYTOSIS

Ninety percent of *Langerhans cell histiocystosis* (LCH), formerly called "histiocytosis X," is present in individuals younger than 40 years of age; the mean age is 19 years.[10] Dagenais et al.[11] were the first to identify and qualify the radiological features of LCH affecting the jaws. They reported that almost all cases affected the posterior sextant of the mandible and in a third of cases also affected the posterior sextant of the maxilla. They found that almost all cases presented as circular or elliptical radiolucencies (Figure 9.7). Although most were well defined, particularly those affecting the alveolus, a cortex or sclerosis was generally observed only in that part of the lesion close to the alveolar crest. Those in the alveolus were multiple, whereas those in the basal process were solitary. The majority of the alveolar lesions were sited about the apices. The lamina dura was absent. Root resorption though common was slight. Although periosteal new bone was observed on occlusal radiographs of lesions affecting the basal process, buccolingual expansion was not observed, except for involvement of the condyle and coronoid processes. Their report, largely consistent with most earlier literature, did not mention "floating teeth," perhaps because almost all their lesions were small, not one case presented with such a feature. Tooth displacement was observed in one-half of cases.[10]

OSTEOPOROSIS

Osteoporosis is a serious disease, but can be treated if detected early, thus avoiding a hip fracture, which, in the elderly, can be fatal.[12] Not only is osteoporosis most frequently observed in postmenopausal women, in which it is most severe, it also affects men.[13] A radiological diagnosis of osteoporosis may be readily determined from examination of the thickness of the lower border of the mandible on a panoramic radiograph. This has been simplified even more and can now be applied to the trabecular pattern on intraoral radiographs.[14] This can be achieved visually.[15] The

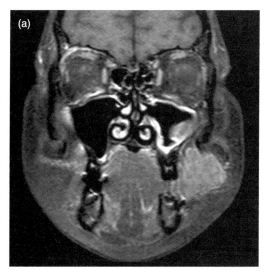
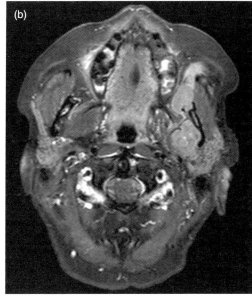
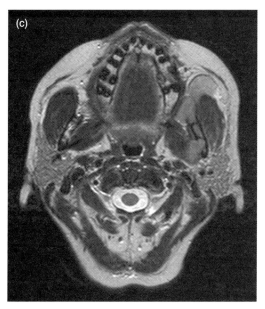

Figure 9.6. *Magnetic resonance imaging* (MRI) of a case of multiple myeloma. See Figure 18.1 for a lateral conventional radiograph of the skull displaying the salt-and-pepper pattern or pepper-pot pattern skull. (a) Coronal T2-weighted magnetic resonance imaging (MRI) displays the neoplasm as a mass within the buccal tissues. It has a heterogeneous hyperintensity within it. The mucosa of the ipsilateral maxillary antrum is very hyperintense, suggestive of sinusitis. The lateral exophytic mass is suggestive of the neoplastic invasion. There is also a hyperintense area within the bone marrow superolateral to the contralateral orbit, the cortex is absent, and the lesion is in continuity with a hyperintense lesion within the orbit. This is suggestive of multiple lesions. (b) Axial T2-weighted magnetic resonance imaging (MRI) displays the neoplasm as a mass within the buccal tissues. It is more hyperintense that the adjacent isointense masseter around whose anterior margin it expands into the deep tissues of the cheek displacing the parotid duct and the subcutaneous tissues outward. The posterior aspect of the lesion abuts the deep pole of the left parotid gland. The neck of the condyle appears substantially eroded. The ipsilateral lateral pterygoid muscle has been substantially infiltrated. (c) Coronal (non–fat-saturated) T1-weighted magnetic resonance imaging (MRI) displays the neoplasm as a mass within the buccal tissues. It appears more isointense in contrast to the more hyperintense fat tissues. Although it is clearly distinguished form the parotid gland the lingual cortex of the base of the neck of the condyle is absent suggesting its infiltration of the marrow space. The attachment of the ipsilateral lateral pterygoid to the neck of the condyle has been completely infiltrated. Figure courtesy of Dr. Montgomery Martin, British Columbia Cancer Agency.

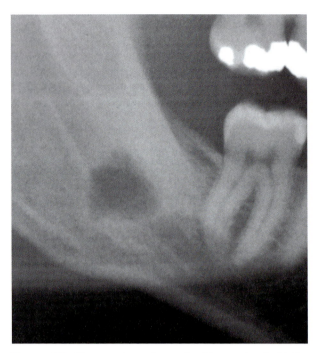

Figure 9.7. Panoramic radiograph of Langerhan's histiocytosis. There are two well-defined radiolucencies in the posterior mandible superimposed upon the mandibular canal. The more distal radiolucency is more translucent and the mandibular canal appears not only widened but does not exhibit the canal's superior or inferior cortex. Its margin is delimited by some sclerosis. The mesial radiolucency's mesial border has the contralateral lower border of the mandible superimposed upon it.

Osteodent research project has recently indicated that dentists now can have a role in the detection and referral of patients with osteoporosis.[12]

HYPERPARATHYROIDISM

The radiology of hyperparathyroidism affecting the jaws, in addition to generalized bone resorption, osteopenia, and osteosclerosis, may reveal localized lesions. These are "brown tumors" and are rarely reported. They may appear either as radiolucencies or leontiasis ossea (discussed further in Chapter 10). A brown tumor presenting as a radiolucency is displayed in Figure 10.9b.

There are three types of hyperparathyroidism, and they all have their effects by disrupting normal calcium homeostasis.[16,17] The most common type is primary hyperparathyroidism, which presents as a disease in middle- to old-aged patients, principally women. Its most common cause is a secreting adenoma arising in one of the 4 parathyroid glands. Occasionally, a secreting carcinoma can cause it.[18]

The secondary type frequently is secondary to *chronic renal insufficiency* (CRI)[16] and familial disease,[17] The third type may arise from neoplasia within the parathyroid glands. These glands' hyperactivity is induced by the above and other diseases and generally persists even after the disease that induced them has been treated. Although hemodialysis has greatly improved the longevity of patients suffering from CRI, its side effect *renal osteodystrophy* (ROD) still remains a real risk, particularly in those patients whose adherence to the recommended diet is poor. ROD manifests itself as calcified deposits. Although Asaumi et al.[19] indicated that these calcified deposits around the jaws can be adequately displayed by conventional radiography and that advanced imaging such as HCT and MRI add little, Chang et al. illustrated that extent and structure of ROD within the hard palate was best displayed by HCT.[16] Calcification within the arteries is discussed further in Chapter 10.

Localized Poorly Defined Radiolucencies

Localized poorly defined radiolucencies suggest both lesions of local origin and of an aggressive nature. Such disease could be either infections or malignant neoplasms. The primary differentiation between them, in the majority of cases, is made upon the clinical findings.

The majority of malignant neoplasms, particularly *squamous cell carcinomas (SCC)*, are poorly defined (Figure 9.8; see also Figure 1.31) and destroy rather than displace bony structures. On occasion, malignancies may present with well-defined margins (Figure 9.9). In such cases other features such as widening of the periodontal ligament space and "floating teeth" and the "spiking" pattern of root resorption (Figure 9.10) may direct the clinician toward consideration of a malignancy.

SCC is the most frequent malignancy of the oral cavity and oropharynx. The World Health Organization's Oral Health Program reveals that its *age-standardized rate* (ASR) is equal to or exceeds 6.9 cases per 100,000 world standard population for males in North America, most of Europe (except Northern Europe), the Indian subcontinent, Australia, Brazil, and Southern and Eastern Africa.[20] For females, the ASR is similar or greater than 6.9 cases per 100,000 in the Indian subcontinent,

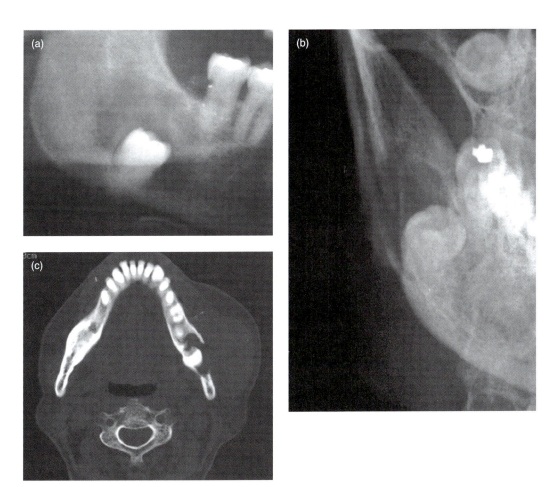

Figure 9.8. Conventional radiographs, computed tomography (CT), and magnetic resonance imaging (MRI) of a carcinoma arising from odontogenic lesion. (a) The panoramic radiograph displays an unerupted third molar with the apex of its completely formed root displaced through the lower border of the mandible. Although a normal follicle space is apparent on the mesial and distal aspects of the crown of this tooth, the occlusal part has been irregularly expanded and largely delineated by a band of sclerosis. Outside this band there is a large area of radiolucency that has a poorly defined periphery. (b) The posterioanterior projection of the mandible does not delineate the buccal cortex, which must now be presumed to have been perforated. The axial (c) and coronal (d) bone window computed tomography confirm the substantial loss of the buccal cortex. The neoplasm has now substantially expanded into the adjacent soft tissue. (e) Fragments of bone are observed at the anterior periphery of the lesion. The precontrast axial soft-tissue window (c) shows that the neoplasm is subjacent to the subcutaneous tissue of the facial skin. (f) The axial contrast section reveals enhancement at the periphery but also as chords extending toward the neoplasm's center indicating an extensive intralesional vasculature. The nonenhanced areas are likely to represent necrosis. (g) The coronal section contrast soft-tissue window displays substantial peripheral enhancement, but the center is substantially necrotic.

Ethiopia, some Southern African states, and New Guinea.[20] Although it primarily presents as a lesion on the oral mucosa, which will be primarily addressed in Chapter 18, some arise within the odontogenic epithelial remnants or "rests," within the jaw bones.

Although SCC far more frequently affects the mucosa rather than the bony jaws, this chapter will focus only on the latter, whereas the former is especially considered in the Chapter 18.

Primary Intraosseous Squamous Cell Carcinomas (ICD-O 9270/3)

Eversole et al. defined the *primary intraosseous squamous cell carcinoma* (PIOSCC) as "a central jaw carcinoma derived from odontogenic epithelial remnants. Subcategories of PIOSCC include (1) a solid tumor that invades marrow spaces and induces osseous resorption, (2) squamous cancer

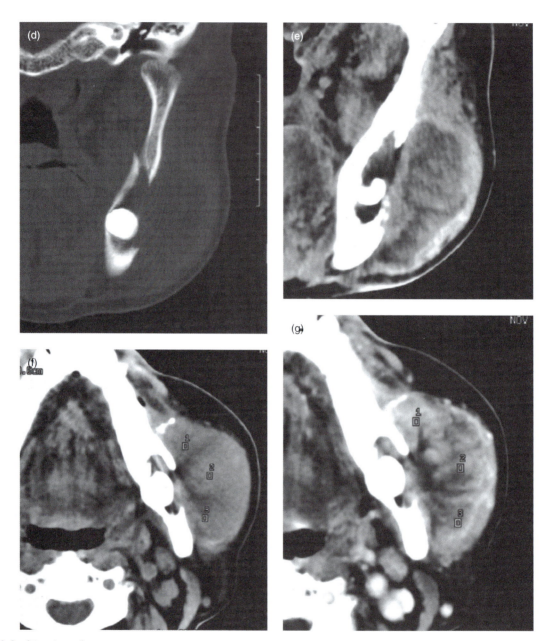

Figure 9.8. (Continued).

arising from the lining of an odontogenic cyst and (3) a squamous cell carcinoma in association with other benign epithelial odontogenic tumors. When the tumor destroys the cortex and merges with the surface mucosa, it may be difficult to distinguish between a PIOSCC and a true carcinoma arising from the oral mucosa. Invasion from an antral primary must also be excluded."[21]

Eversole et al. also report that PIOSCCs are found twice more frequently in males.[21] Although they generally first present with a mean age of 55 years, cases affecting infants have been reported. They are usually found in the posterior sextant of the mandible (Figures 9.8 and 9.11). If they affect the maxilla, they are most frequently observed in the anterior sextant. Although some produce swelling and/or mental paresthesia, the majority are symptom-free and found incidentally on radiographs. Radiographically, early lesions are often indistinguishable from odontogenic cysts

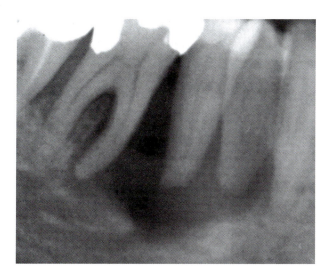

Figure 9.9. Panoramic radiograph of a well-differentiated carcinoma. Although its margin is largely well defined, the widening of the periodontal ligament space on the mesial root of the first molar and the "floating" premolars directs the clinician to consider a malignancy; there is also some resorption of the apices of the premolars. **Note:** This case should be assessed for paresthesia of the lip. The lesion is sited at the mental foramen.

(Figure 9.11). Cortical destruction and invasion of the adjacent soft tissue are features of late lesions (see Figure 9.8).[21]

A PIOSCC associated with an impacted mandibular third molar (see Figure 9.8) is associated with a more favorable prognosis. They are, nevertheless, infrequently reported.[21]

Huang et al.'s recent report on Chinese solid-type PIOSCCs reported them as high-grade malignancies which frequently metastasize to regional lymph nodes.[22] They also have a high recurrence rate (76% at 5 years posttreatment) and thus a high mortality.[22]

Radiolucency of Inflammatory Origin

Radiolucencies of inflammatory origin arise from a dental infection. This most frequently arises from a necrotic pulp secondary to dental caries or trauma. The radiolucencies that result are most frequently sited at the apex of the affected tooth and are unilocular. When observed on a radiograph, these are best simply termed *periapical radiolucencies*. On occasion such a lesion may arise from a lateral canal; it is called a *lateral radicular cyst*, which is entirely different from a lateral periodontal cyst (see later). Periodontal disease can also produce in radiolucencies. These are associated with pockets arising within bone, including loss of bone at the furcation. Radiolucencies associated with inflammation can also arise secondarily in a preexisting lesion such as a cyst or neoplasm. These infected cysts or neoplasms can appear both clinically and radiologically difficult to distinguish from a malignant neoplasm. Infected keratocystic odontogenic tumors affecting the maxillary antrum can present clinically and radiologically as antral malignancies (Figure 11.20).[23]

Periapical Radiolucencies of Inflammatory Origin

The periapical radiolucency is classically juxtapositioned to the apex of a tooth, which has a large carious lesion or large restoration or is fractured and/or is unresponsive to vitality testing, suggesting pulpal necrosis. Nevertheless, such a finding should prompt pulp-vitality testing of affected tooth/teeth because any lesion can present first as a periapical radiolucency, such as osseous dysplasia (Figure 9.12), the already-mentioned PIOSSC,[21,22] and SCC (see Figures 9.9 and 9.10). Other clinically significant lesions that have been reported recently to present as a periapical radiolucency have been the ameloblastomas (Figure 9.13)[24,25] and the giant cell lesion.[26]

The term *periapical radiolucency* is the most appropriate to use when an inflammatory process is suspected because it encompasses the granuloma, the cyst, and the abscess, which differ histologically. Many attempts have been made to correlate the various radiological features of the periapical radiolucency (including its size) with its resultant histopathology. Almost all have substantially failed, other than to report that the periapical radiolucency is more likely to be a cyst if it is large and it has a well-defined margin.

The second edition of the WHO's histological typing of odontogenic tumors describes the radicular cysts as "a cyst arising from the epithelial residues (rests of Malassez) on the periodontal ligament as a consequence of inflammation, usually following the death of the dental pulp."[27] Radicular cysts are the most common cyst found in the jaws. Although they may occur in relation

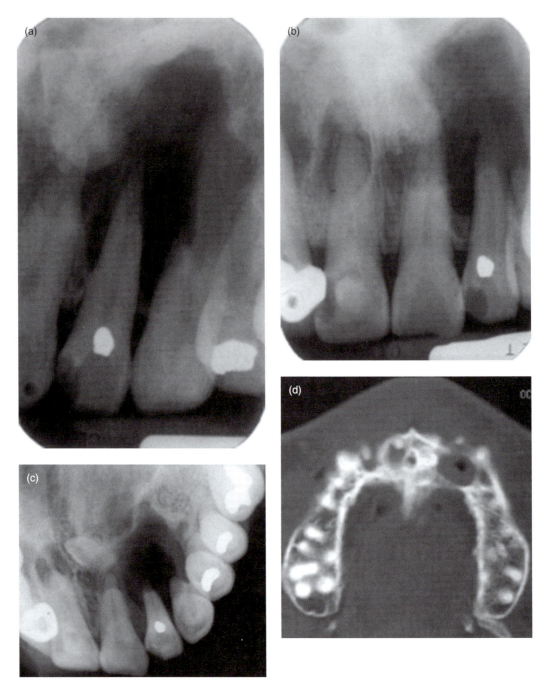

Figure 9.10. Conventional radiographs and computed tomography (CT) of a case of poorly differentiated mucoepidermoid carcinoma. (a) Periapical radiograph exhibiting irregular bone loss (radiolucency) around the left lateral incisor. It is "floating" and its root apex displays spiking pattern resorption. The canine's root has also be resorbed but largely upon its mesial surface. There is some spiking pattern root resorption on the ipsilateral central incisor. A mesiodens is also included within the radiolucency. (b) Periapical radiolucency in addition shows periapical radiolucencies on the contralateral restored incisors. (c) Anterior occlusal displays involvement of the floor of the nose. (d) Axial CT (bone window) reveals radiolucency displacing the left lateral incisor buccally. It has a soft-tissue radiodensity at the periphery and a air-or-gas-filled center.

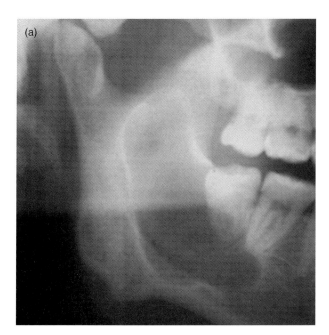
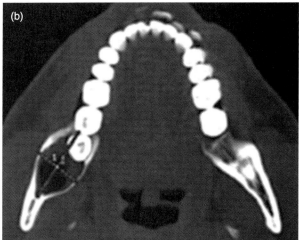
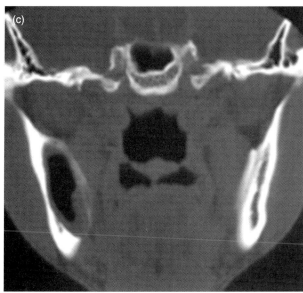

Figure 9.11. Panoramic radiograph and computed tomography (CT) of a primary intraosseous carcinoma. (a) Panoramic radiograph exhibits a well-defined corticated unilocular radiolucency occupying the anterior half of the vertical ramus from the coronoid process to the posterior body of the mandible. The last molar's roots have been displaced mesially. (b) Axial CT (bone window), at the level of the occlusal plane, displaying buccolingual expansion. The content of the lesion is soft tissue at the periphery and air or gas at the center. (c) Coronal CT (bone window), at the level of the condyle, reveals the same features observed in (b).

to any nonvital tooth, they are infrequently associated with deciduous teeth. They are most frequently associated with teeth of the maxillary anterior sextant. Their peak age at first presentation is the third and fourth decades. Males are more frequently affected.[27]

Gundappa et al. compared the images of periapical lesions of *ultrasound* (US) with analogue (film) and digital (charge-couple device) dental radiography.[28] The two dental radiographic technologies did not allow differentiation between cysts and granulomas, whereas US did in each of

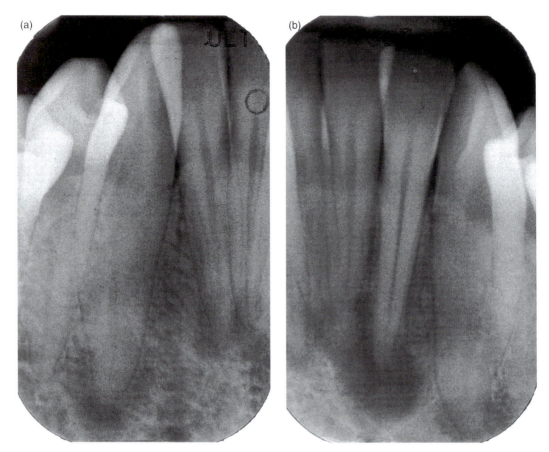

Figure 9.12. The periapical radiographs display periapical radiolucencies associated with noncarious and pristine lower incisors. These are most likely to represent early-stage osseous dysplastic lesions. A positive response to pulp vitality testing is sufficient to confirm their diagnosis.

the 15 cases. They concluded that while dental radiography determines the existence of the periapical disease and its extent, US can determine whether the lesion is or is not cystic if sufficient buccal cortex has been resorbed. The underestimate of the lesion's size by US was attributed to the acoustic shadow cast by the bony edges of the lesion on its lateral walls.[28]

Paradental Cyst

The paradental cyst also called the *buccal bifurcation cyst* (BBC) was defined by the second edition of the WHO histological typing of odontological tumors as "a cyst occurring near to the cervical margin of the lateral aspect of a root as a consequence of an inflammatory process in a periodontal pocket" … "the histological features are … the same as those of the radicular cyst."[29] It arises from the odontogenic epithelium of the periodontal ligament of a vital tooth. Its distinctive feature is its association with the buccal or distal aspects of an erupted mandibular molar tooth. This tooth is most frequently the third molar (Figure 9.14). Similar cysts can arise from the buccal aspect of the first molar tooth in 6- to 8-year-old children.[29]

Pompura et al. reported the largest case series of 44 cases in 31 patients presenting in a children's hospital over 3 years.[30] They all presented with tenderness and discomfort about the time of eruption of the first molar. The eruption of the lingual cusps occurs first, Some buccal swelling may be present. The true occlusal radiograph best displays most of the radiological characteristics of the BBC and is recommended projection to complete the investigation of the BBC. Figure 3 in Pompura et al. is a panoramic radiograph that displays a BBC that looks identical to a dentigerous cyst.[30]

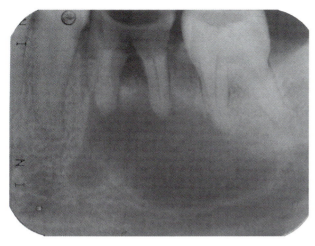

Figure 9.13. A periapical radiolucency of a multilocular ameloblastoma. It presents aa a radiolucency, which has a well-defined inferior margin, but a poorly defined superior margin subjacent to a grossly carious first molar with a large furcation radiolucency. The two clues that this lesion is not primarily of inflammatory origin are its multilocular appearance and the fact that its epicenter is not at or close to the origin of inflammation.

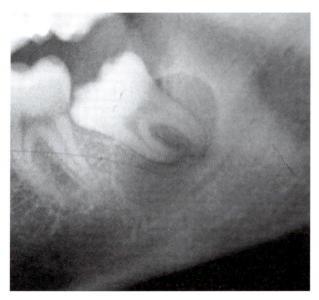

Figure 9.14. Panoramic radiograph of buccal bifurcation cyst. A well-defined partially corticated radiolucency is superimposed upon the root of an unerupted third molar. If has displaced the mandibular canal downward in addition to reducing its diameter.

The rarity of reports other that the preceding indicates the infrequency of this lesion. Iatrou et al. found it in only 9% of all intraosseous lesions of the jaws of children.[31]

Lateral Periodontal Cyst

The lateral periodontal cyst was defined by the second edition of the WHO's classification of odontogenic neoplasms, as "a cyst occurring on the lateral aspect or between the roots of vital teeth and arising from odontogenic epithelial remnants, but not as a result of inflammatory stimuli."[32] According to Shear and Speight, the term *lateral periodontal cyst* should be "confined to those cysts that occur in the lateral periodontal position (the side of the tooth between the apex and cervical margin of the root) and in which inflammatory etiology and diagnosis of collateral OKC (now the keratocystic odontogenic tumor—my parentheses) have been eradicated on clinical and histopathological grounds."[33] In other words a biopsy is required for a definitive diagnosis. The reason for this necessarily rather long and vague definition is that most recent reports on this lesion have been case reports and have not led to a consensus on the pathogenesis of the lesion. Nevertheless, a feasible pathogenesis, suggested by Shear and Speight, is the lateral periodontal cyst's histopathology of a reduced enamel epithelium lining, representing a lateral(-positioned) dentigerous cyst, which was displaced apically as the tooth erupted. The most classical image is Shear and Speight's Figure 6.9.[33] Two other potential origins they discuss are the clear cell rests of the dental lamina and cells of Malassez.[33]

The lateral periodontal cyst accounts for 24 cases referred to Shear's South African pathology service, at a mean annual rate of 0.7 cyst per year.[33] Their average age of 6 reports[34-39] is 47 (37–55) years. Almost all lateral periodontal cysts first present in the fifth to seventh decades, peaking in the sixth. Overall there appears to be equality between the sexes. Rasmussen reports that females present younger than the males.[37] Although the sites vary between the reports, there is a general predilection for anterior sextants.

The clinical presentation ranges from symptomless and discovery incidental to a radiological investigation for another clinical indication, to pain and swelling, which may be fluctuant.[33]

According to Altini and Shear,[34] on radiographs the lateral periodontal cysts present as round or ovoid well-defined radiolucencies with a "sclerotic margin." They are generally smaller than 1 cm. These features differ from those of "botryoid cysts," which are larger, multilocular and extend apical to the periapical area. They do not display root resorption.[34]

All lateral periodontal cysts occur anterior to the molars; Formoso Senanda et al. report 3 mandibular lateral periodontal cysts were sited in the premolar region.[39] Figure 1.26 displays an image that is classical for a lateral periodontal cyst, except that this is of a *keratocystic odontogenic tumor* (KCOT), reinforcing the need to include the latter in the differential diagnosis.

The mandibular premolars are the most frequent site, followed by the maxillary anterior sextant. Furthermore, Formoso Senande et al. noted that all their 8 maxillary lateral periodontal cysts were found between the lateral incisor and the canine.[39] This presentation recalls the globulomaxillary cyst, which, while it is no longer considered to be a true lesion, represents in most cases either a lateral periodontal cyst or a KCOT.

Although the treatment is simple enucleation without sacrifice of the associated tooth, if possible, Shear and Speight advise that those cases that prove on histopathological examination to be of the encapsulated multicystic variety, should be followed up for a number of years because its behavior is not yet completely certain.[33]

Botryoid Odontogenic Cyst

This is microscopically similar to the lateral periodontal cyst, but additionally it is multicystic with thin fibrous connective tissue septa.[40] The name *botryoid* is derived from its bunch-of-grapes–like presentation reflecting its purported multilocular radiological presentation.[41]

Méndez et al. synthesized the literature.[41] Their synthesis revealed that 85% affected the mandible, mainly the premolar-canine region, whereas the maxillary cysts affected the anterior sextant.[41] These sites are identical to those of the lateral periodontal cyst. Two-thirds presented with symptoms; therefore a third would have been detected incidentally. The median age at first presentation was 54 years,[41] older than that for lateral periodontal cysts.

Like the lateral periodontal cyst its margins are well defined.

Although multilocular lesions, hitherto considered an essential characteristic of the botryoid cyst, accounted for only 40% of all botryoid cysts, they were significantly more frequently observed among the 12 cases that recurred. Eleven cases that recurred were multilocular, whereas only 5 of the 21 cases that did not recur were multilocular.

The slight predilection for females was not reflected among the recurrences where there was no gender bias. It was noted that those subjects who had recurrence had a median age of 48 in contrast to a median age of 55 years for those who did not. A third recurred after a median of 8 years.[41]

Perhaps potential markers for a lesion that is more likely to recur are multilocular radiolucencies of 31 mm in contrast to 10 mm first presenting in early middle age (circa 40 years old).[41]

In summary, although this cyst has the broad clinical and radiological appearance of a lateral periodontal cyst and may be considered to be a botryoid variant of the lateral periodontal cyst, it is generally multilocular, presents with symptoms, and tends to recur.[41]

Hemangioma

Waner and Suen substantially reclassified congenital vascular lesions of the head and neck into hemangiomas and vascular malformations. The former are "usually not present at birth, proliferate during the first year and then involute," whereas the latter are "always present at birth, never proliferate and never involute."[42]

Zlotogorski et al.'s synthesis of 86 cases of hemangiomas affecting the jaws revealed predilections for females (56%), the mandible (77%), and the posterior sextants of both jaws (64% for the mandible and 82% for the maxilla). The mean age was 23.3 (0–74) years.[43]

Of the 32% of hemangiomas displaying well-defined borders, none were sited in the maxilla (Figure 9.15). Ninety-six percent of the lesions were radiolucent. Thirty-seven percent of the mandibular lesions and 25% of the maxillary lesions were unilocular. Of the 38 multilocular lesions affecting the mandible, 47%, 34%, and 8% were honeycomb, soap-bubble, and spokelike, respectively. Of the 12 multilocular lesions affecting the maxilla, 42%, 42%, and 8% were honeycomb,

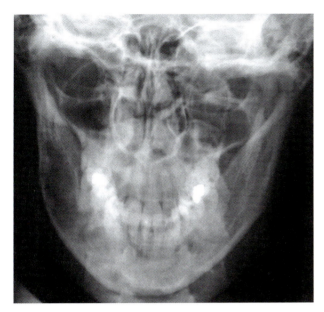

Figure 9.15. Anterioposterior projection of the mandible displaying a hemangioma affecting the buccal half of the posterior body of the mandible. It is well defined with minimal buccolingual displacement. The only hint as to its presence on the accompanying panoramic radiograph was that the trabeculae, but not the teeth, appeared out of focus.

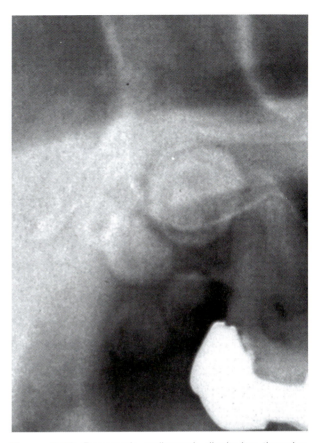

Figure 9.16. Panoramic radiograph displaying thromboliths within a hemangioma. The radiopacities are well defined. The larger one in the center exhibits the alternating concentric radiopaque and radiolucent rings typical of the target pattern, which is almost pathognomic of the thrombolith.

soap-bubble, and spokelike, respectively. Root resorption was reported for 20 cases, 17 in the mandible. Tooth displacement was reported in 14 cases, 11 in the mandible. Of the 41 cases that affected the posterior sextant of the mandible, 10 exhibited involvement of the mandibular canal. This usually presented as an enlargement of the canal's width.[43]

Further clinical and radiological features that may indicate the presence of a hemangioma are port-wine stain on the skin, which may blanch under pressure from a glass microscopic slide. Ascultation may reveal bruits. Teeth may appear larger and have erupted earlier on the affected side. Thromboliths (Figure 9.16) may be observed in hemangiomas that affect soft tissue. The hemangioma enhances on HCT (Figure 4.12) and on MRI (Figure 6.15).

Pericoronal Radiolucencies

Pericoronal radiolucencies, particularly of mandibular third molars, occur frequently. They are particularly important because they not only suggest cystic change within the follicle of the unerupted tooth but may represent perhaps a neoplastic change within that follicle or a secondary envelopment of it by a neoplasm. This neoplastic change is most frequently benign, but it can be malignant (see Figure 9.8). The mandibular third molar site is also the most frequent site for occurrence of many odontogenic neoplasms; some recur if misdiagnosed and inappropriately treated.

Radiolucencies associated with the crown of a unerupted tooth are usually suggestive of dentigerous cysts, but not always. The review by Curran et al. of the over 2,600 pericoronal surgical specimens that had been referred to their pathological service over 6 years revealed that only 67% were not pathological, they were simply normal dental follicles.[44] Twenty-eight percent were dentigerous cysts of which 35 included mucous cells. Of the remaining 5%, 71 were "keratocysts" (see KCOT later), 13 each were ameloblastomas and carcinomas, 6 were calcifying cystic odontogenic tumors, 4 were calcifying epithelial odontogenic tumors, and one was an odontogenic myxoma. The reader should note that the authors considered that only a fraction of all specimens arising from unerupted teeth within their community had actually been referred to them. Furthermore, the specimens sent were considered by the referring surgeon to be potentially pathological. Not only did most come from older patients but also, on subsequent histopathology, those were found most likely to be pathological.

An important feature, which distinguishes dentigerous cysts and follicle spaces from more serious lesions, is the relationship of the cyst to the *cementoenamel junction* (CEJ) (Figure 9.17).

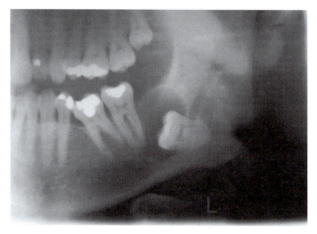

Figure 9.17. This is a panoramic radiograph displaying a classical "central" pattern *dentigerous cyst* (DC) on the left. The coronal radiolucency is attached at the cementoenamel junction at the mesial and distal aspects. The symmetrical enclosure of the crown within the radiolucency is typical of the "central" pattern of radiological presentation of a DC. **Note 1:** The DC has essentially two parts: the cystic cavity itself and the tooth it is attached to. Therefore the DC can affect adjacent structures directly (due to direct contact between the cyst and them) or indirectly (by the tooth it is attached to and them). This DC is affecting the mandibular canal both directly (reducing the mandibular canal's diameter mesially to the unerupted third molar) and indirectly (reduction in the mandibular canal's diameter and its downward displacement by the tooth).

Ikeshima et al. noted that the dentigerous cyst attachment was closer to the CEJ than that of the ameloblastoma.[45] Other important features that tend toward the likelihood of a dentigerous cyst are unilocular radiolucency and an absence of root resorption.

Dentigerous Cyst

The dentigerous cyst, as defined by the second edition of the WHO, is a "cyst which encloses the crown and is attached to the neck of an unerupted tooth. It develops by accumulation of fluid between the reduced enamel epithelium and the crown, or between the layers of the reduced enamel epithelium."[46]

The relationship of the radiographic periphery of the radiolucency surrounding the crown to the "neck" or CEJ is crucially important to a diagnosis of a dentigerous cyst radiologically. Based upon a study by Ikeshima et al.,[45] I conclude that an attachment that is less that 1mm apical to the CEJ is strongly suggestive that the lesion could be a dentigerous cyst.

The global distribution of reports included in the systematic review[47] upon which much of the following is derived is set out in Figure 1.41 and their details in Table 9.1.

The prevalence of the dentigerous cyst could be high at least in one Turkish community. Yildirim et al. reported that pathological changes among symptom-free third molars was 23%, of which two-thirds were dentigerous cysts, the rest were calcifying cystic odontogenic tumors and "odontogenic keratocysts."[48]

The dentigerous cyst is most commonly associated with a mandibular third molar (48%) (see Figure 9.17), a maxillary canine (17%) (Figure 9.19), a mandibular premolar (10%) and a maxillary third molar (4%).[49] It is also associated with supernumeraries; 90% of such cysts are associated with mesiodens (Figures 9.18, 11.26). Kaugars et al. reported that 28% of odontomas had dentigerous cysts.[50]

Dentigerous cysts affect males in 61% of cases of all ethnic groups (Table 9.1). Shear and Speight report it more frequently in South African Whites than in South African Blacks.[49] The mandible is affected in nearly two-thirds of cases (64%). Ninety-two percent are found in the posterior sextant of the mandible. This feature was sig-

Table 9.1. Dentigerous cyst: systematic review

Feature	Total	Western	East Asian	subSaharan	LatinAmer
Male:Female	61%:3%	63%:37%	60%:40%	62%:38%	60%:40%
Mean number per year per report	INA	INA	INA	INA	INA
Mean age	31 years	38 years	INA	18 years	INA
Mean prior awareness	INA	INA	INA	INA	INA
Mand:Max	64%:36%	66%:34%	59%:41%	66%:34%	INA
Mand:Ant:Post	8%:92%	10%:90%	5%:95%	0%:100%*	INA
Max:Ant:Post	59%:41%	59%:41%	62%:38%	0%:100%*	INA
Swelling: Y:N	48%:52%	48%:52%	INA	INA	INA
Pain: Y:N	32%:68%	32%:68%	INA	INA	INA
Incidental: Y:N	24%:76%	24%:76%	INA	INA	INA
Discharge: Y:N	15%:85%	15%:85%	INA	INA	INA
Radiolucent	INA	INA	INA	INA	INA
Uni:Multiloc	INA	INA	INA	INA	INA
Welldefined: Y:N	INA	INA	INA	INA	INA
Cortex: Y:N	INA	INA	INA	INA	INA
Expansion: Y:N	INA	INA	INA	INA	INA
Antrum: Y:N	INA	INA	INA	INA	INA
ToothDispl: Y:N	INA	INA	INA	INA	INA
RootResorp: Y:N	INA	INA	INA	INA	INA

*Advises that the percentages were derived from either one report or from a synthesis of no more that 50 cases. Ant:Post, Anterior:Posterior; INA, Information not available; LatinAmer, Latin American; Mand:Max, Mandible:Maxilla; subSaharan, sub-Saharan African; ToothDispl, Tooth displacement; ToothResorp, Tooth resorption; Western, predominantly Caucasian. Y:N, Yes:No.

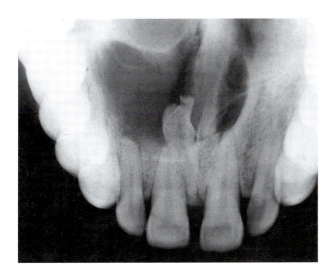

Figure 9.18. This standard anterior occlusal radiograph displays a well-defined radiolucency associated with the crown of an inverted mesiodens as it arises from the cementoenamel junction. This is a "central" pattern *dentigerous cyst* (DC). This DC is also associated with root resorption.

nificantly more likely in the East Asian global group than in the Western. In the maxilla, it is found in the anterior sextant in 59% of cases. It is found as an incidental finding in 24% of cases and with a swelling or pain in 49% and 32%, respectively.[49]

Although the dentigerous cyst is a reasonably common lesion (18% of jaw cysts in a recent report)[51] and has been reasonably frequently reported (see Figure 1.41) its radiology as revealed in a recent systematic review has not yet been subject to a single detailed reported series of consecutive (nonselected) cases affecting a community.[47] The only systematic review-included reports that feature such details are Ioannidou et al.,[52] who reported all dentigerous cysts as well defined, and Ledesma-Montes et al.,[53] who reported them as radiolucent. Clearly these observations do not take us beyond what we already know. Nevertheless, it is generally recognized, in addition to their intimate relationship to CEJ, that all dentigerous cysts present on radiographs as unilocular radiolucencies. A multilocular appearance should direct the

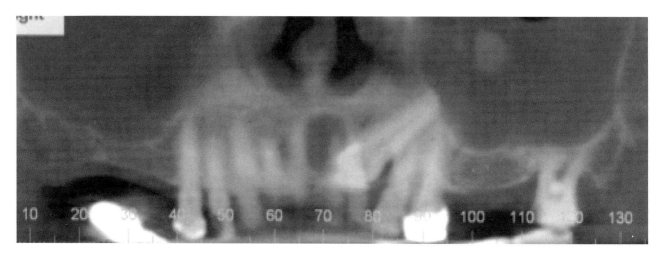
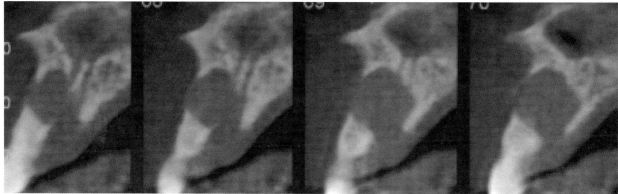

Figure 9.19. This cone-beam computed tomograph (CBCT) of a "lateral" pattern dentigerous cyst (DC) affecting a maxillary canine. The tangential sections display not only buccolingual expansion but also perforation of the labial (buccal) and palatal cortical plates. **Note:** The radiopacity within the left maxillary antrum is most likely to be an antrolith or an exotosis.

clinician straightaway to consider other lesions; these lesions (see Figure 9.4) are locally invasive and are more challenging to treat because they have a tendency to recur. The dentigerous cyst is simply enucleated with the attached tooth, with no recurrence.

The dentigerous cyst presents as one of three radiologic patterns.[49] In the central or "classical" pattern the cyst is attached to both the mesial and distal CEJ and symmetrically envelops the crown of the unerupted tooth (see Figure 9.17). As it expands it can displace the tooth in an apical direction toward the lower border of the mandible or as far as the floor of the orbit in the maxilla. The second pattern, the "lateral" pattern arises from the side of the crown (Figures 9.19 and 9.20). This pattern is frequently associated with teeth, which are mesially or vertically impacted. The third pattern, the "circumferential," is an exaggerated version of the central pattern; the cyst cavity expands down past the level of the CEJ. Its attachment to the root is still at the CEJ. The cyst is separated from the rest of the root by a bony sleeve containing the lamina dura. Frequently this pattern is seen when the tooth cannot be further displaced apically (see Figure 11.26). This pattern of dentigerous cyst needs to be distinguished from the KCOT or ameloblastoma. The last two generally do not appear to be attached at the CEJ.

Trying to distinguish between a dentigerous cyst and a follicle space can be difficult, particularly when the suspected dentigerous cyst is small, about 3 to 4 mm wide. Daley and Wysocki, following their study of this problem, concluded that certainty can be achieved only at the time of surgery.[54]

Struthers and Shear reported that a high percentage of dentigerous cysts in their report were

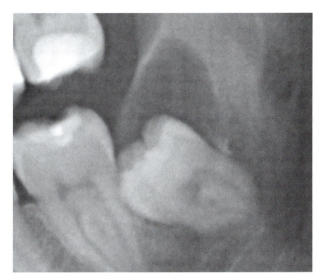

Figure 9.20. The panoramic radiograph displays a dentigerous cyst (DC) associated with the left unerupted third molar tooth. This DC is obviously attached to the distal cementoenamel junction and covers the occlusal surface. The mesial surface of the tooth is in contact with the distal root of the adjacent second molar tooth. This pattern of radiological appearance is "lateral." The root formation is complete. The apices have reduced the diameter of the mandibular canal.

associated with root resorption.[55] This phenomenon does not appear to have been reported elsewhere in the literature. Root resorption of multiple adjacent teeth was observed in an exceptionally large dentigerous cyst (see Figure 11.27).

The dentigerous cyst is unlike most other odontogenic lesions. The dentigerous cyst has both a direct and indirect effect on adjacent structures by virtue that the associated unerupted tooth is an integral part of it. In addition to its direct effect on adjacent teeth, cortices, and mandibular canal, it can exert the same indirectly by its displacement of the attached tooth. Thus, in addition to a complete evaluation of the cyst itself, attention should be paid to the associated tooth and its effects on related structures to ensure minimal injury to them upon its extraction along with the enucleated cyst.

The treatment of choice for the dentigerous cyst is enucleation. Although it does not recur, 15 out of 19 unicystic ameloblastomas were clinically diagnosed as dentigerous cysts.[56] Therefore, the surgical specimen derived from enucleation of such a dentigerous cyst should be submitted for histopathology. Wang et al. also reported that three partial surgical specimens, one derived from a marsupialization and two biopsies, displayed no histopathology suggestive of the ameloblastoma that was subsequently definitively diagnosed after histopathological examination of the entire surgical specimen.[56]

Shear and Speight address a common misapprehension that dentigerous cysts can readily undergo transformation to ameloblastomas which was the definitive diagnosis once the whole lesion had been removed and histopathologically reviewed.[49] Their argument challenging this view is that dentigerous cysts in Shear and Singh's South African study were far more common in Caucasian patients than in Black patients, in which they were uncommon, whereas the ameloblastoma was far more common in the latter, but infrequent in the former.[57]

Ameloblastoma (ICD-O 9310/0)

The WHO 2005 edition defined the ameloblastoma as "a slowly growing, locally invasive epithelial odontogenic tumor of the jaws with a high rate of recurrence if not removed adequately, but with virtually no tendency to metastasize."[58] Although this definition was specifically applied to the intraosseous-sited solid or multilocular ameloblastoma (see Figures 1.18 and 1.30, 11.21), it equally applies to the other three variants. These variants are unicystic (see Figures 1.6, 1.16, 1.34–1.37), desmoplastic, and peripheral ameloblastomas. "The unicystic variant is an ameloblastoma which presents as a cyst. The peripheral variant is an extraosseous counterpart of the intraosseous solid/multilocular ameloblastoma. The last is the desmoplastic ameloblastoma, which exhibits pronounced desmoplasia."[58]

In a case series, 20% of the solid variant was radiologically unilocular.[59,60] As a result, "solid" rather than "multilocular" will be solely used to refer to this variant.

The global distribution of reports included in the systematic review[59,60] upon which much of the following is derived is set out in Figure 1.38 and their details in Table 9.2.

The proportion of the variants varies with the community reported. Some East Asian communities report more unicystic variants,[59,60] whereas the

Table 9.2. Ameloblastoma: systematic review

Feature	Total	Western	East Asian	subSaharan	LatinAmer
Male:Female	56%:44%	56%:44%	56%:44%	60%:40%	46%:54%
Mean number per year per report	4.8	2.3	8.5	8.5	0.8
Mean age#	35.4 years	40.9 years	33.1 years	32.3 years	38.5 years
Mean prior awareness#	2.5 years	1.6 years*	2.7 years	1.8 years*	INA
Mand:Max	91%:9%	85%:15%	92%:8%	93%:7%	93%:7%
Mand:Ant:Post	16%:84%	13%:87%	12%:88%	17%:83%	48%:52%
Max:Ant:Post	18%:82%	8%:92%	24%:76%*	0%:100%*	10%:90%*
Swelling: Y:N	68%:32%	52%:48%*	76%:24%	58%:42%*	INA
Pain: Y:N	18%:82%	44%:56%*	15%:85%*	25%:75%*	18%:82%*
Numb: Y:N	12%:88%	INA	7%:93%*	8%:92%*	INA
Radiolucent	97%:%	86%:14%	99%:1%	98%:2%	100%:0%*
Uni:Multiloc	29%:71%	18%:82%	38%:62%	11%:89%	44%:56%*
Welldefined: Y:N	96%:4%	INA	100%:0%	INA	79%:21%*
Cortex: Y:N	78%:22%	100%:0%*	77%:23%*	INA	INA
Expansion: Y:N	91%:9%	66%:34%	95%:5%	100%:0%*	INA
Antrum: Y:N	75%:25%*	INA	75%:25%*	INA	INA
ToothDispl: Y:N	73%:27%*	INA	73%:27%*	INA	INA
RootResorp: Y:N	59%:41%	56%:44%*	64%:36%	INA	INA
Recurrence: Y:N	14%:86%	16%:84%	15%:85%	9%:91%	4%:96%*

#Advises that this is not the cumulative mean calculated for the published report.
*Advises that the percentages were derived from either one report or from a synthesis of no more that 50 cases. Ant:Post, Anterior:Posterior; INA, Information not available; LatinAmer, Latin American; Mand:Max, Mandible:Maxilla; subSaharan, sub-Saharan African; ToothDispl, Tooth displacement; ToothResorp, Tooth resorption; Uni:Multiloc, Unilocular:Multilocular; Western, predominantly Caucasian; Y:N, Yes:No.

desmoplastic variant may be more frequent in North American and European communities.[61]

The mean number of ameloblastomas per year is globally 4.8. It is significantly greater for both the East Asian and sub-Saharan African global groups, 8.5 ameloblastomas per year for each, than it is for either the Western (2.3) and Latin American (0.8) global groups.

This significantly higher "incidence" is also accompanied by a significantly younger age on first presentation for the East Asian and sub-Saharan African global groups, both about 33 years old. These differences may in part reflect the proportions of the solid, desmoplastic, and unicystic variants. The desmoplastic variant generally has an older mean age. After a synthesis of the literature, Philipsen et al. reported the mean age of the desmoplastic variant was 42 years in contrast to the solid variants' 36 years,[61] whereas, the mean age of the unicystic variant first presents at a significantly younger age.[59,60]

There were significant differences in mean age between the unicystic and solid variants in an East Asian report (Hong Kong Chinese).[59,60] The mean age of the former was 24.6 years of age, whereas the mean age of the latter was 39.0 years of age. The Mainland Chinese report of Luo et al. mirrored the Hong Kong Chinese results closely, except that the proportion of unicystic cases was 31% rather than the latter's 59%.[62] Arotiba et -al.'s subjects were all under 20 years of age. Only 15% were unicystic, whereas the rest were of the solid variant.[63] This clashed with the two Chinese reports, which found the unicystic cases to be significantly younger than the solid variant cases. In Arotiba et al.'s younger case series, if the Chinese results were applied, a markedly larger proportion would have been expected to be unicystic.

The age at first presentation was earliest for Bangladeshis and the Koreans and oldest for the Chinese, among which the earliest, a mean of only 30 years old, was for the Hong Kong Chinese.[59]

This early age of first presentation may have been affected by the very short period of prior awareness of the lesion before presentation in this community. This 0.7 years compared to 4 years overall for the systematic review. This period was even shorter for the younger patients in this Chinese community.[59]

Although overall ameloblastomas are almost equally distributed between sexes, Western, East Asian, and sub-Saharan African reports displayed a predilection for males whereas the Latin Americans displayed a predilection for females, which was significant when compared to the sub-Saharan African global group's greater predilection for males.[59,60]

Overall, in the systematic review,[59,60] 68% present with swelling, 18% with pain, 12% with paresthesia or numbness, 6% with discharge and/or fistulae, and 6% with ulceration. Swelling is significantly more frequent on first presentation in the East Asian global group than in the Western, and pain is more frequent in the Western global group than in the East Asian global group.

The mandible is affected in 91% and the maxilla in 9% of cases. The ameloblastomas display a predilection of the posterior sextants rather than the anterior sextants for both jaws. Although overall these were 84% and 82% for the mandible and maxilla, respectively, the Latin American global group displayed a significant equal distribution between mandibular sextants. The Western global group exhibited a significantly greater predilection for the posterior maxillary sextant than the East Asian global group. Those affecting the posterior sextants, particularly of the mandible in one East Asian report, presented below the age of 25 years and were not only unilocular but unicystic ameloblastomas.[59,60] Luo et al.'s recent report confirms this phenomenon.[62]

Generally the ameloblastoma appears as a well-defined radiolucency (96%) on conventional radiographs[60]; the only exception is those cases of the desmoplastic variant, which appeared as poorly defined.[61] In addition many cases of desmoplastic variant exhibit a mixed radiolucent-radiopaque appearance.[64] This feature, in conjunction with a poorly defined border, "suggested fibro-osseous lesions."[64] Unlike most other ameloblastomas the desmoplastic variant displayed a greater proclivity for the maxilla than the solid variant; the former affects both jaws equally, whereas the latter has an overwhelming predilection for the mandible.[61]

Although the overwhelming majority of ameloblastomas are radiolucent, the Western global group has a significantly higher proportion presenting as radiopacities than those of the East Asian and sub-Saharan African global groups. This may in part reflect the higher proportion of the desmoplastic variant that is purported to be more prevalent in the Western global group.[61]

The East Asian global group displays a significantly higher proportion of unilocular, rather than multilocular, radiolucencies than either the Western or sub-Saharan African global groups. This may represent the higher prevalence of unicystic variant, which presents almost exclusively as a unilocular radiolucency, reported in communities in the East Asian global group.[60,62]

The systematic review reveals that 91% of ameloblastomas are associated with buccolingual expansion (Figure 1.36), this is significantly less in the Western global group than in the East Asian and sub-Saharan African global groups.

Almost all ameloblastomas presenting in the East Asian global group are radiolucent.[60] Ninety-six percent are well defined. Twenty-nine percent of the radiolucencies are unilocular (see Figures 1.6, 1.16, 1.34–1.37) and 71% are multilocular (Figures 9.21 and 9.22a) (see also Figures 1.18 and 1.30). A downward displacement of the lower border of the mandible (see Figure 1.6) and an involvement of the maxillary antrum (see Figure 11.21) occurred in 42% and 75%, respectively, in an East Asian report.[60] Tooth displacement occurred in 73% and root resorption in 59%.[60]

A hitherto previously not observed feature of the ameloblastoma is the displacement of a lateral cortex down past the undisplaced lower border of the mandible (see Figure 1.37). The only other lesion in which this feature has also been observed is the orthokeratinized odontogenic cyst (see Figures 9.34 and 9.35, later in this chapter).

Root resorption, tooth displacement, and unerupted teeth were significantly more associated with the unicystic variant than the other variants in this East Asian (Hong Kong Chinese) report.[60] This means that root resorption affecting ameloblastomas in this community is more likely to be observed in those lesions that first present on or before 25 years old.

Another report did not observe statistical differences between multilocular and unilocular

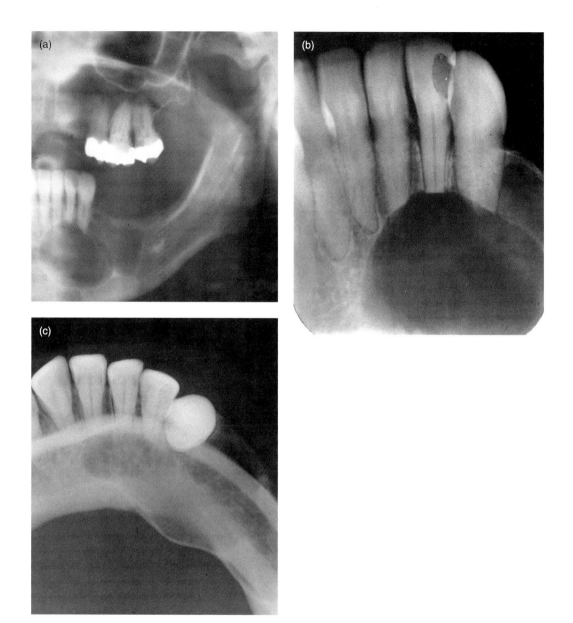

Figure 9.21. The panoramic radiograph (a) exhibits a solid (multilocular) ameloblastoma affecting the anterior teeth and premolar–first molar region. The periapical radiograph (b) displays marked root resorption. It also displays a very thin cortex. The occlusal radiograph (c) exhibits substantial buccolingual expansion, which in conjunction with (a) imparts a beachball-shaped appreciation of the solid ameloblastoma in three dimensions.

radiolucencies with regard to root resorption and unerupted teeth.[63] The differences between these reports may reflect the different ethnic origin, East Asian[60] and sub-Saharan African,[63] and the fact that the latter was derived from selected cases first presenting within a narrow age range (first 2 decades of life).

The effect of accurate diagnosis is necessary for appropriate treatment that minimizes both recurrence and morbidity. This accuracy is not merely confined to determining whether the lesion is an ameloblastoma or not but also to the correct identification of the particular variant of ameloblastoma. The majority of cases are generally of the solid variant, which needs to be resected with a margin. The unicystic variant is generally conservatively treated by enucleation and a cytotoxic agent such as Carnoy's solution.[65] Although this treatment is associated with a 11–16% recurrence rate, which is higher than the 3–4.5 % associated

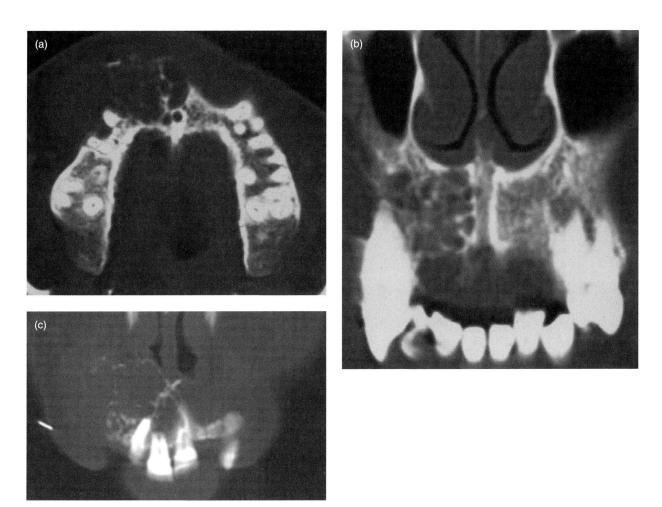

Figure 9.22. Computed tomography (CT) of a solid (multilocular) ameloblastoma affecting the anterior sextant of the maxilla. (a) Axial CT (bone window) displays a multilocular radiolucency affecting the anterior maxilla. It exhibits substantial labial expansion. (b) Coronal CT (bone window) displays a honeycomb appearance extending from the nasopalatine canal to the canine. It has eroded the mesial aspect of the cortex of the floor of the nasal cavity. (c) Coronal CT (bone window) anterior to (b) displays a substantial expansion into the soft tissue of the nose and has expanded into the anterior nares. A multilocular pattern is evident throughout. The pattern is honeycomb distal to the lateral incisor.

with resection, it is much less than the 30% associated with enucleation alone.[66,67] Nevertheless, long-term follow-up of this variant as with the others is recommended, regardless of the mode of treatment.[68]

Conventional radiography reveals that the majority are multilocular cases.[60] This generally ensures that the solid ameloblastoma appears high on the differential diagnosis, particularly if it also displays expansion. The same is not true for the unicystic ameloblastoma, because most present as unilocular radiolucencies, and it may be difficult for many clinicians to distinguish them from dentigerous cysts, which respond to enucleation without recurrence. Much of the lining of a unicystic ameloblastoma may be nonneoplastic and similar to that of a dentigerous cyst, thereby making preoperative biopsy a hit-or-miss affair.[56] Furthermore, Zhang et al.'s histopathological observation of a unicystic ameloblastoma concurrent with a dentigerous cyst supports the notion that occasionally a dentigerous cyst, certainly in this Canadian community may evolve into more serious disease.[69] This is contrary to Speight and Shear's findings within a South African community.[49]

Gardner et al. identified three variants of the unicystic ameloblastoma that are obvious to the histopathologist.[58] These are luminal, intraluminal, and mural (see their Figure 6.22). The cyst lining of the *luminal* variant is composed of ameloblastic

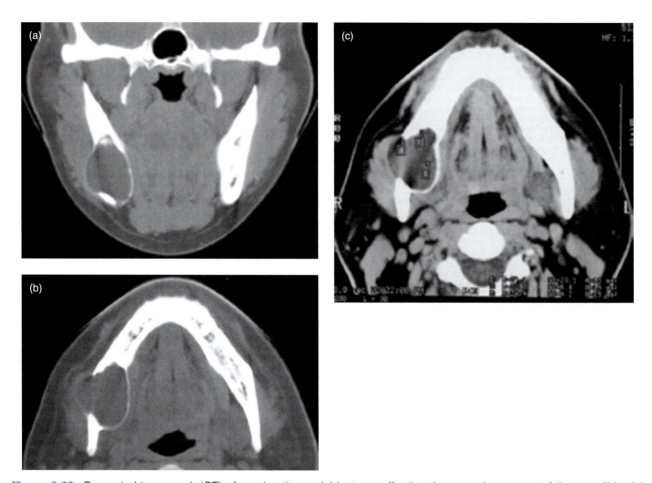

Figure 9.23. Computed tomograph (CT) of a unicystic ameloblastoma affecting the posterior sextant of the mandible. (a) Coronal CT (bone window) displaying buccal and lingual expansion. (b) Axial CT (bone window) displaying buccal and lingual expansion. (c) Axial CT (soft-tissue window) displaying buccal and lingual expansion. The CT number is 22 to 24 Hounsfield units.

cyst epithelium. The *intraluminal* variant displays an ameloblastic mass protruding into the lumen. The *mural* variant exhibits an ameloblastic mass invading the adjacent fibrous tissue wall. Although Li et al. reported that the simple cystic forms of the unicystic ameloblastoma (presumably the luminal variant) are less likely to recur than the other forms,[70] Lee et al. reported that 93% of unicystic ameloblastomas displayed mural invasion of the fibrous tissue wall (the mural variant).[65] This preponderance of the mural variant may explain the still relatively high recurrence rate even after the use of an adjuvant (such as Carnoy's solution) with the enucleation.

There are several subtypes of the unicystic ameloblastoma that are not readily distinguishable radiologically.[60] Although advanced imaging can assist further as a preoperative (including biopsy) investigation, there is a general dearth of such literature for odontogenic neoplasms, particularly with regard to HCT. This may very well indicate that surgeons may consider conventional radiography to be adequate for diagnosis and treatment planning for the majority of ameloblastomas.

Of the 61 consecutive cases of ameloblastoma, only 32 patients were referred for HCT.[71] The unicystic variant (Figure 9.23) was significantly less likely to be referred in comparison to those of the nonunicystic variant (solid variant, Figure 9.24, and desmoplastic variant). This lack of referral for HCT may be due to the fact that cases of unicystic variant are relatively easy to diagnosis on the basis of conventional radiography, they can be conservatively treated (enucleated and Carnoy's solution) with a subsequently good prognosis, and finally that they affect young patients, who are at a higher

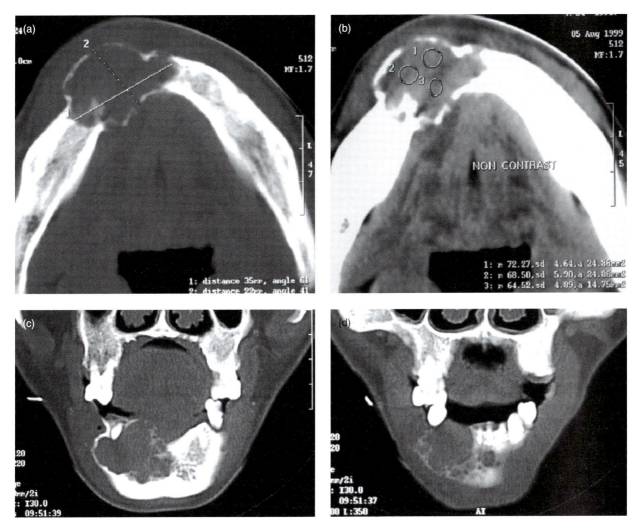

Figure 9.24. Computed tomography (CT) of a solid (multilocular) ameloblastoma affecting the anterior sextant of the mandible. (a) Axial CT (bone window) displaying buccal expansion with some perforation of the cortex. This lesion, unlike Figure 9.23, is more oval in shape. (b) Axial CT (soft-tissue window) displaying a higher CT number than Figure 9.23. It is 64 to 72 Hounsfield units. (c,d) Coronal CT (bone window) displaying a multilocular pattern, which ranges from large soap-bubble to honeycomb.

risk of radiation-induced disease. The other most significant feature of those patients referred was that they were older males. It is possible that the desire to avoid increasing the radiation dose in general to young patients and more particularly to females of reproductive age inhibited the referral of the younger patients for HCT.[71]

Nevertheless, HCT of certain cases can reveal and allow better assessment of soft-tissue involvement following perforation Figure 9.25). Furthermore, it allows a fuller evaluation of the extent of ameloblastoma affecting the anterior maxilla (see Figure 9.21). Although the multilocular pattern does not vary greatly between solid ameloblastomas, in the few cases reported it appears that those affecting the anterior maxilla present with the honeycomb pattern.

Asaumi reported that MRI can distinguish between the ameloblastoma and the odontogenic myxoma on the basis of their dynamic behaviors.[72] The ameloblastoma enhances rapidly, within 45 to 60 seconds, whereas the odontogenic myxoma enhances after 500–600 seconds. Asaumi suggested that MRI may better assist, because of its better spatial resolution and multiplanar features.[73] The mural nodule is easier to identify with MRI. Furthermore, dynamic contrast can reflect the intratumor angiogenesis. The essential cystic

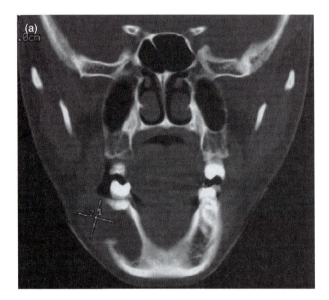

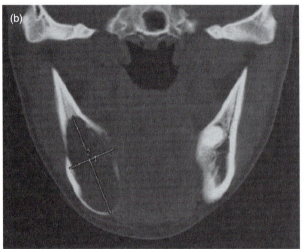

Figure 9.25. These coronal computed tomographs (bone window) display an ameloblastoma affecting the mandible (a), which has resorbed the tooth root and expanded the adjacent buccal cortex and obturated the buccal sulcus (measurements are 17 × 18 mm). (b) It has expanded the body of the mandible buccolingually and vertically (measurements are 48 × 23 mm). The lower border has been displaced downward.

nature of the unicystic ameloblastoma can be distinguished from the solid variant by a homogeneous hyperintense signal on T2-weighting. The follicular histopathology of the solid ameloblastoma can be predicted by the multiple cystic spaces on T2-weighting.

The "contrast index" is computed by subtracting the precontrast plotting from the postcontrast plotting.[73] Although the MRI allows the potential to determine differences in the contrast index for different lesions, unfortunately no differences have been observed between primary and recurrent ameloblastoma or between the ameloblastoma[73] and the glandular odontogenic cyst. The last displays similar clinical and radiological features and postoperative behavior.[74]

The desmoplastic variant, despite its poorly defined margins that underscores both its lack of a capsule and its infiltrative microscopic character and frequent perforation of cortices, appears to have a reduced tendency to recur. In spite of this, Philipsen at al. advise that treatment of this variant should "follow the same radical treatment modality ... as that of the 'classical' ameloblastoma."[61]

The systematic review's overall recurrence rate was 14%.[59,60] This recurrence rate will be affected by the period and quality of follow-up and the proportion of the unicystic variant, which is less likely to be resected. The Hong Kong case series exhibits these. After 5 years follow-up 12% recurred, of which 86% were conservatively treated unicystic ameloblastomas.

Although Hong et al. suggested that preoperative biopsy had no significant effect on recurrence, the particular histopathological pattern could be important; the follicular pattern had the most recurrences.[66] One-half recurred after conservative treatment.

In addition to recurrence, the clinician is concerned with pulmonary metastasis, of otherwise benign-appearing ameloblastoma cells, because, if untreated, it can lead to death. This clinical phenomenon was called a "malignant ameloblastoma." The WHO's 2005 edition has renamed it the *metastasizing ameloblastoma*.[75] Its ICD-O code is 9310/3. Unlike squamous cell carcinoma, these metastasized ameloblastoma cells are bloodborne (hematogeneous spread). Their emboli generally impact in the pulmonary capillaries.[76] Although such metastasis is generally associated with multiple operations,[77] Hong reported a case occurring 8 years postoperatively.[66] Although they have metastasized, these cells display the same benign histology as the local ameloblastoma at the primary site. This is the main point that distinguishes the malignant ameloblastoma from ameloblastic carcinoma (ICD-O 9270/3).[75] The latter is a rare primary malignant neoplasm. It may arise *de novo*, as a primary lesion, or arise secondarily by dedifferentiation of a preceding benign odontogenic neoplasm, such as an ameloblastoma. Microscopically it displays the features of the ameloblastoma, but with atypia. It, like other malignant neoplasms, can metastasize. Five-year survival is 67%.[78] So far 60 cases have been reported, many from China.[75]

Odontogenic Myxoma (ICD-O 9320/0)

The odontogenic myxoma is defined by the WHO's 2005 edition as "an intraosseous neoplasm characterized by stellate and spindle-shaped cells embedded in an abundant myxoid or mucoid extracellular matrix. When a relatively greater amount of collagen is evident, the term myxofibroma may be used."[79] The odontogenic myxoma is the fourth most common odontogenic neoplasm,[79] after odontomas, ameloblastomas, and KCOTs.

The global distribution of reports included in a recent update of the sole systematic review on odontogenic myxomas published in 2002[80] is set out in Figure 1.39 and their details in Table 9.3. This update not only addressed the many hiatuses, particularly with regard to sub-Saharan African and Latin American global groups, revealed by the original systematic review,[80] but it also revealed significant differences between the global groups, which were not apparent in the original systematic review.[80]

The mean number of odontogenic myxomas per year is globally 0.9 (Table 9.3). Although not

Table 9.3. Odontogenic myxoma: systematic review, updated (March 2010)

Feature	Total	Western	East Asian	subSaharan	LatinAmer
Male:Female	40%:60%	40%:60%	49%:51%	36%:64%	36%:64%
Mean number per year per report	0.9	0.8	1.0	1.3	0.9
Mean age#	30.5 years	31.0 years	29.0 years	27.8 years	28.3 years
Mean prior awareness#	1.9 years	1.8 years	2.2 years	INA	1.8 years*
Mand:Max	56%:44%	55%:45%	54%:46%	63%:37%	53%:47%
Mand:Ant:Post	19%:81%	23%:77%	5%:95%	28%:72%*	27%:73%
Max:Ant:Post	25%:75%	45%:55%	4%:96%	32%:68%*	14%:86%*
Swelling: Y:N	56%:44%	68%:32%	49%:51%*	80%:20%*	58%:42%
Pain: Y:N	28%:72%	14%:86%	72%:28%*	14%:86%*	30%:70%
Incidental: Y:N	14%:86%	12%:88%*	6%:94%*	INA	19%:81%
Loose teeth: Y:N	19%:80%	20%:80%*	20%:80%*	INA	15%:85%
Displ teeth: Y:N	28%:72%	34%:66%	50%:50%*	19%:81%*	INA
Numb: Y:N	8%:92%	7%:93%*	10%:90%*	8%:92%	INA
Radiolucent	83%:17%	98%:2%	100%:0%*	71%:29%	97%:3%
Uni:Multiloc	32%:68%	43%:55%*	27%:73%	23%:77%	40%:60%*
Welldefined: Y:N	51%:49%	53%:47%	50%:50%	39%:61%*	85%:15%*
Cortex: Y:N	45%:55%*	38%:62%*	100%:0%*	64%:36%*	INA
Expansion: Y:N	84%:16%	97%:3%*	80%:20%*	95%:5%*	70%:30%*
LBMd: Y:N	30%:70%*	0%:100%*	50%:50%*	INA	INA
Antrum: Y:N	91%:9%*	90%:10%*	86%:14%*	100%:0%*	INA
ToothDispl: Y:N	47%:53%	39%:61%*	80%:20%*	INA	57%:43%*
RootResorp: Y:N	27%:73%	18%:82%	35%:65%*	52%:48%*	0%:100%*
Recurrence: Y:N	18%:82%	20%:80%	0%:100%*	13%:87%*	29%:71%*

#Advises that this is not the cumulative means calculated for the published report.
*Advises that the percentages were derived from either one report or from a synthesis of no more that 50 cases. Ant:Post, Anterior:Posterior; Displ teeth, symptom of displaced teeth; INA, Information not available; LatinAmer, Latin American; LBMd, downward expansion of the lower border of the mandible; Mand:Max, Mandible:Maxilla; subSaharan, sub-Saharan African; ToothDispl, Tooth displacement; ToothResorp, Tooth resorption; Uni:Multiloc, Unilocular:Multilocular; Western, predominantly Caucasian; Y:N, Yes:No.

significant, it is greater for the sub-Saharan African global group than for the Western and Latin American global groups (Table 9.3).

The mean age for first presentation is about 30 years for all global groups (Table 9.3). The patients may first become aware of their disease a mean of 1.9 years before presenting.[80]

Although the odontogenic myxoma globally displays a predilection for females (60%), the East Asian global group significantly approximates to equality (Table 9.3).

Overall swellings present in 56% of cases (Table 9.3). Swellings present significantly less in the East Asian global group. Those cases first presenting with pain present only in 28% overall, whereas 70% of East Asian cases significantly report pain (Table 9.3). About 8% of cases in all global groups present with numbness. Twenty-eight percent present with displaced teeth and 19% with loose teeth. Only 14% are discovered as incidental findings.

The mandible is most frequently affected (56%) in all global groups (Table 9.3). Although 81% of the mandibular cases and 75% of maxillary cases overall affect the posterior sextants, this predilection was significantly greatest in the East Asian global group for both jaws (Table 9.3).

Odontogenic myxomas on conventional radiography in the original systematic review generally appeared as radiolucencies (98%).[80] The update of this systematic review, displayed in Table 9.3, reveals that a higher proportion of lesions were described as "radiopacities."[81,82] The Western and Latin American global groups significantly display a higher proportion of radiolucencies than the East Asian and sub-Saharan African global groups (Table 9.3). Two reports[81,82] reveal that the radiopaque pattern ranges from ground glass to a mixed appearance. These together with a poorly defined margin appear like fibrous dysplasia.[81] Furthermore, 3 of Zhang et al.'s cases had a "moth-eaten appearance," which could not be distinguished from a malignant lesion.[81]

The poorly defined margins (51%) and the general lack of a cortex or sclerosis when well defined (these present in only 51% of all cases with well-defined margins; Table 9.3) are consistent with the gelatinous nonencapsulated histopathology and the high recidivist nature of this lesion.[80]

Odontogenic myxomas display buccolingual expansion in 84% of cases (Table 9.3). Buccolingual expansion is significantly more frequently observed in Western and sub-Saharan African global groups than in East Asian and Latin American global groups (Table 9.3).

Odontogenic myxomas, if large enough, do expand the lesion in all directions. This expansion out with the maxillary antrum (Figure 11.28) is not the balloonlike expansion frequently observed in ameloblastomas or most odontogenic cysts. Instead, the pattern of expansion of the odontogenic myxoma is very similar to that observed for fibrous dysplasia; it is almost fusiform (Figure 9.26) (see also Figure 1.19).[80,83] This shape reflects the infiltrative nature of this lesion, which incidentally, like fibrous dysplasia, "shows little encapsulation."[84] This lack of a capsule reflects the degree of definition of the lesion-normal adjacent bone interface. As already mentioned, not only is this margin poorly defined in half of all cases, but also half of the well-defined margins have neither a cortex or are sclerotic. The radiologically apparent "normal" trabeculae immediately adjacent to the lesion are encased in tumor (see Buchner and Odell's Figure 6.64[79]), because "some odontogenic myxomas permeate into the marrow spaces in a pseudomalignant pattern."[79] Therefore, in order to ensure that all the neoplastic cells have been removed, the adjacent "normal-appearing" bone should also be removed with the lesion. In other words the odontogenic myxoma should be routinely resected.

In addition to buccolingual expansion, odontogenic myxomas displace the lower border of the mandible downward in nearly a quarter of the few reported cases that reach the lower border. Table 9.3 reveals that nearly every lesion subjacent to the maxillary antrum involves it. Zhang et al. reported that such antral lesions displayed a "reticular appearance (see their Figure 4) at the anterior part and a unilocular lesion in the posterior part on a panoramic radiograph."[81] This reticular (netlike) pattern could be observed in Figure 9.27), whereas Figure 9.28 displayed, in part, a honeycomb pattern.

In the original systematic review,[80] odontogenic myxomas were almost evenly divided between unilocular and multilocular radiolucencies on their presentation on conventional radiography. On the contrary, the update of this systematic review includes a larger proportion of multilocular lesions (Table 9.3); this is now 68% in comparison to the former's 53%.[80] This dramatic shift in presentation within such as short space of time may

Chapter 9: Radiolucencies **123**

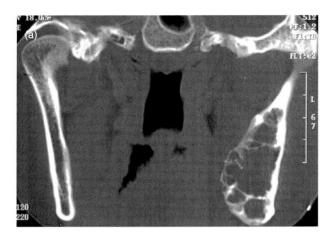
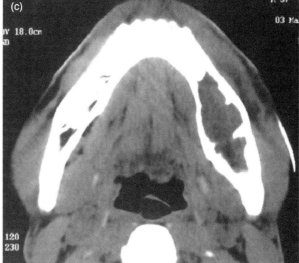
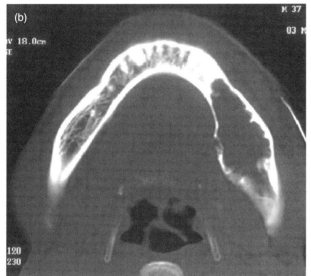

Figure 9.26. Computed tomography (CT) of a odontogenic myxoma affecting the posterior sextant of the mandible. These coronal (a) and axial (b) computed tomographs (bone window) display the peripheral arrangement of the septa and cells around a center devoid of septa. (c) Axial computed tomograph (CT) (soft-tissue window) is the precontrast image that corresponds to Figure 1.1c., which displays enhancement of a biopsy site. The biopsy had been taken prior to the CT. Figure (a) reprinted with permission from MacDonald-Jankowski DS, Yeung R, Li TK, Lee KM. Computed tomography of odontogenic myxoma. *Clinical Radiology* 2004;59:281–287.

be in part explained by the increased use of advanced imaging[83] that would have made more obvious the presence of septae.[85] The multilocular cases are evenly distributed between coarse (see Figures 1.19 and 9.26) and fine (Figure 9.27) septa.[80]

A quarter of odontogenic myxomas in a Black South African report[82] and a third in a Japanese report[86] presented with the tennis-racket pattern (Figure 1.19), which, although it is pathognomonic for the odontogenic myxoma, was generally otherwise infrequently seen in the original systematic review.[80] The sunburst or sunray appearance typical of the osteogenic sarcoma (Figure 11.28a)[81,82] was also displayed on CT images of large odontogenic myxomas affecting a Chinese community.[83] A suggestion for this pattern is that it may represent the honeycomb pattern in profile.[83]

The multilocular lesions reported by Noffke et al.[82] and by Martinez-Mata et al.[87] were significantly larger than the unilocular lesions, suggesting that the multilocularity is a feature of the larger lesion. Noffke et al. also found that relationship between the multilocular lesion and the age of the

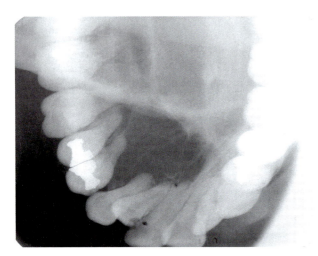

Figure 9.27. This upper occlusal radiograph displays a well-defined radiolucency between the canine and first premolar. The multilocular presentation excluded consideration of the lateral periodontal cyst. This was an odontogenic myxoma. Reprinted with permission from MacDonald-Jankowski DS, Yeung R, Lee KM, Li TK. Odontogenic myxomas in the Hong Kong Chinese: clinico-radiological presentation and systematic review. *Dentomaxillofacial Radiology* 2002;31:71–83.

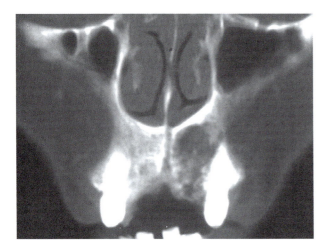

Figure 9.28. This coronal tomograph displays a multilocular radiolucency within the anterior maxilla. This is an odontogenic myxoma. The multilocular pattern is honeycomb and made up of small cells. The closeness of the radiodense septa may occasionally lead to radiopaque appearance. Reprinted with permission from MacDonald-Jankowski DS, Yeung R, Li TK, Lee KM. Computed tomography of odontogenic myxoma. *Clinical Radiology* 2004;59:281–287.

female (but not male) patient was also significant.[82] Hisatomi et al. revealed that the smaller lesion, displacing adjacent roots, presented as a nonspecific radiolucency.[88] They presented almost identical images of ameloblastoma, keratocyst odontgenic tumors, simple bone cysts, and odontogenic myxomas (see Figure 1.26). To this list can be added the lateral periodontal cyst.

The few cases reporting the radiology of the odontogenic myxoma affecting the anterior maxilla suggest that the honeycomb pattern is frequent (see Figure 9.28). Figure 9.29 is a flowchart for lesions that most frequently present as radiolucencies in the anterior maxilla.

The multilocular pattern observed on the conventional radiograph appears entirely different on HCT (see Figures 1.19 and 9.26). Instead of dividing the entire lesion into locules, the locules are confined to the periphery of the lesion leaving the center of the lesion, the atrium (my term), almost completely devoid of septae.[83]

Displacement of teeth in the systematic review is 47%, but it is significantly more marked in the sole East Asian global report than in the Western global group. Root resorption occurs in 27% of lesions, but it is significantly more frequent in a single report,[89] representing the entire sub-Saharan African global group, than in the Western global group. These two radiologically apparent features appear to relate well with the proportion of cases first presenting with displaced (53%) and loose teeth (27%), respectively (Table 9.3).

Li et al. reported a few cases with an association with unerupted teeth.[90]

As already indicated, the increased use of cross-sectional imaging of the odontogenic myxoma has already transformed a recent systematic review. It displayed a higher proportion of odontogenic myxomas presenting as multilocular radiolucencies. Although CBCT produces images of a better spatial resolution necessary for high detailed images of the bone[85] it is not recommended for infiltrative benign odontogenic neoplasms (including the ameloblastoma and KCOT), because it cannot show the soft tissue. If advanced imaging is indicated, it would be better to refer for an HCT so that its soft-tissue window could permit determination of infiltration of the adjacent soft tissue, particularly if the cortex has been perforated or extensively eroded.

Koseki et al. reported that two-thirds of their HCT cases displayed interruption of their corti-

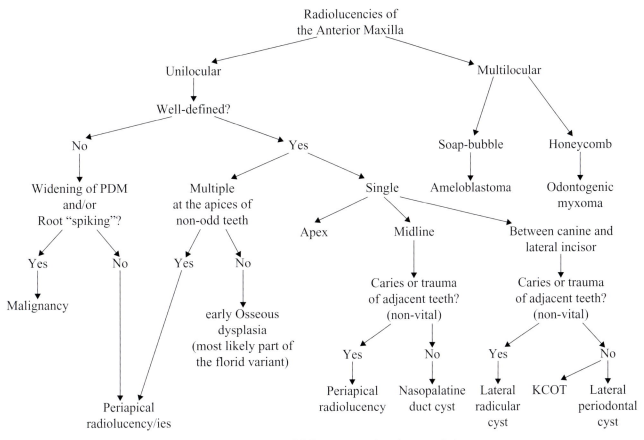

Figure 9.29. Radiolucencies of the anterior maxilla. KCOT, keratocystic odontogenic tumor.

ces.[86] The margins of the now extraosseous lesions to surrounding soft tissue were clearly recognized and smooth on the HCT images, even without cortical continuity.[86] Nevertheless, in order to determine whether the adjacent soft tissue has been infiltrated, intravenous contrast medium (see Chapter 4) should be routinely used as it also should be for the ameloblastoma. Infiltration may be indicated by "enhancement."

Although the contrast medium "enhances" the peripheral vascular supplying the odontogenic myxoma,[83] it is infrequent to find enhancement in the center of the lesion unless the lesion had been recently biopsied prior to the HCT (see Figure 1.2c).[83]

When MRI is used, the bulk of the lesion T1-weighted MRI presents as a hypointense homogeneous signal,[79] but on T2-weighted MRI it presents as a hyperintense homogeneous signal.[88] Asaumi et al. reported that the odontogenic myxoma and the ameloblastoma can be distinguished on the basis of their dynamic behavior on MRI. The odontogenic myxoma enhances after 500–600 seconds in contrast to the ameloblastoma's 45–60 seconds.[73]

Resection is necessary because the systematic review on odontogenic myxoma revealed an overall recurrence rate of 18% even after resection. This is higher than the 14% of the ameloblastoma. The Western and Latin American global groups exhibited significantly greater recurrence rates than that observed for the East Asian global group (Table 9.3). Li et al. reiterated the need for radical treatment of this lesion; after follow-up the only odontogenic myxoma that recurred had only been enucleated.[90]

Keratocystic Odontogenic Tumor (ICD-O 9270/0)

The "odontogenic keratocyst" has recently been renamed the *keratocystic odontogenic tumor* (KCOT) and reclassified as an odontogenic neoplasm in the WHO's 2005 edition of its histological classification of odontogenic tumors.[91] According to this edition, the KCOT has been defined as "A benign uni- or multicystic intraosseous tumor of odontogenic origin, with a characteristic lining of parakeratinised stratified squamous epithelium and potentially aggressive, infiltrative behaviour. It may be solitary or multiple. The latter is usually one of the stigmata of the inherited *nevoid basal cell carcinoma syndrome* (NBCCS)."[91] Furthermore, to emphasis the essential parakeratotic feature of this new "tumor," Philipsen adds that "Cystic jaw lesions that are lined by orthokeratinizing epithelium do not form part of the spectrum of a ... KCOT."[91]

The reader should be aware that leading world authorities on cysts of the jaws, Shear and Speight, do not agree with this change in nomenclature. Although Shear and Speight acknowledge that KCOT is a neoplasm, they prefer the older name of odontogenic keratocyst.[92] As Philipsen's definition of the KCOT is an essentially parakeratotic lesion,[91] Wright's[93] orthokeratotic variant, which was still recognized by the WHO's second edition,[94] has now become an entirely separate lesion, the orthokeratinized odontogenic cyst. These developments essentially exclude the mixed parakeratotic and orthokeratotic variant from either lesion, which account for an average 10% of formerly named odontogenic keratocysts.[95] This was significantly greater in the Western global group (11.%) than in the East Asian (8%) and Latin American (8%) global groups. Although many reports have identified the number of cases, which are mixed (7%), very few have reported on their clinical and radiological features and treatment outcomes. Nevertheless, because of their parakeratinized component they should be treated and followed up as if they were KCOTs.[91]

The KCOT is now the third most common odontogenic neoplasm after the odontoma and ameloblastoma. The global distribution of reports included in the systematic review,[95] upon which much of the following is derived is set out in Figure 1.40 and their details in Table 9.4.

The mean number of KCOTs per year is 6 globally. Although, globally, 6% of cases are syndromic. This rises to 14% among Latin Americans, which is significantly greater than those of the East Asian (5%) and Western (6%) global groups.[95]

The clinical and radiological features described below particularly pertain to the nonsyndromic KCOT. The syndromic cases, which are more prone to recur, first present at a significantly younger age in an East Asian report.[96]

The mean age at first presentation is 38 years old. Although the mean ages for the global groups range between 34 years old for Latin Americans to 43 years old for East Asians, the differences are not significant. The most frequent decade is the third (29%).[95] Females predominate in the first decade (91%), whereas males generally predominate in subsequent decades.

The majority (60%) of nonsyndromic patients are males. The predilection for males in East Asian and Western global groups is significantly greater than that for the Latin American group.

Swelling is the most common presenting symptom; it occurs in 58% of cases. It is significantly less frequent in the Latin American global group than in the East Asian and Western global groups. Pain presents in 32% and presents significantly more frequently in the East Asian global group than in the Western global group. Those patients first presenting with pain were significantly older than those that did not.[95] Twenty-one percent of nonsyndromic KCOTs are found as incidental findings; The East Asian, Latin American, and Western groups differ significantly. Although numbness is infrequent (2%), it may be more prevalent among Latin Americans (21%), but their sample is small. A purulent discharge presented in 12%; it presented significantly more in the Western (22%) than in the East Asian global group (9%).[95]

The mandible overall is affected in 72% of cases, particularly among East Asian and Latin American global groups. It was significantly less for the Western global group.[95] KCOTs in the Hong Kong Chinese, affecting the maxilla, first presented significantly earlier (22 years old) than those of the mandible (36 years old);[96] this was contrary to that in the ameloblastoma affecting the same community.[60] The posterior sextants are more frequently affected for both jaws: 88% for the mandible and 69% for the maxilla. Almost all cases confined to the posterior sextant, like that already reported for the ameloblastoma[60] with regard to the posterior sextant of the mandible in the same community, were 25 years old or younger.[96] The association between involvement of the posterior sextant and younger presentation was also a feature of the maxillary cases in the Hong Kong Chinese report.[96] Conversely, the maxillary ameloblastomas affecting the Hong Kong Chinese exhibit a predilection for the anterior sextant, which may extend to the

Table 9.4. Keratocystic odontogenic tumor: systematic review

Feature	Total	Western	East Asian	subSaharan	LatinAmer
Male:Female	60%:40%	60%:40%	61%:39%	50%:50%*	53%:47%
Mean number per year per report	5.8	6.5	6.3	0.0*	3.7
Mean age	38 years	40 years	35 years	43 years*	34 years
Mean prior awareness	0.7 years	INA	0.6 years	2.1 years*	INA
Mand:Max	72%:28%	65%:35%	77%:23%	100%:0%*	73%:27%
Mand:Ant:Post	12%:88%	11%:89%	12%:88%	0%:100%*	0%:100%*
Max:Ant:Post	31%:69%	37%:63%*	31%:69%	INA	INA
Swelling: Y:N	58%:42%	68%:32%	60%:40%	100%:0%*	37%:63%
Pain: Y:N	32%:68%	17%:83%	41%:39%	100%:0%*	0%:100%*
Incidental: Y:N	21%:79%	26%:74%	11%:89%	0%:100%*	58%:42%
Numb: Y:N	2%:98%	INA	1%:99%	INA	21%:79%*
Discharge: Y:N	12%:88%	22%:78%	9%:91%	67%:33%*	28%:72%*
Radiolucent	100%:0%	100%:0%	100%:0%	100%:0%*	100%:0%
Uni:Multiloc	72%:28%	80%:20%	63%:37%	0%:100%*	76%:24%
Welldefined: Y:N	64%:36%	45%:55%	100%:0%*	INA	INA
Cortex: Y:N	60%:40%	26%:74%*	88%:12%*	INA	INA
Expansion: Y:N	62%:38%*	14%:86%*	82%:18%*	INA	INA
LBMd: Y:N	71%:29%*	INA	71%:29%*	INA	INA
Antrum: Y:N	100%:0%*	INA	100%:0%*	INA	INA
ToothDispl: Y:N	69%:31%*	INA	69%:31%*	INA	INA
RootResorp: Y:N	23%:77%	8%:92%*	41%:59%*	INA	INA
Unerupted: Y:N	35%:65%	44%:56%	31%:69%	INA	20%:80%
Syndromic: Y:N	6%:94%	6%:94%	5%:95%	25%:75%*	14%:86%
Mixed(Ex): Y:N	7%:93%	8%:92%	9%:91%	11%:89%	2%:98%
Orthok(Ex): Y:N	10%:90%	11%:89%	8%:92%	8%:92%	8%:92%
Recurrence: Y:N	28%:72%	33%:67%	24%:76%	50%:50%*	17%:83%*

*Advises that the percentages were derived from either one report or from a synthesis of no more that 50 cases. Ant:Post, Anterior:Posterior; INA, Information not available; LatinAmer, Latin American; LBMd, downward expansion of the lower border of the mandible; Mand:Max, Mandible:Maxilla; subSaharan, sub-Saharan African; ToothDispl, Tooth displacement; ToothResorp, Tooth resorption; Uni:Multiloc, Unilocular:Multilocular; Western, predominantly Caucasian; Y:N, Yes:No.

anterior aspect of the posterior sextant. Maxillary ameloblastomas also first present significantly later in life.[60]

All KCOTs are radiolucent and nearly three-quarters (72%) are unilocular (Figure 9.30).[95] The unilocular KCOTs (24 years old) presented significantly earlier than the multilocular KCOTs (36 years old).[96] The mesiodistal length of the unilocular KCOTs in "dental units" is 4.22 in comparison to the multilocular KCOT's 5.70; this was not significant.[96] Nevertheless, in the mandible, the unilocular KCOTs (3.20) were much smaller than the multilocular KCOTs (6.17); this tended to significance.[96]

Sixty-four percent are well defined, of which 60% are corticated.[95] East Asian global group's 100% well-defined cases were significant compared to 45% in the Western global group. Also, East Asian global group 88% cortication was significantly greater than the Western global group's 26%.[95] Although buccolingual expansion was observed in 62% of KCOTs, this assumed a pattern similar to the fusiform expansion of fibrous dysplasia (Figure 9.30b). It was significantly more prevalent in the East Asian global group (82%) than in the Western global group (14%).[95] This pattern of expansion is most frequently observed in those KCOTs involving the body of the mandible.[95] Therefore, a low buccolingual:mesiodistal ratio of a radiolucency in the body of the mandible, should suggest a KCOT (and also an odontogenic

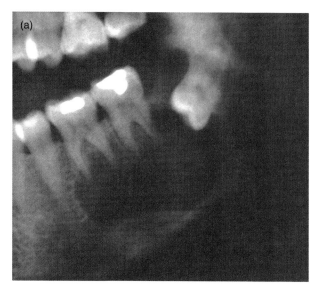
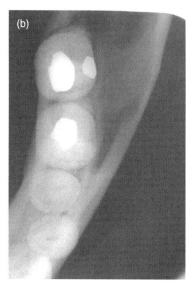

Figure 9.30. The panoramic radiograph (a) and the true occlusal radiograph (b) are of a keratocystic odontogenic tumor. It presents as a radiolucency associated with the crown of an unerupted third molar. Although it is "attached" to the tooth's mesial (or inferior) cementoenamel junction (CEJ), its distal (superior) "attachment" is to the root distant from the CEJ. It almost extends through the entire vertical height of the body of the mandible without expanding it. It also minimally expands the mandible buccolingually. **Note:** 'The body of the hyoid is superimposed upon the inferior anterior aspect of the lesion in (a). Reprinted with permission from MacDonald-Jankowski DS, Li TK. Keratocystic odontogenic tumor in a Hong Kong community; the clinical and radiological presentations and the outcomes of treatment and follow-up. *Dentomaxillofacial Radiology* 2010;39:167–175.

myxoma and simple bone cyst), whereas a high buccolingual:mesiodistal ratio would suggest an ameloblastoma or, as a lead onto the next lesion, an orthokeratinized odontogenic cyst.

Tooth displacement was recorded in one report, an East Asian; it occurred in 69% of cases.[96] Contrary to a statement in the WHO's 2005 edition that root resorption is very rare,[91] it occurred in 41% and 9% of cases, in an East Asian[96] and a Western[97] report, respectively. By virtue of these two reports, root resorption occurred in 23% of cases in the systematic review.[95] Although 41% in KCOTs of the Hong Kong Chinese[96] was less than the ameloblastoma's 59% in the same community,[59,60] this was not significant.[96] Furthermore, the presentation of Hong Kong Chinese KCOTs with root resorption in older-aged patients[96] differs from that in Hong Kong Chinese ameloblastomas' first presentation in younger patients; this phenomenon is significantly more frequent in earlier presenting unicystic variant lesions than in the nonunicystic (predominantly solid variant) and predominantly radiologically multilocular lesions, which first present later in life.[59,60]

An association with unerupted teeth occurred in a third of cases overall.[95] It differed significantly between the East Asian, Latin American, and Western global groups. The surprise that defies easy explanation was that one of the Western reports was American, in which prophylactic removal of unerupted teeth has been practiced for decades. Those KCOTs in the Hong Kong Chinese presenting with associated unerupted teeth were significantly younger than those that did not.[96]

All nonsyndromic KCOTs affecting the posterior maxilla affected the maxillary antrum, expanding up into it in a balloonlike fashion. Figure 11.23 displays such an example.[96]

KCOTs affecting the anterior maxilla appear to have some predilection for the canine-lateral incisor site (Figures 9.31 and 9.32).

Treatment of the nonsyndromic KCOT is usually conservative; in a recent report 79% of cases were treated by enucleation and Carnoy's solution.[96]

The global recurrence rate for nonsyndromic KCOTs is 28%.[95] It is significantly greater in the Western global group (32%) than in the East Asians (24%). This recurrence rate is far higher than that of the ameloblastoma (14%) and the odontogenic myxoma (18%), emphasizing the need for long-term follow-up.[96] Although a reason

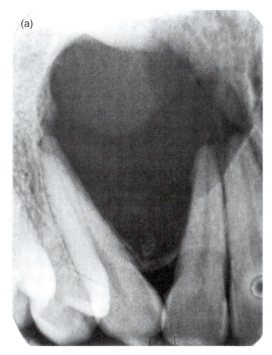
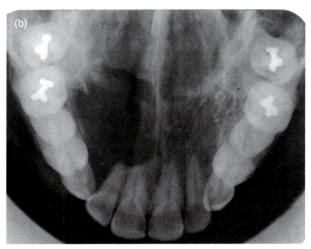

Figure 9.31. Conventional images of keratocystic odontogenic tumor affecting the anterior maxilla. In contrast to the periapical radiograph (a), the standard anterior occlusal (b) reveals the full vertical extend of the lesion that caused this well-defined radiolucency, displacing the lateral incisor and the canine. (a) This image reveals the absence of the root resorption, which is suspected on (b). **Note:** The tip of the nose is obvious in the upper half of the radiolucency in Figure 9.31 (a).

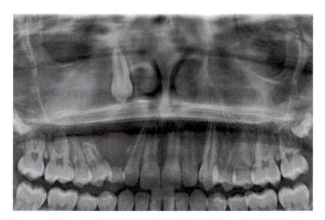

Figure 9.32. This panoramic radiograph depicts a radiolucent lesion in the canine site splaying the lateral incisor and first premolar. The canine is unerupted and its root completely formed. This phenomenon earlier would have invoked consideration of the now-defunct globulomaxillary cyst. Now this presentation invokes a lateral radicular cyst, lateral periodontal cyst, and keratocystic odontogenic tumor. This is a keratocystic odontogenic tumor.

for this high recurrence with regard to KCOT is the general conservative approach to its treatment, both the degree of follow-up and the recurrence rates for both KCOT[96] and ameloblastoma[60] in a Hong Kong Chinese community were equally high and low, respectively. Follow-up for KCOTs and ameloblastomas were 73% (for a mean of 5 years) and 89% (for a mean of 5 years), respectively. The recurrence rates for KCOTs and ameloblastomas were 9%[96] and 11%[60], respectively. The reasons for this degree of follow-up for both lesions are both the surgeons' vigorous follow-up protocol and the relative stability of the Hong Kong community as a whole.[60]

KCOTs (frequently multiple), along with calcified falx cerebri and basal cell carcinomas, are major features of NBCCS, which generally manifest in the patient's teens.[98] The syndromic cases first presented at a significantly younger age than the nonsyndromic cases in a recent East Asian report.[96] This syndrome is a multisystem disease.[98,99] Lo Muzio revealed that 5–10% of NBCCS patients develop a brain malignancy, the medulloblastoma, which can cause early death.[98] The recurrence rate of the syndromic KCOTs is higher than for the nonsyndromic KCOTs; they recur in the former in 60%[98] of cases in contrast to the latter's 28%.[95]

KCOTs, particularly the nonsyndromic cases, do not appear to prompt clinicians to refer them for advanced imaging. Nevertheless, Lam et al. advise that "there should be early and frequent monitoring of NBCCS patients for the development of KOTs [their abbreviation for KCOT] in youth and adolescence, and that CT imaging should play a role in these investigations."[100] Both conventional radiography and CT (both HCT and CBCT) were equally effective in displaying the majority of KCOTs in an association with dental follicles of unerupted teeth. CT was significantly more effective at demonstrating endosteal scalloping of the cortex, which presumably assists in differentiating them from dentigerous cysts.[100] Conventional radiography displays tooth displacement significantly more frequently. Lam et al. are very clear that CT should not displace conventional radiography as the primary imaging modality.[100] A case of syndromic KCOT investigated by HCT is displayed in Figure 11.24.

Orthokeratinized Odontogenic Cyst

The *orthokeratinized odontogenic cyst* (OOC) was first clearly identified as the orthokeratotic variant of the odontogenic keratocyst by Wright in 1981,[93] due to its different histopathology and reduced likelihood to recur. The last was confirmed in a systematic review.[101] Although the first two editions of the WHO's histological classification of odontogenic tumors recognized that "cases with orthokeratosis are seen,"[94] the WHO's 2005 edition, expressly excluded it from its definition of a KCOT.[91]

Although Shear and Speight agree that on the basis of mounting evidence, the orthokeratotic variant of the KCOT should be considered a separate entity,[92] those mixed lesions that contain a substantial component of parakeratotic epithelium should be regarded with caution. As already mentioned, such mixed cases (Figure 9.33), which account for 10% of all cases formerly considered to be odontogenic keratocysts, should be managed as if they were KCOTs.[101]

The global distribution of reports included in the systematic review[101] upon which much of the following is derived is set out in Figure 1.42 and their details in Table 9.5.

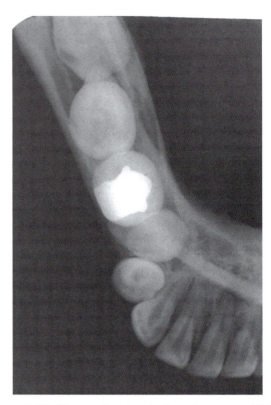

Figure 9.33. True occlusal projection of a case exhibiting a mixture of parakeratin and orthokeratin. Although extensive in its mesiodistal extent it displays no buccolingual expansion.

The mean number of OOCs per year is globally 0.8. The OOC accounts for 10% of all nonsyndromic former odontogenic keratocysts.[101] Their mean age at first presentation is 35 years old. They have a very short period of first awareness of their lesions. Although it is 0.9 years for the East Asian global group alone, that of the Hong Kong study was significantly shorter than that of the Malaysian report.[101] Most (32%) first present in the third decade.

Forty-eight percent are found incidentally, the rest present with symptoms. Swelling occurs in 41%, significantly most frequently in the East Asian global group (58%). Pain and discharge occur in 24% and 16%, respectively. The mandible is affected in 71% of cases. The posterior mandibular sextant is affected (92%) more frequently than the anterior sextant. The anterior and posterior maxillary sextants are affected equally.[101]

Table 9.5. Orthokeratinized odontogenic cyst: systematic review

Feature	Total	Western	East Asian	subSaharan	LatinAmer
Male:Female	66%:34%	67%:33%	65%:35%	INA	67%:35%*
Mean number per year per report	0.8	1.1	0.8	INA	0.4
Mean age	35 years	35 years	35 years	INA	37 years*
Mean prior awareness	0.9 years*	INA	0.9 years*	INA	INA
Mand:Max	71%:29%	69%:31%	79%:21%	INA	50%:50%*
Mand:Ant:Post	8%:92%	10%:90%*	4%:96%*	INA	INA
Max:Ant:Post	50%:50%*	50%:50%*	50%:50%*	INA	INA
Swelling: Y:N	41%:59%	23%:77%*	58%:42%*	INA	67%:33%*
Pain: Y:N	24%:76%	23%:77%*	22%:78%	INA	0%:100%*
Incidental: Y:N	48%:52%	53%:47%*	31%:69%*	INA	INA
Infected: Y:N	16%:84%*	11%:89%*	31%:69%*	INA	33%:67%*
Radiolucent	100%:0%	100%:0%	100%:0%	INA	100%:0%*
Uni:Multiloc	93%:7%	95%:%	89%:11%	INA	100%:0%*
Welldefined Y:N	100%:0%*	INA	100%:0%*	INA	INA
Cortex Y:N	100%:0%*	INA	100%:0%*	INA	INA
Expansion Y:N	100%:0%*	INA	100%:0%*	INA	INA
LBMd: Y:N	25%:75%*	INA	25%:75%*	INA	INA
Antrum Y:N	100%:0%*	INA	100%:0%*	INA	INA
ToothDispl: Y:N	80%:20%*	INA	80%:20%*	INA	50%:50%*
RootResorp: Y:N	0%:100%*	INA	0%:100%*	INA	INA
Unerupted: Y:N	69%:31%	76%:24%*	63%:37%*	INA	62%:38%*
Recurrence: Y:N	4%:96%	5%:95%	0%:100%*	INA	0%:100%*

*Advises that the percentages were derived from either one report or from a synthesis of no more that 50 cases. Ant:Post, Anterior:Posterior; INA, Information not available; LatinAmer, Latin American; LBMd, downward expansion of the lower border of the mandible; Mand:Max, Mandible:Maxilla; subSaharan, sub-Saharan African; ToothDispl, Tooth displacement; ToothResorp, Tooth resorption; Uni:Multiloc, Unilocular:Multilocular; Western, predominantly Caucasian; Y:N, Yes:No.

All OOCs present as radiolucencies (Figure 9.34). Ninety-three percent are unilocular (Figure 9.34), but some are multilocular (Figure 9.35). Marginal definition is recorded for only two reports included in the systematic review; OOCs are well defined with corticated margins. Sixty-nine percent are associated with unerupted teeth, the majority of which are third molars.[101] One report indicates that they may affect the maxillary antrum and displace teeth.[102]

Two out of three consecutive mandibular OOCs displayed the downward displacement of a lateral cortex past the otherwise undisplaced lower border of the mandible (Figures 9.34b and 9.35),[102] hitherto a feature observed only in ameloblastomas.[60]

Although the overall recurrence rate is 4%,[101] this may be an overestimate because it is very likely that a few "mixed" cases displaying both parakeratotic and orthokeratotic epithelium may have been simply assigned the diagnosis of orthokeratotic variant of OKC or OOC because no alternative was available.

The OOC differs significantly from the KCOT by being observed as an incidental finding with a well-defined margin and an association with an unerupted tooth.[101]

Glandular Odontogenic Cyst

The *glandular odontogenic cyst* (GOC), also known as the sialo-odontogenic cyst, was first clearly identified as a separate entity by Gardner et al. in 1988.[103] The WHO's second edition histological classification of odontogenic tumors in 1992 recognized it as "a cyst arising in the tooth-bearing areas of the jaws and characterized by an epithelial lining with cuboidal or columnar cells both at the surface and lining crypts or cyst-like spaces within the thickness of the epithelium."[104] Slootweg added

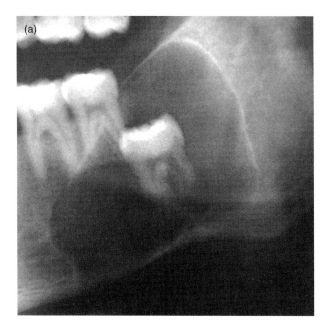
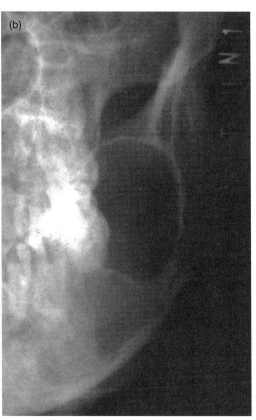

Figure 9.34. The panoramic radiograph (a) and the posterioanterior projection of the mandible displays an orthokeratinized odontogenic cyst (OOC). This presents as a well-defined corticated unilocular radiolucency, which not only exhibits buccolingual expansion but also downward expansion of either lateral cortex past the undisplaced lower border of the mandible anteriorly. The associated unerupted tooth, with which the OOC has no obvious relationship to the cementoenamel junction, in (a) has been displaced lingually in (b). The mandibular canal has been displaced to the lower border of the mandible. Reprinted with permission from MacDonald-Jankowski DS, Li TK. Orthokeratinising odontogenic cyst in a Hong Kong community; the clinical and radiological presentations and the outcomes of treatment and follow-up. *Dentomaxillofacial Radiology* 2010;39:238–243.

that this complex epithelium is also partly nonkeratinized and that mucus-producing cells may be present. The latter contributes to the GOC's histopathological resemblance to a well-differentiated mucoepidermoid carcinoma.[105]

The global distribution of reports included in the systematic review[106] upon which much of the following is derived is set out in Figure 1.43 and their details in Table 9.6.

The mean number of GOCs per year globally is 0.4. The sub-Saharan African global group had the highest mean number, which was significantly greater than that for the Latin American global group with the least.[106]

The GOC has a predilection for males (66%).[106] Its mean age on first presentation is 45 years old, ranging significantly between the global groups from 35 (sub-Saharan African) and 37 (East Asian), through 46 (Western), to 52 (Latin-American). The decade with the most frequent first presentations is the fifth (26%). The period of prior awareness was 10 years for two small European reports.[106]

Of the vast majority present with symptoms, only 10% are found incidentally. The most frequent symptom was swelling (88%), followed by pain (37%) and numbness (15%).[106]

The mandible is affected in 80% of cases. The anterior sextants are more frequently affected; 78% and 100% for the mandible and maxilla, respectively.[106]

Radiographically, 98% are radiolucent.[106] Fifty-eight percent are unilocular (Figure 9.36a),

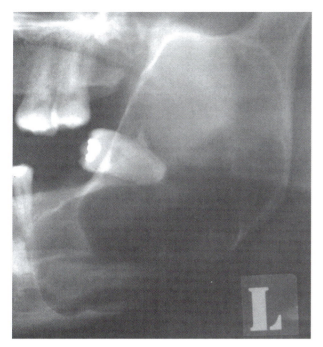

Figure 9.35. Panoramic radiograph of an orthokeratinized odontogenic cyst affecting the posterior sextant of the mandible. It exhibits a well-defined multilocular radiolucency with a corticated margin. It extends from the distal aspect of the second premolar back to the base of the sigmoid notch. It expands the anterior margin of the vertical ramus and displaces a lateral cortex down past the lower border of the mandible. Reprinted with permission from MacDonald-Jankowski DS, Li TK. Orthokeratinising odontogenic cyst in a Hong Kong community; the clinical and radiological presentations and the outcomes of treatment and follow-up. *Dentomaxillofacial Radiology* 2010;39: 238–243.

and the rest are multilocular (Figure 9.36b). Ninety-six percent are well defined, of which 30% are corticated. Eighty-two percent display buccolingual expansion. In a small report, when subjacent to the maxillary antrum, 3 out of 4 cases involve it. One-half of cases displace teeth, 30% resorb roots, and 11% are associated with a unerupted tooth. The prevalence of tooth displacement varies significantly between the global groups: most frequent for the sub-Saharan African and least frequent for the East Asian global group. Also, root resorption significantly occurs least in the sub-Saharan African and most in the Western global group. With regard to the unerupted teeth, although the unilocular lesions appear like dentigerous cysts, most infrequently affect third molars because of their anterior location.[106]

Eighteen percent recur after treatment.[106] This is similar to that of odontogenic myxoma (18%).[95]

Squamous Odontogenic Tumor (ICD-O 9312/0)

Reichart defined the *squamous odontogenic tumor* as "a locally infiltrative neoplasm consisting of islands of well-differentiated squamous epithelium in a fibrous stroma."[107] Less than 50 cases have been reported arising in a very wide age range of patients; the mean age is 39 years old. It classically develops between the roots of vital teeth. Although its etiology is still unknown it is likely to be of periodontal ligament origin.[107]

It may present with pain, swelling, and/or mobile teeth. The small lesion is unilocular or triangular, whereas the larger lesion is more likely to be multilocular.[107] Although it generally responds well to conservative treatment, the rare recurrence may be due to incomplete removal.[107]

Central Giant Cell Lesion

Jundt defined the central giant cell lesion as "a localized benign but sometimes aggressive osteolytic proliferation consisting of fibrous tissue with hemorrhage and hemosiderin deposits, presence of osteoclastlike giant cells and reactive bone formation."[108]

Sixty-four percent of the cases in de Lange et al.'s synthesis of the literature were female,[109] which is in broad agreement with Stavropoulos and Katz' case series in which 55% were female.[110] The mean age of Stavropoulos and Katz's case series was 32 years old.[110] The mandible was affected in 61% of cases,[109] but it was 85% in Stavropoulos and Katz' own case series.[110] The only clear clinical feature expressed in Stavropoulos and Katz' systematic review was paresthesia in 10% of their cases and in 6% of their systematic review.[110]

Stavropoulos and Katz, in reporting their own case series, observed that 81% were anterior to the molars. There was no association between size (some were over 6 centimeters although the mean was 3.6) and age.[110]

Radiologically, Stavropoulos and Katz' Table 2 revealed that 45% presented as radiolucencies

Table 9.6. Glandular odontogenic cyst: systematic review

Feature	Total	Western	East Asian	subSaharan	LatinAmer
Male:Female	64%:36%	70%:30%*	67%:33%*	44%:56%*	57%:43%*
Mean number per year per report	0.4	0.5	0.4	0.7	0.2
Mean age	45 years	46 years	38 years*	35 years *	52 years*
Mean prior awareness	10 years*	10 years*	INA	INA	INA
Mand:Max	80%:20%	82%:18%*	58%:42%*	67%:33%*	91%:9%*
Mand:Ant:Post	78%:22%*	84%:16%*	100%:0%*	0%:100%*	50%:50%*
Max:Ant:Post	100%:0%*	100%:0%*	INA	100%:0%*	100%:0%*
Swelling: Y:N	88%:12%*	81%:19%*	92%:8%*	INA	100%:0%*
Pain: Y:N	37%:63%*	23%:77%*	25%:75%*	INA	40%:60%*
Incidental: N:Y	10%:90%*	14%:86%*	8%:92%*	INA	0%:100%*
Numb: Y:N	15%:85%*	14%:86%*	0%:100%*	INA	INA
Radiolucent: Y:N	98%:2%	95%:5%*	100%:0%*	100%:0%*	100%:0%*
Uni:Multiloc	58%:42%	65%:35%*	58%:42%*	67%:33%*	20%:80%*
Welldefined: Y:N	96%:4%*	100%:0%*	100%:0%*	87%:13%*	INA
Cortex: Y:N	30%:70%*	36%:64%*	INA	22%:78%*	INA
Perforation: Y:N	52%:48%*	57%:43%*	INA	44%:56%*	INA
Expansion: Y:N	82%:18%*	57%:43%*	INA	100%:0%*	100%:0%*
LBMd: Y:N	INA	INA	INA	INA	INA
Antrum: Y:N	75%:25%	75%:25%	INA	INA	INA
ToothDispl: Y:N	50%:50%*	50%:50%*	17%:83%*	100%:0%*	INA
RootResorp: Y:N	32%:68%*	54%:46%*	33%:67%*	0%:100%*	INA
Unerupted: Y:N	11%:89%*	14%:86%*	17%:83%*	0%:100%*	INA
Recurrence: Y:N	18%:82%	25%:75%*	10%:90%*	0%:100%*	20%:80%*

*Advises that the percentages were derived from either one report or from a synthesis of no more that 50 cases. Ant:Post, Anterior:Posterior; INA, Information not available; LatinAmer, Latin American; LBMd, downward expansion of the lower border of the mandible; Mand:Max, Mandible:Maxilla; subSaharan, sub-Saharan African; ToothDispl, Tooth displacement; ToothResorp, Tooth resorption; Uni:Multiloc, Unilocular:Multilocular; Western, predominantly Caucasian; Y:N, Yes:No.

and 55% were "mixed radiolucent-radiopaque," Their systematic review revealed that the lesions were wholly well defined in 34% of cases, whereas their own report found all wholly or partly were well defined. Forty-six percent in their systematic review were unilocular, whereas 45% were multilocular in their study. Fifty percent displayed "trabeculae" within the lesions. Scalloping between teeth, also in their study, was observed in 53%.[110]

Stavropoulos and Katz' own study alone reported that 45% displayed buccolingual expansion and 50% displayed perforation of the cortex, whereas their systematic review revealed 51% and 38% for these features, respectively.[110]

Stavropoulos and Katz' own study alone reported the following features: Tooth and follicle displacement occurred in 45% (Figure 9.37). The mandibular canal was inferiorly displaced in 5%. The lamina dura was absent in 94% of cases (Figure 9.37).[110]

Root resorption is 37% in Stavropoulos and Katz' own study and 33% in their systematic review.[110]

Ameloblastic Fibroma (ICD-O 9330/0)

Slootweg defined the ameloblastic fibroma (also known as a fibrodentinoma) as consisting "of odontogenic ectomesenchyme resembling the dental papilla and epithelial strands and nests resembling dental lamina and enamel organ. No dental and hard tissues are present."[111] If there is dentine formation or dentine and enamel, the lesion is referred to as *ameloblastic fibrodentinoma*

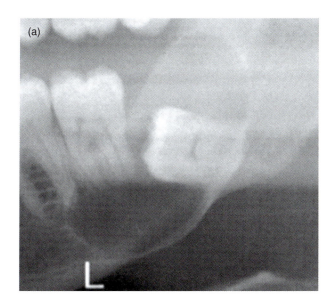

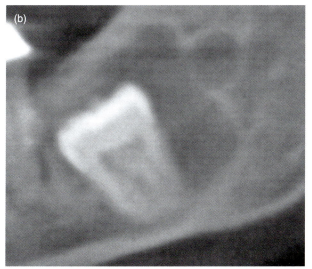

Figure 9.36. Panoramic radiographs of two separate glandular odontogenic cysts (GOC). (a) A unilocular GOC has enveloped the crown and root of an unerupted third molar. The GOC has no attachment with the cementoenamel junction. The tooth has been displaced to the lower boder of the mandible. The latter and the mandibular canal have been displaced downward. (b) A multilocular GOC associated with the distal side of the crown and root of an unerupted third molar. The mandibular canal has been displaced downward.

(ICD-O 9271/0) and *ameloblastic fibro-odontoma* (ICD-O 9290/0), respectively.[111] Because these are likely to present with radiopacities (dentine and enamel), these will not be considered further in this chapter.

Chen et al. recently synthesized the literature on ameloblastic fibroma.[112] They reported 123 cases in 55 reports of which 8 were case series. Sixty-eight percent were male. The mean age was 15.9 years. Eighty percent affected the mandible. Seventy-four percent of cases were found in the posterior sextants of the mandible. The ameloblastic fibroma presented as a hard swelling in 72% and as an incidental finding in 23%. The mean period of awareness of the lesion prior to presentation was 1.2 (0.1–6.0) years. The mean size at first presentation was 4.0 (0.7–16) cm. The 38 cases reported some radiography. All were well defined. Sixty percent were unilocular and 40 percent multilocular. Those cases that were symptom-free were unilocular, whereas those with symptoms were multilocular; this difference was significant.[112]

Although the histopathology leads to an expectation that these lesions will be radiolucent, this is not so in every case. The marginal cortex may be so sclerotic as to confer a "mixed" presentation and therefore cause at least some cases to be first considered as radiopacities.[112] This mixed presentation is clearly displayed in Figure 9.38 and the sclerotic marginal cortex (in another case) in Figure 9.39.

Ninety percent were conservatively treated.[112] A third recurred, 4% and 69% after 5 and 10 years, respectively. Resected cases recurred after a slightly longer interval. Furthermore, 11% underwent malignant transformation, 10% and 22% for 5 and 10 years, respectively. The cases first presenting younger than 22 years were less likely to undergo malignant transformation. This suggests that the ameloblastic fibroma should be subjected to long-term follow-up.[112]

Cherubism

Jundt recently defined cherubism as "an autosomal dominant inherited disease that is characterized by a symmetrical distension of the jaws, often leading to a typical facial expression. The histology is indistinguishable from a central giant cell lesion."[113] "Although the histology is not specific, the combination of clinical appearance, radiology and central giant-cell lesionlike histology is diagnostic."[113]

Two reports of multiple kindred published in the last decade are Von Wowern's 18 cases from 6 Danish kindred[114] and Meng et al.'s 14 from 6 Chinese kindred.[115] The latter also included 10

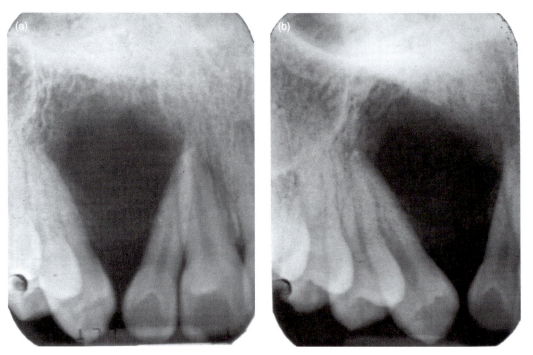

Figure 9.37. Periapical radiographs of a giant cell granuloma affecting the anterior sextant of the maxilla. This case appears similar to Figure 9.31 in the way it displays the roots of the maxillary canine and lateral incisor.

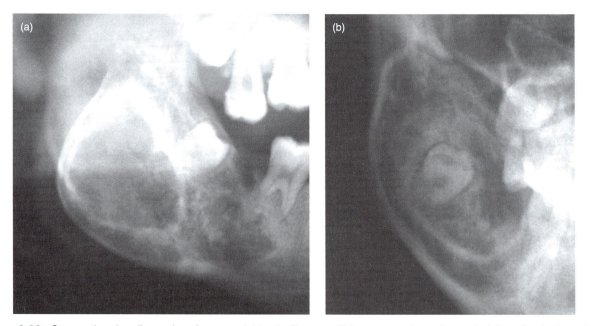

Figure 9.38. Conventional radiography of an ameloblastic fibroma. This panoramic radiograph (a) and anteroposterior projection of the mandible (b) displays an ameloblastic fibroma sited within the angle of the mandible and posterior body of the mandible. This lesion presents with buccolingual expansion. Its margin is well defined and sclerotic. Its internal structure is complex. Although largely translucent, it contains a separate radiolucency well defined by a sclerotic margin. The lesion also encompasses an unerupted third molar tooth, whose normal-sized follicle space is still patent. The distal root of the second molar has been resorbed.

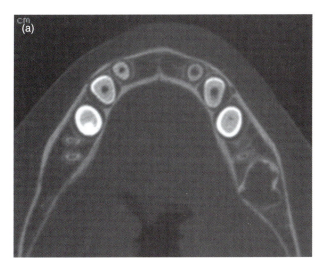

Figure 9.39. Computed tomography of an ameloblastic fibroma. (a) Axial CT (bone window) displaying a well-defined radiolucency with a sclerotic margin. (b) Coronal CT (bone window) displaying a well-defined radiolucency with a sclerotic margin. It exhibits lingual expansion obliterating the submandibular fossa.

nonfamilial cases.[115] Von Wowern followed up 18 individuals for 36 years.[114] Both studies[114,115] used variations of the Seward and Hankey grading system

Grade 1. Involvement of the bilateral mandible vertical ramus and posterior body.
Grade 2. In addition to grade 1, the maxillary tuberosities are involved.
Grade 3. Massive involvement of the entire maxilla and mandible except the condyles.

Grade 1 accounted for 79% in Von Wowern's[114] and 33% in Meng et al.'s[115] reports.

Although both reports revealed that, proportionally, males rather than females within a kindred are more likely to be affected, this was not so in one of Von Wowern's kindred.[114] Nevertheless, males were more likely to be affected by the more severe grades.

Although Von Wowern reported, on the basis of conventional radiography, that the lingual cortex was not expanded,[114] lingual expansion may be apparent on computed tomography.

Meng et al. did report 1 case that affected the condyles,[115] but all of the condyles of Von Wowern's cases were unaffected.[114] Meng et al. reported that all cases displayed tooth displacement, one-half aplasia, one-third noneruption, and one-quarter root resorption.[115] Furthermore, 3 of Meng et al.'s cases exceptionally displayed an increase in serum phosphorous and alkaline phosphatase.[115] Screening of the entire skeleton did not reveal the involvement of any other bone outside the jaws.

In those few cases of asymmetrical cherubism, Roginsky et al., reporting a Russian series, advised that minor features such as tooth malposition, premature tooth loss, or rare or previously unreported features such as gingival hyperplasia and enamel hypoplasia should suggest cherubism.[116]

Von Wowern's long-term follow-up revealed that although the lesions of most of her cases became apparent by 7 years of age[114] (only 62% of Meng et al.'s cases became apparent by 10 years of age[115]), they achieved maximum buccal expansion by 12 years of age. The 15 patients with grade 1 returned to normality by their late teens, and grade 2 by their mid- to late twenties.[114] The grade 3 patients still showed a modest expansion by 23 years of age and a mild expansion at the end of the reported period of follow-up.[114]

During the active stage of the disease, the lesions presented, on conventional radiography, as multilocular radiolucencies (Figure 9.40a, c).[114] Upon long-term follow-up, the outline of the initial multilocular lesion in the mandible may be apparent as a "sketched" outline (Figure 9.40c), and the radiolucency is filled in with normal bone structure. On the other hand, the maxillary radiolucency now appears completely replaced by a normal bone structure. If followed up long enough some mandibular cases may show "normal mandibular bone without signs of earlier cherubic lesions." There appears to be a transition stage between the initial radiolucent and "sketched" stages. The osseous infilling initially assumes a ground-glass appearance arising from the small tightly compressed trabecular pattern.[114]

Not only is the degree of buccal expansion proportional to the grade (Figure 9.40b), but so also is the degree of aplasia (Figure 9.40a).[114] Furthermore, two patients with the most severe

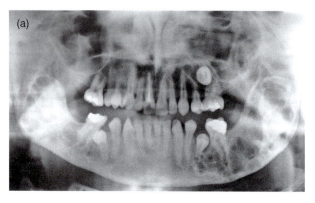
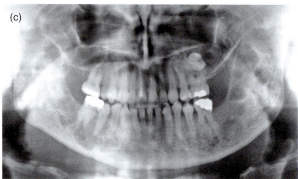
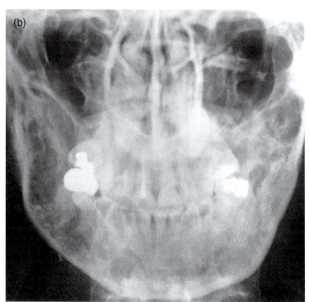

Figure 9.40. Conventional images of a case of cherubism observed at an earlier (a,b) and a later (c) stage (c) (a) Panoramic radiograph displays bilateral and bimaxillary radiolucencies confined to the posterior sextants. This is cherubism. It also displays absence (agenesis) of some molars and tooth displacement. (b) The posterioanterior projection reveals substantial buccolingual expansion of the vertical rami. The pattern of the radiolucencies is multilocular. (c) The later panoramic radiograph reveals regression of the lesions. They are now more radiodense and normal bone has infilled the earlier radiolucencies. The sketched outlines of these earlier radiolucencies can still be discerned.

grade displayed root resorption of the remaining mandibular teeth resulting in total mandibular tooth loss before 24 years of age in one patient.[114] Root malformations of the lower first molar were observed in two-thirds of patients.[114]

Although conventional radiography was clearly adequate for both Von Wowern's[114] and Meng et al.'s[115] reports, the latter advised that CT (both HCT and CBCT) may be indicated for the rare severe cases of grade 3.[115] Other than better delineation of the extent of the lesions, their expansion and the lack of a periosteal reaction, advanced imaging contributed little to our understanding of grades 1 and 2 cherubism.[117] Occasionally advanced imaging, particularly by HCT, of very extensive cases complemented the conventional radiography.[118] This was especially so in those exceptionally rare cases involving the orbit.[119–121] Nevertheless, a report by Beaman et al. functions as a reasonable pictorial essay of HCT and MRI imaging.[117] MRI has been performed for a few cases, but so far there is no clear role for this modality in the management of cherubism. Scintigraphy has little role in the diagnosis of cherubism because it is most frequently negative.[122]

In view of the self-regressive nature of most cases of cherubism after the onset of puberty and the likelihood of postsurgical complications, Von Wowern advises that "conservative management is appropriate until functional or emotional disturbances demand surgical intervention."[114]

Simple Bone Cyst

The simple bone cyst (SBC) has been defined by Jundt as "an intraosseous pseudocyst devoid of an epithelial lining, either empty or filled with serous or sanguineous fluid."[123]

The SBC has a number of synonyms, which have been set out by Suei et al.[124] The possible etiologies have been recently reviewed by Harnet et al.[125] The SBC is not a common lesion.[126] In a recent UK report, it accounted for 1.2 lesions per year.[51] Although there is a view that they rarely are detected after 25 years of age,[127] another report indicates that this is not true for every community.[128]

The SBC of the jaws affects both sexes equally.[129] Suei et al.[124] suggest that a proportion of the female patients with SBC are afflicted by osseous dysplasia (Figure 1.20), which can itself be accompanied by SBC.[130]

Therefore, it may be difficult to determine in some cases whether a particular lesion is primarily an SBC or an osseous dysplasia.

Because the SBC affecting the jaw is rarely symptomatic, it is frequently discovered incidentally on a radiograph prescribed for another purpose.[129] Although SBCs may be associated intimately with teeth, these teeth are almost always vital. SBCs predominantly affect the mandible (94%).[131]

On the conventional radiograph a frequent presentation is as a unilocular radiolucency with no or minimal buccolingual expansion. To distinguish it from the KCOT, which has a similar presentation, Ferreira et al. reported that the SBC is significantly more likely to display scalloping between the roots of teeth (Figure 1.17).[132]

The degree of marginal definition can vary from corticated to poorly defined.[128] Ferreira et al. reported that the SBC may also be distinguished from the KCOT because the latter's border was frequently more corticated.[132]

Expansion was present in 40% of Suei et al. case series[133] and in 21% of MacDonald-Jankowski's case series.[128] Although the interior dental canal was displaced in 54% of Matsumura et al. cases,[131] this was only observed one case in MacDonald-Jankowski's report.[128] Furthermore, although that report observed the constriction of the canal in one case and absence of its outline in 2 cases, the canal was unaffected although superimposed upon the SBC in another 8 cases.[128] Root resorption or tooth displacement were not observed.[128,131]

Although the majority present as radiolucencies, at least one report included a high proportion of cases that include central radiopacities (Figure 1.20).[131] Matsumura et al. divided their 53 simple bone cysts into 2 groups.[131] Type A has a connective tissue lining, whereas Type B has a partially thickened wall with dysplastic bone formation. Recurrence was more associated with Type B.[131]

Suei et al.[129] and Eriksson et al.[134] could not substantiate, by HCT and MRI, respectively, the frequent observation that, at the time of surgery, SBCs were air-filled.

Suei et al., in their recent report of their case series and synthesis of the literature, reported that 26% of 132 cases recurred.[129] This is much higher than hitherto reported. Furthermore, this recurrence rate was even higher for cases with multiple cysts and osseous dysplasia, 71% and 75%, respectively.[129] Healing occurred within 1 to 1.5 years and recurrence up to 2.5 years. They recommend that follow-up is continued until healing has been radiographically confirmed, particularly with regard to those cases with multiple cysts or osseous dysplasia.[129] MacDonald-Jankowski reported a case that recurred at least 6 times in 9 years.[128] The initially four separate lesions coalesced into a lesion that occupied the entire body of the mandible (Figure 1.28).[128] The need for follow-up until complete healing occurred was emphasized, because frequently some bone regeneration was apparent before it regressed.[128]

Aneurysmal Bone Cyst

Jundt defined the aneurysmal cone cyst (ABC) as "an expansile osteolytic lesion often multilocular, with blood filled spaces separated by fibrous septa containing osteoclast-type giant cell and reactive bone." ABCs affecting the jaws are rare.[135]

Sun et al. synthesized the literature capturing 92 cases out of 75 reports.[136] Nearly three-quarters first present in the first 2 decades. Two-thirds affect the mandible, of which 85% affect the posterior body and vertical ramus. A bony swelling with or without pain is the most common clinical presentation. Kaffe et al. additionally reported that paresthesia occurred in 9%.[137]

Ninety-four of Sun et al.'s 49 cases, for which radiology could be determined, present as radiolucencies of which two-thirds are multilocular; 2 maxillary cases are radiopaque and 1 mandibular case is mixed.[136] Kaffe et al.'s synthesis of 64 cases reported 87% as radiolucent (53% multilocular); 1 maxillary case was radiopaque and 6 (4 mandible and 2 maxilla) were "mixed."[137] The mixed and radiopaque cases may reflect reactive bone formation within the ABC.[137]

Both Sun et al.[136] and Kaffe et al.'s[137] findings differ from that by Motamedi et al.'s Iranian case series,[138] the largest and most recently reported case series of ABCs affecting the jaws. They displayed similar age and anatomical distribution to reported by Sun et al.[136] except that there was more equal distribution between the mandibular sextants. All were radiolucent.[138]

Sixty-five percent of Kaffe et al.'s multilocular ABCs were greater than 4 cm, whereas 60% of unilocular ABCs were smaller than 4 cm.[137] This may suggest that multilocularity may depend on the lesion's size.

Thirty-nine percent of Kaffe et al.'s 64 cases have a cortex, 33% of the rest are well defined, and 28% poorly defined.[137]

Fifty-five percent of Kaffe et al.'s cases displayed expansion. Twenty-six percent of ABCs in dentate areas displayed tooth displacement and 14% root resorption.[137]

Five of Kaffe et al.'s cases were also associated with unerupted teeth,[137] which may reflect the ABC's predilection for the posterior sextants and the child and adolescent.

Fifteen percent of Sun et al.'s ABCs are considered to arise secondarily to other lesions; the majority are fibrous dysplasia.[136] It is an indication of the ABC's rarity that no clear mention of ABCs being secondary to fibrous dysplasia appeared in any reported case series on fibrous dysplasia. The overall recurrence rate is 13% and is not influenced by resection or curettage. Motamedi et al. reported a similar recurrence rate of 16% after follow-up ranging from 2 to 30 years.[138]

Nasopalatine (Duct) Cyst

The WHO's second edition defines the nasopalatine duct cyst as "a cyst arising from the epithelial residues in the nasopalatine (incisive) canal."[139]

According to Shear and Speight, the nasopalatine duct cyst is the most common nonodontogenic cyst.[140] Its definition should also include the former median palatine cyst and the cyst of the palatine papilla.[140]

Shear and Speight reported that the nasopalatine duct cyst accounted for 12% of all jaw cysts referred to Shear's South African histopathological service.[140] The majority were detected in the fourth decade. The nasopalatine duct cyst was found two times more frequently in males; there was no difference between the White and Black South Africans in this regard.[140]

The most common symptom is swelling.[140] This swelling may be both palatal and buccal. Symptoms are generally not severe and are often ignored by the patient for several years prior to presentation.[140] Nevertheless, Nortje and Farman report that the symptoms may be severe in Black South Africans.[141] Hertzamu et al. (1985) reported that all cysts over 3 cm were in Black South Africans.[142] Tooth displacement was observed often.[140]

Radiographically, a diagnosis of nasopalatine duct cyst may be considered if the width of the canal exceeds 4.79 mm (standard deviation 1.33 mm) and if the anterior-posterior length exceeds 10.19 mm (standard deviation 3.34 mm).[140]

Figure 15.13 is normal, whereas Figures 9.41 and 15.14 are likely to represent nasopalatine duct cysts at the earliest stages. Figures 9.42 and 9.43 are nasopalatine duct cysts.

Lingual Bone Defect

The lingual bone defect was formerly known as "Stafne's bone cyst." It represented a well-defined ovoid radiolucency classically sited between the mandibular canal and the lower border of the mandible (Figure 9.44). It is not a true cyst, but a concavity within the mandible arising from the lingual surface. Philipsen et al. have since defined four subtypes of this phenomenon.[143] The lingual posterior constitutes the majority. The lingual anterior is sited lingual to the incisors, canines, and premolars. Because it lies above the mylohyoid muscle it may be superimposed upon the apices of these teeth and mimic a periapical radiolucency. The border of this radiolucency may less distinct than those of the lingual posterior subtype. The buccal ramus and finally the lingual ramus are sited just below the neck of the condyle. The incidence of the most frequent, the lingual posterior, is 0.10 to 0.48% of panoramic radiographs. Its radiographic visibility requires a reduction in the mineralized tissue by 12%.[143]

The lingual bone defects are rarely observed before 40 years of age, suggesting that the process that results in their creation is not merely the passive development of the mandible around a submandibular gland as was once thought; it is more active (Figure 9.45).[143] Furthermore, they are overwhelmingly reported in males. Philipsen et al. discuss the possible etiologies, clearly favoring the

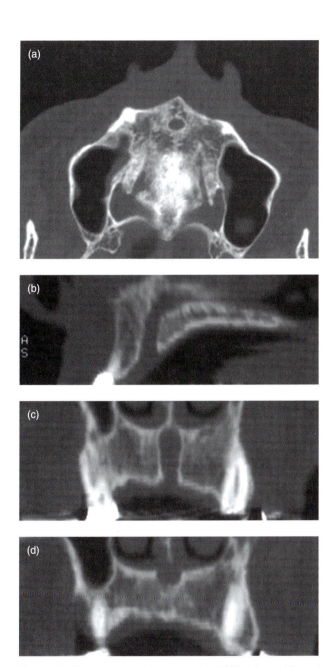

Figure 9.41. Computed tomography (CT) of a widened nasopalatine canal. (a) Axial CT (bone window) displaying the widened canal. (b) Sagital CT (bone window). (c) Coronal CT (bone window). (d) Coronal CT (bone window).

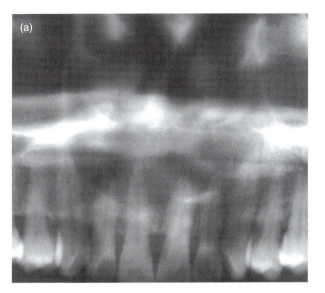

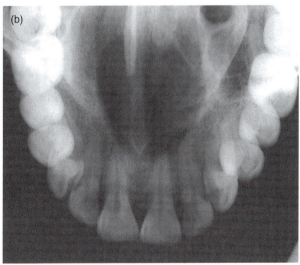

Figure 9.42. Conventional radiographs of a nasopalatine duct cyst. (a) Panoramic radiograph reveals a well-defined radiolucency in the anterior maxilla. (b) Anterior standard occlusal revealed that there was a well-defined radiolucency occupying the midline of the hard palate. If adjacent teeth were carious or restored, pulp vitality testing may be required to distinguish between the nasopalatine duct cyst and a periapical radiolucency of inflammatory origin.

"glandular" hypothesis.[143] This suggests the presence of a hyperplastic or hypertrophic lobe of the submandibular (lingual posterior), sublingual (lingual anterior), and parotid (lingual or buccal ramus) glands putting pressure on the adjacent bone causing focal bone atrophy or resorption.[143]

This clinical phenomenon has been reported globally. Shields and Mann discuss its reduced prevalence and equal sex ratio in African Americans.[144]

Shimizu et al. reported two patterns among their 32 cases on panoramic radiographs of what

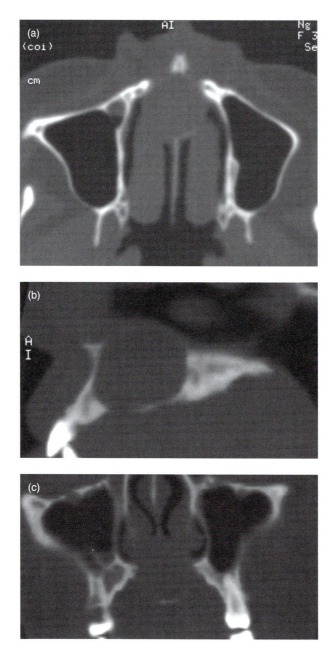

Figure 9.43. Computed tomography (CT) of a large nasopalatinal duct cyst. (a) Axial CT (bone window), above the floor of the nasal cavity, displays round shape and thin cortex in places. (b) Sagital CT (bone window) upward displacement of the nasal floor. (c) Coronal CT (bone window), at level of premolars exhibiting substantial expansion up into the nasal cavity and downward in to the oral cavity.

appears to be, from their images, the lingual posterior subtype.[145] These are "Stafne type" (Figure 9.46) or "Cyst type" (Figure 9.47); the former shows a connection to the lower border of the mandible, whereas the latter does not.[145] Shimizu et al.'s HCT study revealed that the "Cyst type" included only fat and that the "Stafne type" contained other soft tissues. The submandibular glands were more anteriorly positioned in contrast to their contralateral controls. In addition, the glands associated with the "Stafne type" were more laterally placed.[145]

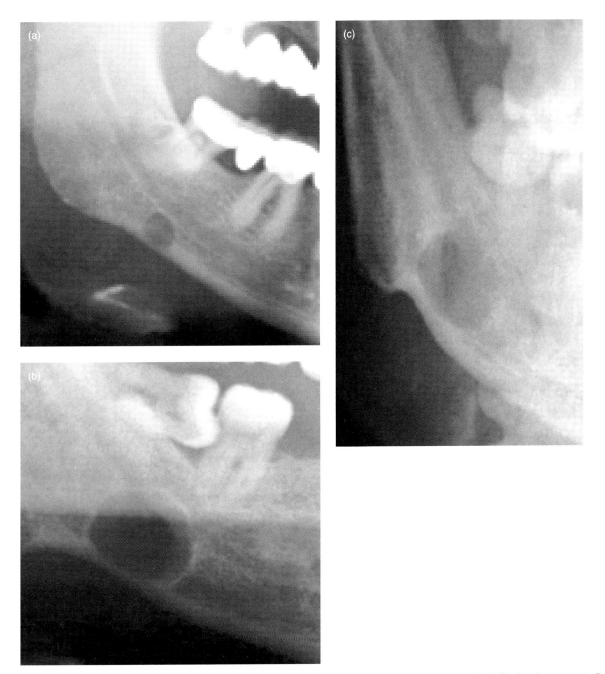

Figure 9.44. Conventional radiography of lingual bone defects (LBD). They were formerly called "Stafne bone cysts." (a) Panoramic radiograph exhibiting a classical LBD. This is a round or oval-shaped well-defined radiolucency, lying anterior to the gonial notch, between the lower border of the mandible and mandibular canal. (b) Panoramic radiograph of another case showing extension above the mandibular canal. (c) Posteroanterior projection of the mandible of the case in (b).

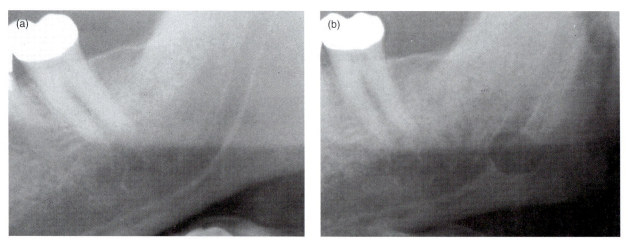

Figure 9.45. Panoramic radiographs taken of the same area of a lingual bone defect (LBD) separated by a few years. (a) A well-defined radiolucency partly superimposed upon the inferior aspect of the inferior dental canal and the lower border of the mandible. A panoramic radiograph taken earlier (b) does not display it at all. LBD generally appear in later life. Although related to the contact between the submandibular gland and mandible, the mechanism is not yet fully understood.

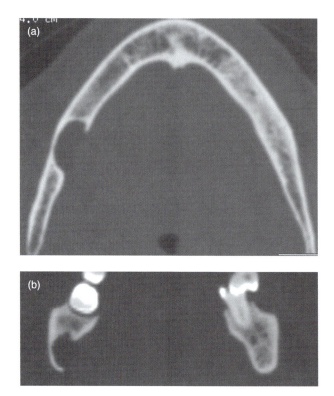

Figure 9.46. Computed tomography (CT) of a lingual bone defect (LBD) in the posterior sextant of a edentulous mandible. (a) Axial CT displays the LBD reaching the inside of the buccal cortex and extending mesially into the bone marrow past the most anterior aspect of the lingual cortical defect. (b) Coronal CT reveals that it extends fully to the lower border of the mandible and extends upward past its most superior aspect of the lingual cortical defect.

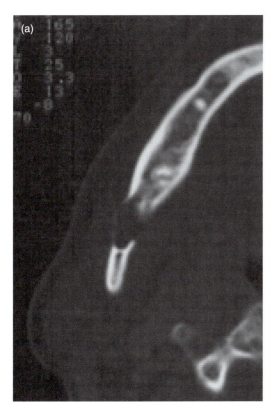 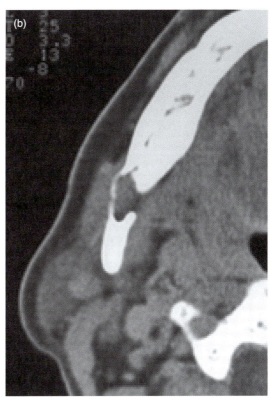

Figure 9.47. The axial computed tomography (CT) displays a lingual bone defect. The bone window (a) depicts the lingual osteum of the defect as being narrower than the mesiodistal extend of the defect within the mandible. The intraosseous defect is lined by a cortex. The buccal cortex is eroded and slightly expanded. The soft-tissue window (b) reveals soft-tissue density. **Note:** The radiolucent band separating the soft-tissue with the defect from that lingual to it is due to beam-hardening caused by the adjacent lingual cortex bordering the defect.

Smith et al. reported a case of bilateral lingual anterior subtype.[146] The use of MRI confirmed the presence of sublingual glands within the defects obviating the need for a biopsy.[146]

References

1. Hazza'a AM, Al-Jamal G. Dental development in subjects with thalassemia major. *J Contemp Dent Pract* 2006;7:63–70.
2. Hazza'a AM, Al-Jamal G. Radiographic features of the jaws and teeth in thalassaemia major. *Dentomaxillofac Radiol* 2006;35:283–388.
3. Tyler PA, Madani G, Chaudhuri R, Wilson LF, Dick EA. The radiological appearances of thalassaemia. *Clin Radiol* 2006;61:40–52.
4. White SC, Cohen JM, Mourshed FA. Digital analysis of trabecular pattern in jaws of patients with sickle cell anemia. *Dentomaxillofac Radiol* 2000;29:119–124.
5. Faber TD, Yoon DC, White SC. Fourier analysis reveals increased trabecular spacing in sickle cell anemia. *J Dent Res* 2002;81:214–218.
6. Roodman GD. Skeletal imaging and management of bone disease. *Hematology Am Soc Hematol Educ Program* 2008;2008:313–319.
7. Lae ME, Vencio EF, Inwards CY, Unni KK, Nascimento AG. Myeloma of the jaw bones: a clinicopathologic study of 33 cases. *Head Neck* 2003;25:373–381.
8. Witt C, Borges AC, Klein K, Neumann HJ. Radiographic manifestations of multiple myeloma in the mandible: a retrospective study of 77 patients. *J Oral Maxillofac Surg* 1997;55:450–453; discussion 454–455.
9. Pisano JJ, Coupland R, Chen SY, Miller AS. Plasmacytoma of the oral cavity and jaws: a clinicopathologic study of 13 cases. *Oral Surg Oral Med Oral Pathol Oral Radiol Endod* 1997;83:265–271.
10. Hicks J, Flaitz CM. Langerhans cell histiocytosis: current insights in a molecular age with emphasis on clinical oral and maxillofacial pathology practice. *Oral Surg Oral Med Oral Pathol Oral Radiol Endod* 2005; 100:S42–66.

11. Dagenais M, Pharoah MJ, Sikorski PA. The radiographic characteristics of histiocytosis X. A study of 29 cases that involve the jaws. *Oral Surg Oral Med Oral Pathol* 1992;74:230–236.
12. Devlin H, Allen P, Graham J, Jacobs R, Nicopoulou-Karayianni K, Lindh C, Marjanovic E, Adams J, Pavitt S, van der Stelt P, Horner K. The role of the dental surgeon in detecting osteoporosis: the OSTEODENT study. *Br Dent J* 2008;204:E16; discussion 560–561.
13. Liu H, Paige NM, Goldzweig CL, Wong E, Zhou A, Suttorp MJ, Munjas B, Orwoll E, Shekelle P. Screening for osteoporosis in men: a systematic review for an American College of Physicians guideline. *Ann Intern Med* 2008;148:685–701. Review. Summary for patients in *Ann Intern Med* 2008;148:I35.
14. Geraets WG, Verheij JG, van der Stelt PF, Horner K, Lindh C, Nicopoulou-Karayianni K, Jacobs R, Marjanovic EJ, Adams JE, Devlin H. Selecting regions of interest on intraoral radiographs for the prediction of bone mineral density. *Dentomaxillofac Radiol* 2008;37:375–379.
15. Lindh C, Horner K, Jonasson G, Olsson P, Rohlin M, Jacobs R, Karayianni K, van der Stelt P, Adams J, Marjanovic E, Pavitt S, Devlin H. The use of visual assessment of dental radiographs for identifying women at risk of having osteoporosis: the OSTEODENT project. *Oral Surg Oral Med Oral Pathol Oral Radiol Endod* 2008;106:285–293.
16. Chang JI, Som PM, Lawson W. Unique imaging findings in the facial bones of renal osteodystrophy. *AJNR Am J Neuroradiol* 2007;28:608–609.
17. Stålberg P, Carling T. Familial parathyroid tumors: diagnosis and management. *World J Surg* 2009;33:2234–2243.
18. DeLellis RA, Mazzaglia P, Mangray S. Primary hyperparathyroidism: a current perspective. *Arch Pathol Lab Med* 2008;132:1251–1262.
19. Asaumi J, Aiga H, Hisatomi M, Shigehara H, Kishi K. Advanced imaging in renal osteodystrophy of the oral and maxillofacial region. *Dentomaxillofac Radiol* 2001;30:59–62.
20. World Health Organization (Oral health Program). Global data on incidence of oral cancer. 2005. http://www.who.int/oral_health/publications/oral_cancer_brochure.pdf
21. Eversole LR, Siar CH, van der Waal I. Primary intraosseous squamous cell carcinomas. *The Clinical Outline of Oral Pathology Diagnosis and Treatment*. Lea & Febiger, Philadelphia 1992: pp 290–291.
22. Huang JW, Luo HY, Li O, Li TJ. Primary intraosseus squamous cell carcinoma of the jaws. Clinicopathological presentation and prognostic factors. *Arch Pathol Lab Med* 2009;133:1834–1840.
23. MacDonald-Jankowski DS. The involvement of the maxillary antrum by odontogenic keratocysts. *Clin Radiol* 1992;45:31–33.
24. Faitaroni LA, Bueno MR, De Carvalhosa AA, Bruehmueller Ale KA, Estrela C. Ameloblastoma suggesting large apical periodontitis. *J Endod* 2008;34:216–219.
25. Cunha EM, Fernandes AV, Versiani MA, Loyola AM. Unicystic ameloblastoma: a possible pitfall in periapical diagnosis. *Int Endod J* 2005;38:334–340.
26. Lombardi T, Bischof M, Nedir R, Vergain D, Galgano C, Samson J, Küüffer R. Periapical central giant cell granuloma misdiagnosed as odontogenic cyst. *Int Endod J* 2006;39:510–515.
27. Kramer IRH, Pindborg JJ, Shear M. *Histological Typing of Odontogenic Tumours*, 2nd ed. Springer-Verlag, London 1992: pp 40–41.
28. Gundappa M, Ng SY, Whaites EJ. Comparison of ultrasound, digital and conventional radiography in differentiating periapical lesions. *Dentomaxillofac Radiol* 2006;35:326–333.
29. Kramer IRH, Pindborg JJ, Shear M. *Histological Typing of Odontogenic Tumours*, 2nd ed. Springer-Verlag, London 1992: pp 42.
30. Pompura JR, Sándor GK, Stoneman DW. The buccal bifurcation cyst: a prospective study of treatment outcomes in 44 sites. *Oral Surg Oral Med Oral Pathol Oral Radiol Endod* 1997;83:215–221.
31. Iatrou I, Theologie-Lygidakis N, Leventis M. Intraosseous cystic lesions of the jaws in children: a retrospective analysis of 47 consecutive cases. *Oral Surg Oral Med Oral Pathol Oral Radiol Endod* 2009;107:485–492.
32. Kramer IRH, Pindborg JJ, Shear M. *Histological Typing of Odontogenic Tumours*, 2nd ed. Springer-Verlag, London 1992: p 37.
33. Shear M, Speight P. *Cysts of the Oral and Maxillofacial Regions*, 4th ed. Blackwell, Munksgaard 2006: pp 59–79.
34. Altini M, Shear M. The lateral periodontal cyst: an update. *J Oral Pathol Med* 1992;21:245–250.
35. Wysocki GP, Brannon RB, Gardner DG, Sapp P. Histogenesis of the lateral periodontal cyst and the gingival cyst of the adult. *Oral Surg Oral Med Oral Pathol* 1980;50:327–334.
36. Cohen DA, Neville BW, Damm DD, White DK. The lateral periodontal cyst. A report of 37 cases. *J Periodontol* 1984;55:230–234.
37. Rasmusson LG, Magnusson BC, Borrman H. The lateral periodontal cyst. A histopathological and radiographic study of 32 cases. *Br J Oral Maxillofac Surg* 1991;29:54–57.
38. Jones AV, Franklin CD. An analysis of oral and maxillofacial pathology found in adults over a 30-year period. *J Oral Pathol Med* 2006;35:392–401.
39. Formoso Senande MF, Figueiredo R, Berini Aytés L, Gay Escoda C. Lateral periodontal cysts: a retrospective study of 11 cases. *Med Oral Patol Oral Cir Bucal* 2008;13:E313–317.

40. Ramer M, Valauri D. Multicystic lateral periodontal cyst and botryoid odontogenic cyst. Multifactorial analysis of previously unreported series and review of literature. *N Y State Dent J* 2005;71:47–51.
41. Méndez P, Junquera L, Gallego L, Baladrón J. Botryoid odontogenic cyst: clinical and pathological analysis in relation to recurrence. *Med Oral Patol Oral Cir Bucal.* 2007;12:E594–598.
42. Waner M, Suen JY. Management of congenital vascular lesions of the head and neck. *Oncology* (Williston Park) 1995;9:989–994, 997; discussion 998 passim.
43. Zlotogorski A, Buchner A, Kaffe I, Schwartz-Arad D. Radiological features of central haemangioma of the jaws. *Dentomaxillofac Radiol* 2005;34:292–296.
44. Curran AE, Damm DD, Drummond JF. Pathologically significant pericoronal lesions in adults: Histopathologic evaluation. *J Oral Maxillofac Surg* 2002;60:613–617; discussion 618. Comments by Slater and Flick in *J Oral Maxillofac Surg* 2003;61:149–150; author reply 150.
45. Ikeshima A, Tamura Y. Differential diagnosis between dentigerous cyst and benign tumor with an embedded tooth. *J Oral Sci* 2002;44:13–17.
46. Kramer IRH, Pindborg JJ, Shear M. *Histological Typing of Odontogenic Tumours*, 2nd ed. Springer-Verlag, London 1992: p 36.
47. MacDonald-Jankowski DS, Chan KC. Clinical presentation of dentigerous cysts: systematic review. *Asian J Oral Maxillofac Surg* 2005;15:109–120.
48. Yildirim G, Ataoğlu H, Mihmanli A, Kiziloğlu D, Avunduk MC. Pathologic changes in soft tissues associated with asymptomatic impacted third molars. *Oral Surg Oral Med Oral Pathol Oral Radiol Endod* 2008;106:14–18.
49. Shear M, Speight P. *Cysts of the Oral and Maxillofacial Regions*, 4th ed. Blackwell, Munksgaard 2006: pp 59–79.
50. Kaugars GE, Miller ME, Abbey LM. Odontomas. *Oral Surg Oral Med Oral Pathol* 1989;67:172–176.
51. Jones AV, Craig GT, Franklin CD. Range and demographics of odontogenic cysts diagnosed in a UK population over a 30-year period. *J Oral Pathol Med* 2006;35:500–507.
52. Ioannidou F, Mustafa B, Seferiadou-Mavropoulou T. [Odontogenic cysts of the jaws. A clinicostatistical study] *Stomatologia* (Athenai) 1989;46:81–90. (Greek, Modern)
53. Ledesma-Montes C, Hernández-Guerrero JC, Garcés-Ortíz M. Clinico-pathologic study of odontogenic cysts in a Mexican sample population. *Arch Med Res* 2000;31:373–376.
54. Daley TD, Wysocki GP. The small dentigerous cyst. A diagnostic dilemma. *Oral Surg Oral Med Oral Pathol Oral Radiol Endod* 1995;79:77–81.
55. Struthers P, Shear M. Root resorption by ameloblastomas and cysts of the jaws. *Int J Oral Surg* 1976;5:128–132.
56. Wang JT. Unicystic ameloblastoma: a clinicopathological appraisal. *Taiwan Yi Xue Hui Za Zhi* 1985;84:1363–1370.
57. Shear M, Singh S. Age-standardized incidence rates of ameloblastoma and dentigerous cyst on the Witwatersrand, South Africa. *Community Dent Oral Epidemiol* 1978;6:195–199.
58. Gardner DG, Heikinheimo K, Shear M, Philipsen HP, Coleman H. Ameloblastomas. Barnes L, Eveson J, Reichert P, Sidransky D, eds. *WHO Classification of Tumours, Pathology and Genetics of Tumours of the Head and Neck*. International Agency for Research on Cancer (IARC), Lyon 2005: pp 296–300.
59. MacDonald-Jankowski DS, Yeung R, Lee KM, Li TK. Ameloblastoma in the Hong Kong Chinese. Part 1: systematic review and clinical presentation. *Dentomaxillofac Radiol* 2004;33:71–82.
60. MacDonald-Jankowski DS, Yeung R, Lee KM, Li TK. Ameloblastoma in the Hong Kong Chinese. Part 2: systematic review and radiological presentation. *Dentomaxillofac Radiol* 2004;33:141–151.
61. Philipsen HP, Reichart PA, Takata T. Desmoplastic ameloblastoma (including "hybrid" lesion of ameloblastoma). Biological profile based on 100 cases from the literature and own files. *Oral Oncol* 2001;37:455–460.
62. Luo HY, Li TJ. Odontogenic tumors: A study of 1309 cases in a Chinese population. *Oral Oncol* 2009 (Jan 13) [Epub ahead of print].
63. Arotiba GT, Ladeinde AL, Arotiba JT, Ajike SO, Ugboko VI, Ajayi OF. Ameloblastoma in Nigerian children and adolescents: a review of 79 cases. *J Oral Maxillofac Surg* 2005;63:747–751.
64. Kaffe I, Buchner A, Taicher S. Radiologic features of desmoplastic variant of ameloblastoma. *Oral Surg Oral Med Oral Pathol* 1993;76:525–529.
65. Lee PK, Samman N, Ng IO. Unicystic ameloblastoma—use of Carnoy's solution after enucleation. *Int J Oral Maxillofac Surg* 2004;33:263–267.
66. Hong J, Yun PY, Chung IH, Myoung H, Suh JD, Seo BM, Lee JH, Choung PH. Long-term follow up on recurrence of 305 ameloblastoma cases. *Int J Oral Maxillofac Surg* 2007;36:283–288.
67. Lau SL, Samman N. Recurrence related to treatment modalities of unicystic ameloblastoma: a systematic review. *Int J Oral Maxillofac Surg* 2006;35:681–690.
68. Philipsen HP, Reichart PA. Unicystic ameloblastoma. A review of 193 cases from the literature. *Oral Oncol* 1998;34:317–325.
69. Zhang LL, Yang R, Zhang L, Li W, MacDonald-Jankowski D, Poh CF. Dentigerous cyst: a retrospective clinicopathological analysis of 2082 dentigerous cysts in British Columbia, Canada. *Int J Oral Maxillofac Surg* 2010;39:878–882.
70. Li TJ, Wu YT, Yu SF, Yu GY. Unicystic ameloblastoma: a clinicopathologic study of 33 Chinese patients. *Am J Surg Pathol* 2000;24:1385–1392.

71. MacDonald-Jankowski DS, Li TK. Computed tomography of ameloblastomas affecting a Hong Kong Chinese community. Part presented at the 17th International Congress of Dental and Maxillofacial Radiology in Amsterdam. June 28–July 2, 2009.
72. Asaumi J, Matsuzaki H, Hisatomi M, Konouchi H, Shigehara H, Kishi K. Application of dynamic MRI to differentiating odontogenic myxomas from ameloblastomas. *Eur J Radiol* 2002;43:37–41.
73. Asaumi J, Hisatomi M, Yanagi Y, Matsuzaki H, Choi YS, Kawai N, Konouchi H, Kishi K. Assessment of ameloblastomas using MRI and dynamic contrast-enhanced MRI. *Eur J Radiol* 2005;56:25–30.
74. Hisatomi M, Asaumi J, Konouchi H, Yanagi Y, Kishi K. A case of glandular odontogenic cyst associated with ameloblastoma: correlation of diagnostic imaging with histopathological features. *Dentomaxillofac Radiol* 2000;29:249–253.
75. Sciubba JJ, Eversole LR, Slootweg PJ. Odontogenic/ameloblastic carcinomas. Barnes L, Eveson J, Reichert P, Sidransky D, eds. *WHO Classification of Tumours, Pathology and Genetics of Tumours of the Head and Neck*. International Agency for Research on Cancer (IARC), Lyon 2005: pp 287–189.
76. Senra GS, Pereira AC, Murilo dos Santos L, Carvalho YR, Brandão AA. Malignant ameloblastoma metastasis to the lung: a case report. *Oral Surg Oral Med Oral Pathol Oral Radiol Endod* 2008;105:e42–46.
77. Henderson JM, Sonnet JR, Schlesinger C, Ord RA. Pulmonary metastasis of ameloblastoma: case report and review of the literature. *Oral Surg Oral Med Oral Pathol Oral Radiol Endod* 1999;88:170–176.
78. Benlyazid A, Lacroix-Triki M, Aziza R, Gomez-Brouchet A, Guichard M, Sarini J. Ameloblastic carcinoma of the maxilla: case report and review of the literature. *Oral Surg Oral Med Oral Pathol Oral Radiol Endod* 2007;104:e17–24.
79. Buchner A, Odell EW. Odontogenic myxoma/myxofibroma. Barnes L, Eveson J, Reichert P, Sidransky D, eds. *WHO Classification of Tumours, Pathology and Genetics of Tumours of the Head and Neck*. International Agency for Research on Cancer (IARC), Lyon 2005: pp 316–317.
80. MacDonald-Jankowski DS, Yeung R, Lee KM, Li TK. Odontogenic myxomas in the Hong Kong Chinese: clinico-radiological presentation and systematic review. *Dentomaxillofac Radiol* 2002;31:71–83.
81. Zhang J, Wang H, He X, Niu Y, Li X. Radiographic examination of 41 cases of odontogenic myxomas on the basis of conventional radiographs. *Dentomaxillofac Radiol* 2007;36:160–167.
82. Noffke CE, Raubenheimer EJ, Chabikuli NJ, Bouckaert MM. Odontogenic myxoma: review of the literature and report of 30 cases from South Africa. *Oral Surg Oral Med Oral Pathol Oral Radiol Endod* 2007;104:101–119.
83. MacDonald-Jankowski DS, Yeung R, Lee KM, Li TK. Computed tomography of odontogenic myxoma. *Clin Radiol* 2004;59:281–287.
84. Kramer IRH, Pindborg JJ, Shear M. *Histological Typing of Odontogenic Tumours*, 2nd ed. Springer-Verlag, London 1992: p 23.
85. Araki M, Kameoka S, Mastumoto N, Komiyama K. Usefulness of cone beam computed tomography for odontogenic myxoma. *Dentomaxillofac Radiol* 2007;36:423–427.
86. Koseki T, Kobayashi K, Hashimoto K, Ariji Y, Tsuchimochi M, Toyama M, Araki M, Igarashi C, Koseki Y, Ariji E. Computed tomography of odontogenic myxoma. *Dentomaxillofac Radiol* 2003;32:160–165.
87. Martínez-Mata G, Mosqueda-Taylor A, Carlos-Bregni R, de Almeida OP, Contreras-Vidaurre E, Vargas PA, Cano-Valdéz AM, Domínguez-Malagón H. Odontogenic myxoma: clinico-pathological, immunohistochemical and ultrastructural findings of a multicentric series. *Oral Oncol* 2008;44:601–607.
88. Hisatomi M, Asaumi J, Konouchi H, Yanagi Y, Matsuzaki H, Kishi K. Comparison of radiographic and MRI features of a root-diverging odontogenic myxoma, with discussion of the differential diagnosis of lesions likely to move roots. *Oral Dis* 2003;9:152–157.
89. Simon EN, Merkx MA, Vuhahula E, Ngassapa D, Stoelinga PJ. Odontogenic myxoma: a clinicopathological study of 33 cases. *Int J Oral Maxillofac Surg* 2004;33:333–337.
90. Li TJ, Sun LS, Luo HY. Odontogenic myxoma: a clinicopathologic study of 25 cases. *Arch Pathol Lab Med* 2006;130:1799–1806.
91. Philipsen HP. Keratocystic odontogenic tumour. Barnes L, Eveson J, Reichert P, Sidransky D, eds. *WHO Classification of Tumours, Pathology and Genetics of Tumours of the Head and Neck*. International Agency for Research on Cancer (IARC), Lyon 2005: pp 306–307.
92. Shear M, Speight P. *Cysts of the Oral and Maxillofacial Regions*, 4th ed. Blackwell, Munksgaard 2006: pp 6–58.
93. Wright JM. The odontogenic keratocyst: orthokeratinized variant. *Oral Surg Oral Med Oral Pathol* 1981;51:609–618.
94. Kramer IRH, Pindborg JJ, Shear M. *Histological Typing of Odontogenic Tumours*, 2nd ed. Springer-Verlag, London 1992: p 35.
95. MacDonald-Jankowski DS. Keratocystic odontogenic tumour; a systematic review. *Dentomaxillofac Radiol* 2011;40:1–23.
96. MacDonald-Jankowski DS. Keratocystic odontogenic tumour in a Hong Kong community; the clinical and radiological presentations and the outcomes of treatment and follow-up. *Dentomaxillofac Radiol* 2010;39:167–175.

97. Haring JI, Van Dis ML. Odontogenic keratocysts: a clinical, radiographic, and histopathologic study. *Oral Surg Oral Med Oral Pathol* 1988;66:145–153.
98. Lo Muzio L. Nevoid basal cell carcinoma syndrome (Gorlin syndrome). *Orphanet J Rare Dis* 2008;3:32.
99. Kimonis VE, Mehta SG, Digiovanna JJ, Bale SJ, Pastakia B. Radiological features in 82 patients with nevoid basal cell carcinoma (NBCC or Gorlin) syndrome. *Genet Med* 2004;6:495–502.
100. Lam EW, Lee L, Perschbacher SE, Pharoah MJ. The occurrence of keratocystic odontogenic tumours in nevoid basal cell carcinoma syndrome. *Dentomaxillofac Radiol* 2009;38:475–479.
101. MacDonald-Jankowski DS, Li TKL. Orthokeratinized odontogenic cyst in a Hong Kong community; the clinical and radiological presentations and the outcomes of treatment and follow-up. *Dentomaxillofac Radiol* 2010;39:240–245.
102. MacDonald-Jankowski DS. Orthokeratinized odontogenic cyst; systematic review. *Dentomaxillofac Radiol* 2010;39:455–467.
103. Gardner DG, Kessler HP, Morency R, Schaffner DL. The glandular odontogenic cyst: an apparent entity. *J Oral Pathol* 1988;17:359–366.
104. Kramer IRH, Pindborg JJ, Shear M. *Histological Typing of Odontogenic Tumours*, 2nd ed. Springer-Verlag, London 1992: p 38.
105. Slootweg PJ. Lesions of the jaws. *Histopathology* 2009; 54:401–418.
106. MacDonald-Jankowski DS. Glandular odontogenic cyst: a systematic review. *Dentomaxillofac Radiol* 2010;39:127–139.
107. Reichart PA. Squamous odontogenic tumour. In Barnes L, Eveson J, Reichert P, Sidransky D, eds. *WHO Classification of Tumours, Pathology and Genetics of Tumours of the Head and Neck*. International Agency for Research on Cancer (IARC), Lyon 2005: p 301.
108. Jundt G. Central giant cell lesion. In Barnes L, Eveson J, Reichert P, Sidransky D, eds. *WHO Classification of Tumours, Pathology and Genetics of Tumours of the Head and Neck*. International Agency for Research on Cancer (IARC), Lyon 2005: p 324.
109. de Lange J, van den Akker HP, van den Berg H. Central giant cell granuloma of the jaw: a review of the literature with emphasis on therapy options. *Oral Surg Oral Med Oral Pathol Oral Radiol Endod* 2007; 104:603–615.
110. Stavropoulos F, Katz J. Central giant cell granulomas: a systematic review of the radiographic characteristics with the addition of 20 new cases. *Dentomaxillofac Radiol* 2002;31:213–217. Review. Erratum in *Dentomaxillofac Radiol* 2002;31:394.
111. Slootweg PJ. Ameloblastic fibroma. In Barnes L, Eveson J, Reichert P, Sidransky D, eds. *WHO Classification of Tumours, Pathology and Genetics of Tumours of the Head and Neck*. International Agency for Research on Cancer (IARC), Lyon 2005: p 308.
112. Chen Y, Wang JM, Li TJ. Ameloblastic fibroma: a review of published studies with special reference to its nature and biological behavior. *Oral Oncol* 2007;43:960–969.
113. Jundt G. Cherubism. In Barnes L, Eveson J, Reichert P, Sidransky D, eds. *WHO Classification of Tumours, Pathology and Genetics of Tumours of the Head and Neck*. International Agency for Research on Cancer (IARC), Lyon 2005: p 325.
114. Von Wowern N. Cherubism: a 36-year long-term follow-up of 2 generations in different families and review of the literature. *Oral Surg Oral Med Oral Pathol Oral Radiol Endod* 2000;90:765–772.
115. Meng XM, Yu SF, Yu GY. Clinicopathologic study of 24 cases of cherubism. *Int J Oral Maxillofac Surg* 2005;34: 350–356.
116. Roginsky VV, Ivanov AL, Ovtchinnikov IA, Khonsari RH. Familial cherubism: the experience of the Moscow Central Institute for Stomatology and Maxillo-Facial Surgery. *Int J Oral Maxillofac Surg* 2009;38:218–223.
117. Beaman FD, Bancroft LW, Peterson JJ, Kransdorf MJ, Murphey MD, Menke DM. Imaging characteristics of cherubism. *AJR Am J Roentgenol* 2004;182:1051–1054.
118. Pontes FS, Ferreira AC, Kato AM, Pontes HA, Almeida DS, Rodini CO, Pinto DS Jr. Aggressive case of cherubism: 17-year follow-up. *Int J Pediatr Otorhinolaryngol* 2007;71:831–835.
119. Carroll AL, Sullivan TJ. Orbital involvement in cherubism. *Clin Experiment Ophthalmol* 2001;29: 38–40.
120. Ozkan Y, Varol A, Turker N, Aksakalli N, Basa S. Clinical and radiological evaluation of cherubism: a sporadic case report and review of the literature. *Int J Pediatr Otorhinolaryngol* 2003;67:1005–1012.
121. Elfahsi A, Oujilal A, Lahlou M, Lazrak A, Kzadri M. [An ophthalmological complication of cherubism] *Rev Stomatol Chir Maxillofac* 2007;108:58–60. (in French)
122. Von Wowern N, Hjørting-Hansen E, Edeling CJ. Bone scintigraphy of benign jaw lesions. *Int J Oral Surg* 1978;7:528–533.
123. Jundt G. Simple bone cyst. Barnes L, Eveson J, Reichert P, Sidransky D, eds. *WHO Classification of Tumours, Pathology and Genetics of Tumours of the Head and Neck*. International Agency for Research on Cancer (IARC), Lyon 2005: p 327.
124. Suei Y, Taguchi A, Tanimoto K. A comparative study of simple bone cysts of the jaw and extracranial bones. *Dentomaxillofac Radiol* 2007;36:125–129.
125. Harnet JC, Lombardi T, Klewansky P, Rieger J, Tempe MH, Clavert JM. Solitary bone cyst of the jaws: a review of the etiopathogenic hypotheses. *J Oral Maxillofac Surg* 2008;66:2345–2348.
126. Shear M, Speight P. *Cysts of the Oral and Maxillofacial Regions*, 4th ed. Blackwell, Munksgaard 2006: pp 156–161.

127. Beasley JD 3rd. Traumatic cyst of the jaws: report of 30 cases. *J Am Dent Assoc* 1976;92:145–152.
128. MacDonald-Jankowski DS. Traumatic bone cysts in the jaws of a Hong Kong Chinese population. *Clin Radiol* 1995;50:787–791.
129. Suei Y, Taguchi A, Tanimoto K. Simple bone cyst of the jaws: evaluation of treatment outcome by review of 132 cases. *J Oral Maxillofac Surg* 2007;65:918–923.
130. Melrose RJ, Abrams AM, Mills BG. Florid osseous dysplasia. A clinical-pathologic study of thirty-four cases. *Oral Surg Oral Med Oral Pathol* 1976;41:62–82.
131. Matsumura S, Murakami S, Kakimoto N, Furukawa S, Kishino M, Ishida T, Fuchihata H. Histopathologic and radiographic findings of the simple bone cyst. *Oral Surg Oral Med Oral Pathol Oral Radiol Endod* 1998;85: 619–625.
132. Ferreira Júnior O, Damante JH, Lauris JR. Simple bone cyst versus odontogenic keratocyst: differential diagnosis by digitized panoramic radiography. *Dentomaxillofac Radiol* 2004;33:373–378.
133. Suei Y, Taguchi A, Kurabayashi T, Kobayashi F, Nojiri M, Tanimoto K. Simple bone cyst: investigation on the presence of gas in the cavity using computed tomography—review of 52 cases. *Oral Surg Oral Med Oral Pathol Oral Radiol Endod* 1998;86:592–594.
134. Eriksson L, Hansson LG, Akesson L, Ståhlberg F. Simple bone cyst: a discrepancy between magnetic resonance imaging and surgical observations. *Oral Surg Oral Med Oral Pathol Oral Radiol Endod* 2001; 92:694–698.
135. Jundt G. Aneurysmal bone cyst. Barnes L, Eveson J, Reichert P, Sidransky D, eds. *WHO Classification of Tumours, Pathology and Genetics of Tumours of the Head and Neck*. International Agency for Research on Cancer (IARC), Lyon 2005: p 326.
136. Sun ZJ, Sun HL, Yang RL, Zwahlen RA, Zhao YF. Aneurysmal bone cysts of the jaws. *Int J Surg Pathol* 2009 (Feb 19) [Epub ahead of print].
137. Kaffe I, Naor H, Calderon S, Buchner A. Radiological and clinical features of aneurysmal bone cyst of the jaws. *Dentomaxillofac Radiol* 1999;28:167–172.
138. Motamedi MH, Navi F, Eshkevari PS, Jafari SM, Shams MG, Taheri M, Abbas FM, Motahhari P. Variable presentations of aneurysmal bone cysts of the jaws: 51 cases treated during a 30-year period. *J Oral Maxillofac Surg* 2008;66:2098–2103.
139. Kramer IRH, Pindborg JJ, Shear M. *Histological Typing of Odontogenic Tumours*, 2nd ed. Springer-Verlag, London 1992: p 39.
140. Shear M, Speight P. *Cysts of the Oral and Maxillofacial Regions*, 4th ed. Blackwell, Munksgaard 2006: pp 108–118.
141. Nortjé CJ, Farman AG. Nasopalatine duct cyst. An aggressive condition in adolescent Negroes from South Africa? *Int J Oral Surg* 1978;7:65–72.
142. Hertzamu Y, Cohen M, Mendelsohn DB. Nasopalatine duct cyst. *Clin Radiol* 1985;36:153–158.
143. Philipsen HP, Takata T, Reichart PA, Sato S, Suei Y. Lingual and buccal mandibular bone depressions: a review based on 583 cases from a world-wide literature survey, including 69 new cases from Japan. *Dentomaxillofac Radiol* 2002;31:281–290.
144. Shields ED, Mann RW. Salivary glands and human selection: a hypothesis. *J Craniofac Genet Dev Biol* 1996;16:126–136.
145. Shimizu M, Osa N, Okamura K, Yoshiura K. CT analysis of the Stafne's bone defects of the mandible. *Dentomaxillofac Radiol* 2006;35:95–102.
146. Smith MH, Brooks SL, Eldevik OP, Helman JI. Anterior mandibular lingual salivary gland defect: a report of a case diagnosed with cone-beam computed tomography and magnetic resonance imaging. *Oral Surg Oral Med Oral Pathol Oral Radiol Endod* 2007;103:e71–78.

Chapter 10
Radiopacities

Introduction

A radiopacity is the "white" area on a conventional radiograph. it represents a tissue or a structure within the patient, which attenuates the primary beam of X-rays more than adjacent tissue or structures. In the normal patient presenting to the oral and maxillofacial practitioner, the normal radiopaque structures are anatomical: the teeth, the bones of the jaws (including the middle-third of the face and nasal bones), the stylohyoid complex (including the hyoid bone), the skull base, and cervical vertebrae. Although the term *radiopacity* can be used for any such tissue or structure it is frequently applied only to those, which suggest a lesion or disease process. These radiopacities are due to deposition of mineralized tissue. This deposition reflects two different processes. Deposition either by bone cells in dysplastic and neoplastic lesions or by nonbone cells in dystrophic processes. The last arise usually in chronic inflammatory lesions or those with multiple episodes of inflammation. Dystrophic calcification is commonly observed in the soft tissues of the jaws as calcification of the tonsils (now usually secondary to tonsillitis) or cervical lymph nodes. It also occasionally presents in atherosclerotic plaques in blood vessel walls.

In addition to *conventional radiography*, advanced imaging modalities such as *computed tomography* (CT) and *magnetic resonance* (MRI) are frequently used to investigate jaw lesions. Although calcified lesions and structures are still white on images made by either *helical computed tomography* (HCT, see Chapter 4) or *cone-beam computed tomography* (CBCT, see Chapter 5), they appear black on *MRI* as do the air-filled spaces and blood vessels (see Chapter 6).

The term *significant* will be used only when the feature it is qualifying is $P < 0.05$. Table 10.1 overviews the causes of radiopacities of the jaws. Because almost all of these have been addressed in other texts, they will not be pursued further here. For radiopacities within the bony jaws, refer to Figures 10.1 and 10.2.

Radiopacities Outside the Bony Jaws

CALCIFICATIONS OF THE STYLOHYOID COMPLEX

The 12 most common patterns of calcification of the stylohyoid complex are set out in Figure 10.3.[1] These are based on the four developmental regions of this complex: the tympanohyal (skull base), stylohyal (majority of the styloid process), ceratohyal (contributes to elongated styloid process and stylohyoid ligament), and hypohyal (lesser horn of hyoid component of stylohyoid ligament). This system of patterns was introduced for clinical use because traditional measurement from panoramic radiographs is subject to substantial magnification and distortion, especially in the region posterior to the alveolar processes.[1]

The following paragraphs summarize the different prevalences of the more common patterns in two world communities, Hong Kong and London. The following percentages are cited in that order. A normal styloid process (patterns a to d; 84% to 73%) and pattern d on its own (67% to 40%) were significantly more prevalent in the Hong Kong Chinese (Figures 10.4, 10.5), whereas a calcified stylohyoid ligament (patterns f to k) were significantly more frequent in Londoners (4% to 16%) (Figure 10.6). Segmentation was significantly more frequent in Londoners (6% to 23%). Bilateral symmetry, with regard to pattern, was significantly more frequent in the Chinese (100% to 93%) (Figures 10.5, 10.7). There was no significant

Oral and Maxillofacial Radiology: A Diagnostic Approach,
David MacDonald. © 2011 David MacDonald

Table 10.1. Radiopacities of the jaws

Step	Cause of Radiopacity	Step	Cause of Radiopacity

1.	Development Artifact
	"White metal-like spots"
	Fixer on undeveloped film
		5.	Radiopacity/ies in the soft tissues
			Round
			  Neck; in vertical line (dystrophic from behind ramus to thoracic inlet
			    Lymph nodes
			      a. TB in the elderly or Indian subcontinent patients
			      b. Sarcoidosis
			      c. Previous radiotherapy treated lymphoma

2.	Inadequate Patient Preparation Jewelry (earrings, necklace, facial, and tongue piercing) Hair ornaments Hearing aid Removable dentures and orthodontic appliances		

3.	Normal Anatomy and Variants
	Radiopacity over roots of mandibular premolars (on panoramic radiograph), perhaps bilateral
	  Torus mandibularis
	Linear or ovoid opacities in near vertical axis behind or superimposed on angle of mandible
	  Stylohoid complex
	A vertical fingerlike radiopacity usually with a reticular or honeycomb pattern below greater horn of the hyoid
	  Thyroid; superior horn
			Neck; broadly in horizontal line from:
			  a. Behind ramus to 2nd maxillary molar
			    Parotid sialoliths
			  b. Below gonial notch
			    Submandibular sialoliths
			Group centered about mandibular foramen
			  Tonsilloliths
			Single, anywhere
			  Acne scar (obvious skin scar)
			Multiple, anywhere
			  Gardner's Syndrome
			Target and round; anywhere with perhaps discoloration of overlying skin
			  Hemangioma

4.	Metal-like/Iatrogenic
	Restorative material
	  Overextended root filling
	  Amalgam in socket
	  Amalgam retrograde with apicectomy
	Broken instrument
	  Endodontic
	  Elevator
	Surgical packs.
	Bone plates/wires
	Implants
			Sheets
			  a. Generalized
			    Scleroderma
			  b. Localized
			    Myositis ossificans
			Bars
			  Cystercicosis (parasite)
			Parallel lines
			  Arteriosclerotic blood vessels (diabetes OR chronic renal disease)

		6.	Radiopacities within the Bony Jaws

difference between Hong Kong and London for an elongated styloid process (pattern e; 9% and 8%, respectively) (Figure 10.7).

Using the same table, Okabe et al.[2] observed similarity in pattern distribution between their 80-year-old Japanese and the above Hong Kong Chinese. They suggested that this was perhaps due to similarities in East Asian genotypes and phenotypes. The major difference was that their sample displayed a substantially greater proportion of pattern e (35%). As the Okabe et al.[2] study was exclusively 80-year-olds, this difference was considered to be a phenomenon of aging. This was supported by a Brazilian report measuring directly from panoramic radiographs.[3] Okabe et al.[2] also reported that the elongated styloid process (Figure 10.7) correlated significantly with increased serum calcium concentration and heel bone density. It also correlated significantly with the patient's height and weight. Therefore, these findings "may provide potentially life-saving information about elderly people."[2]

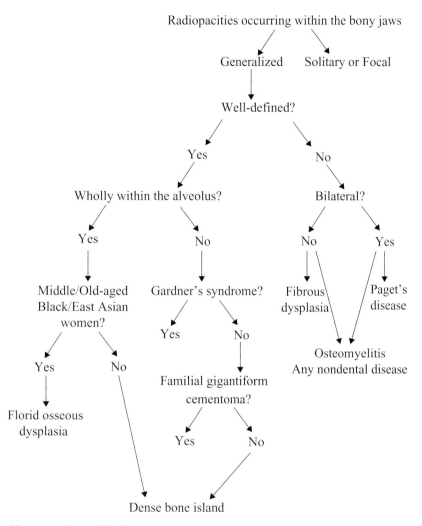

Figure 10.1. Radiopacities occurring within the bony jaws.

The styloid process also runs in an anterior and medial direction to the hyoid bone (Figures 10.7a and b, respectively); the inferior horn and upper half of the body of the hyoid bone form the caudal component of the stylohyoid complex.

The upper half of the hyoid bone was derived in common with the cranial components of the stylohyoid complex from the second pharyngeal arch or Reichert's cartilage. The rest of the hyoid is derived from the third pharyngeal arch. The hyoid bone arises from 6 centers of ossification, 2 for the body and 1 for each horn.[4] The lesser horn may not fuse with the body (see Figure 1.13b), but it is attached by fibrous tissue to the greater horn, which in turn articulates with the body by a diarthrodial synovial joint (Figure 10.8). This last feature is important, because it is clearly observed on panoramic and lateral cephalometric radiographs, particularly in children. It should not be mistaken for a hyoid bone fracture.[5]

The stylohyoid complex infrequently causes difficulty in its recognition, except perhaps pattern j in which no other landmarks of the stylohyoid complex are available. This could be mistaken for calcified carotid artery atheroma. Occasionally, a long styloid process may cause Eagle's syndrome, which among other features may cause atypical pain.[6]

CALCIFIED CAROTID ARTERY ATHEROMA

There are two types of calcification of the arteries depending upon whether the tunica media or

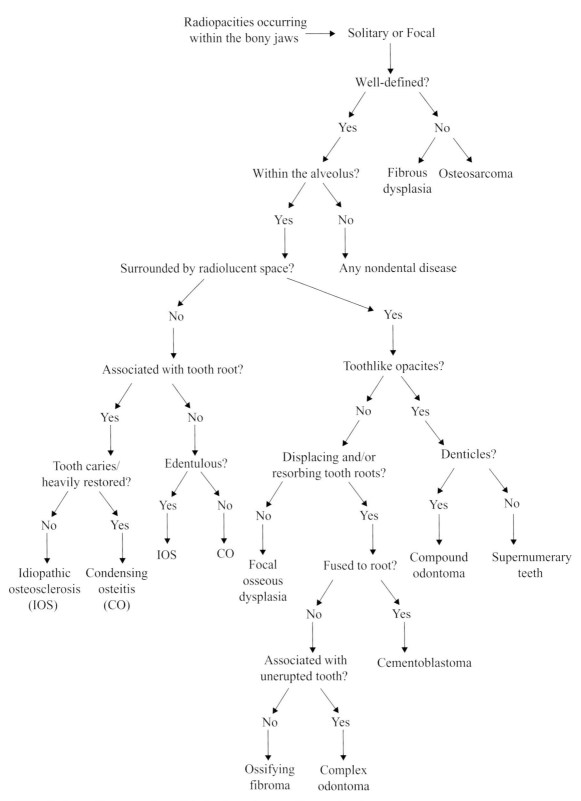

Figure 10.2. Radiopacities occurring within the bony jaws: solitary or focal.

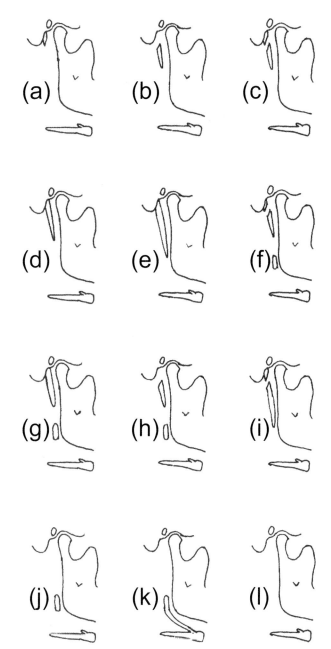

Figure 10.3. The 12 patterns of calcification of stylohyoid complex used by MacDonald-Jankowski (2001). Pattern: (a) Region 1 = tympanohyal alone; (b) Region 2 = stylohyal alone; (c) Region 1 and 2, separate; (d) Regions 1 and 2 continuous; (e) Regions 1, 2, and 3 continuous; (f) Regions 1, 2, and 3 separate; (g) Regions 1 and 2 continuous, but separate from 3; (h) Regions 2 and 3 separate; (i) Regions 2 and 3 continuous, but separate from 1; (j) Region 3 alone; (k) Region 3 and 4 continuous (may include calcification in one other region); (l) No styloid process visible. **Note 1:** Regions 1, 2, 3, and 4 coincide with the 4 centers of ossification of the stylohyoid complex, tympanohyal, stylohyal, ceratohyal, and hypohyal. **Note 2:** Patterns (a) to (d) are normal styloid processes, pattern (e) is an elongated styloid process, and patterns (f) to (k) are calcified stylohyoid ligaments.

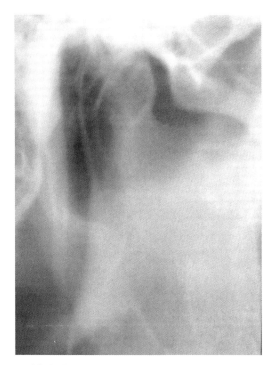

Figure 10.4. Panoramic radiograph displaying a normal styloid process. A normal styloid process can extend as inferior as the middle of the mandibular foramen. This example is of a continuous bone and is comprised of the tympanohyal and stylohyal components (pattern [d]).

tunica intima is involved. The former is termed *medial calcific sclerosis* or *Monckeberg's arteriosclerosis*, whereas the latter, because of its narrowing of the lumen and atheroma formation is termed *calcified carotid artery atheroma* (CCAA) (Figure 10.9a). The former's synonym, Monckeberg's arteriosclerosis, implies vessel hardening and may be radiologically apparent as a tramline or pipe-stem pattern on conventional radiography and computed tomographic sections (Figure 10.9). Although supposedly benign in comparison to CCAA in outcome, Monckeberg's arteriosclerosis can be associated with medical conditions such as parathyroidism and osteoporosis. Most commonly observed in the limbs, it infrequently presents in the head and neck. Although the definitive diagnosis for calcification of the tunica intima or tunica media rests on histopathology, biopsy of an artery has its own obvious risks.

CCAA presents as a round radiopacity initially and becomes linear as it becomes larger; it frequently presents as 2 parallel vertical lines. It is sited at or below the intervertebral space between the third and fourth cervical vertebrae.[7] The CCAA observed on panoramic radiographs was first reported by Friedlander and Landes in 1981.[8] Since

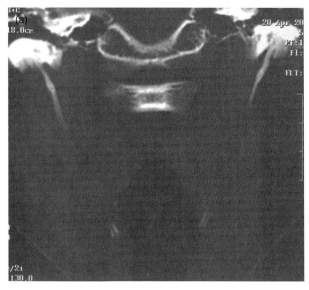
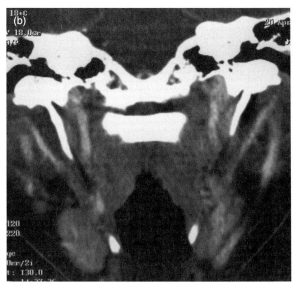

Figure 10.5. Coronal computed tomography (CT) displays the styloid process within the bone (a) and soft-tissue (b) windows. The coronal section is at the level of the anterior arch of C1. The styloid processes on these coronal sections are angled medially, running almost straight (in part tracing out the stylohyoid ligament) toward the junction between the body and the greater horn of the hyoid bone. The greater horn on each side are observed in cross-section.

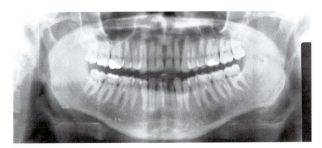

Figure 10.6. Panoramic radiograph displayed a bilateral and symmetrical calcified stylohyoid complex.

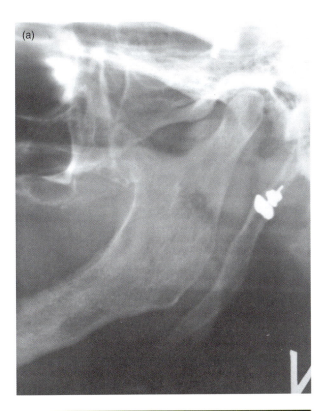

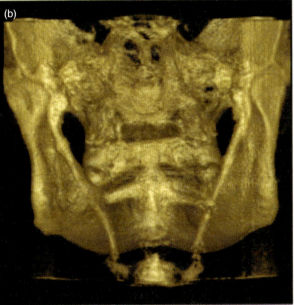

Figure 10.7. The anterior (a) and medial (a) path of the calcified stylohyoid ligament. (a) Panoramic radiograph showing a long styloid process extending inferiorly to as far as the hyoid bone. A long styloid process includes the tympanohyal and stylohyal components of the normal styloid process, but also includes the ceratohyal component. (b) Cone-beam computed tomography (CBCT) of a completely calcified stylohyoid ligament. Figure courtesy of Dr. Alexandre Khairallah, University of Lebanon.

then many reports have indicated that this phenomenon is widespread, affecting almost every global group (see Chapter 1 for their definitions). So far there does not appear to be a report from the sub-Saharan Africa. Nevertheless, the clinical significance of the CCAA and the utility of the panoramic radiograph as a screening tool for the CCAA are very controversial. Their "results suggest that the presence of carotid artery calcifications on panoramic radiographs may be related to the history of past vascular diseases; however, this is not a useful marker for subsequent vascular diseases and related death among 80-year-olds."[9] This was supported by a systematic review, which concluded that clinical guidelines based upon the hypothesis that CCAA, detectable on panoramic radiographs, is associated with an increased risk of stroke ... cannot be established on the basis of the current evidence."[10] Furthermore, Madden et al.[11] reported that panoramic radiography, when compared to ultrasonography, is not a reliable means to detect CCAA or stenosis. This poor assessment of the panoramic radiograph was qualified by Damaskos et al.[12] Furthermore, Friedlander and Cohen maintain that "incidental finding of a CCAA ... portents significant risk of a future adverse vascular event."[13] They observed that 15% of such cases had occult metabolic disease, formerly known as insulin-resistant syndrome. This syndrome is composed of increased abdominal obesity, raised triglycerides, reduced high density lipids and cholesterol, hypertension, and insulin resistance. A referral in such cases is necessary "because aggressive management may preclude a stroke."[13] Finally, developing Maddens et al.'s conclusions, Farman states that patients for whom CCAA have been detected should be further screened by ultrasonography.[14]

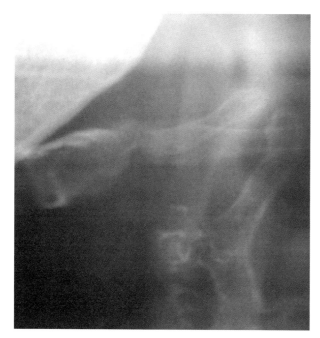

Figure 10.8. Panoramic radiograph displays the hyoid bone immediately below the angle of the mandible. Below the hyoid bone is the superior horn and partially calcified lateral lamina of the thyroid cartilage. In reality, in contrast to its traditional display in many anatomical texts, the hyoid bone infrequently exists as a complete unified bone, except in the oldest patients. The various parts are more frequently observed radiologically as separate bones joined by radiolucent spaces, representing the diarthrodial synovial joints. In this figure the anterior rectangular bone represents the body of the hyoid, whereas the small round bone superimposed and or in contact with its superiodistal margin is the lesser horn. Immediately distal to the above joint is a horizontal V whose upper limb represents the image of the ipsilateral greater horn. The lower, longer, almost horizontal image represents the contralateral greater horn. Calcification of the thyroid cartilage is encountered increasingly in the older patient. Nevertheless, as is obvious here, it can coexist with a hyoid bone displaying all of its components as still separate bony entities.

TRITICEOUS CARTILAGE (TRITICAE CARTILAGO)

The thyroid cartilage complex can undergo calcification (Figures 10.8, 10.9a) with increasing age. Calcification of the thyroid cartilage itself is infrequently mistaken for CCAA. On the other hand, the triticeous cartilage is more frequently mistaken for CCAA, particularly by students. Their shape, outline, and location assist in distinguishing them.

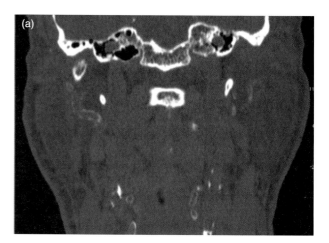

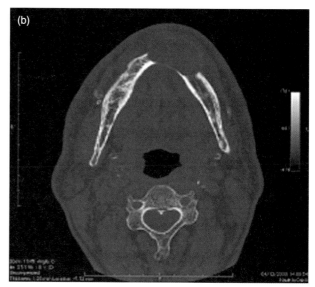

Figure 10.9. Computed tomography (bone windows and no contrast media used) of parathyroidism secondary or tertiary to chronic kidney disease displaying calcification. Monckeberg's arteriosclerosis of the branches of the external carotid artery. The classical stempipe or railtrack pattern of calcification of the arteries tunica media is evident. (a) A coronal section (bone window), at the level of the anterior arch of the first cervical vertebra, displays the rostral part of the external carotid artery. (b) An axial section (bone window), at the level of the second cervical vertebra, displays the "brown tumor," as a expansile unilocular radiolucency, which was considered on clinical and conventional radiological examination to be an ameloblastoma. Monckeberg's arteriosclerosis plots the torturous course of the facial artery medial and lateral to the mandible. There is a short arc of calcification within the internal carotid artery. Figures courtesy of Dr. Lewei Zhang, Oral and Maxillofacial Pathology, Faculty of Dentistry, UBC.

The triticeous cartilage lies within the lateral thyrohyoid ligament between the superior horn of the thyroid (if apparent) and the distal end of the greater horn of the hyoid. It is round and of homogeneous density, whereas the CCAA is round initially and becomes linear as it becomes larger, frequently as 2 parallel vertical lines. The triticeous cartilage is an intrinsic component of the larynx and serves as an attachment of the vocal cords and the musculature responsible for phonation. It undergoes calcification more frequently in females (12%) than in males (8%).[7]

TONSILLOLITHS

Tonsilloliths affecting the palatine tonsil are frequently observed on panoramic radiographs and are now becoming increasingly observed on CBCT (Figure 10.10). In the former they are most commonly observed superimposed upon the mandibular foramen. Formerly they were associated with tubercular lesions in the older patient, but are now observed in younger patients, presumably secondary to multiple episodes of tonsillitis earlier in life. These tonsilloliths may be associated with halitosis.[15] At least one has caused dysphagia.[16] They also present on CT[17] and MRI.[18]

OTHER NONPATHOLOGICAL CAUSES OF RADIOPACITIES

Other structures that can appear as radiopacities, but are not included in Table 10.1, are ironically soft-tissue structures. These include not only normal anatomical structures such as the tongue, the soft palate, and the pharynx clearly apparent on the panoramic radiograph, but they also include soft-tissue lesions within the maxillary antrum. All these appear radiopaque by virtue of being contrasted (silhouetted) against the air-filled space. The reason for this phenomenon is discussed in Chapter 11.

Artifacts causing radiopacities arise from three main sources: image development, inadequate preparation, and earlier treatment (Table 10.1). Although image development artifacts in the traditional chemistry-based technology (film) and the strategies to avoid them are well known, the recent advent of the digital imaging technologies have the potential for different artifacts. The photostimulable phosphor plates are easy to damage, resulting in white scratches and bite marks.[19]

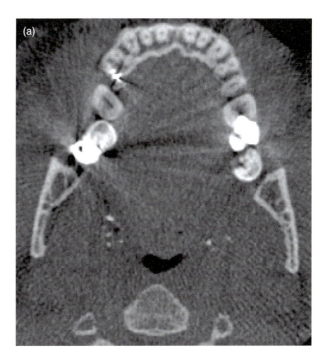

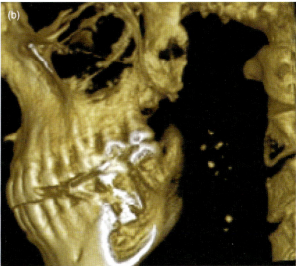

Figure 10.10. Cone-beam computed tomography (CBCT) of tonsilloliths. (a) Axial CBCT showing tonsilloliths adjacent to the oropharynx just medial to the mandibular foramen. (b) Three-dimensional reconstruction, cutting away the posterior body and vertical ramus of the mandible to display the tonsilloliths. Figures courtesy of Dr. Alexandre Khairallah, University of Lebanon.

Although artifacts caused by metal restorations (amalgam restorations, crowns, and bridges) in conventional imaging are extremely infrequent, even in panoramic radiography with its secondary imaging of the contralateral jaws (except for long or large earrings), this is not true for advanced imaging. In spite of the development of *metal artifact reduction* (MAR) software metal dental restorations pose significant problems for HCT (see Figure 4.9), CBCT (see Figures 5.2a, 5.4), and MRI (see Figure 6.6).

Radiopacities Occurring within the Bony Jaws

The flowcharts in Figures 10.1 and 10.2 generally flow from the most important clinical and radiological findings, addressing systemic lesions and malignancies first. Multiple radiopacities, particularly if they are distributed throughout the jaws, suggest a systemic cause, whereas the single radiopacity suggests a local cause.

The degree of marginal definition is crucially important to determining potentially serious disease. If it is well defined, the radiopacity is more likely to be benign, whereas a poorly defined radiopacity, in addition to inflammation or fibrous dysplasia, could represent a malignancy. The radiopacity's relationship to the mandibular canal or the image of the hard palate (on panoramic or cephalometric radiographs) indicates whether it is likely to be of odontogenic origin. A radiopacity occurring above the mandibular canal or below the image of the hard palate is within the dental alveolus and therefore could be of odontogenic origin.

If the radiopacity is sited within the alveolus, its relationship to teeth is important, in order to refine further the differential diagnosis. If it is associated with the root of an erupted tooth, which has a large carious lesion or a large restoration, suggesting the possibility of a necrotic pulp, inflammation is a likely cause. If the radiopacity is associated with the crown of an unerupted tooth, an odontogenic lesion, most likely a neoplasm, should be considered.

The effect of the radiopacity on the tooth or adjacent structures is manifested by either displacement or erosion. The latter when applied to teeth, particularly their roots, is termed *root resorption*. Although all lesions presenting as radiopacities may in due course cause root resorption, this would appear to be a particular feature of certain odontogenic neoplasms. Displacement of teeth and buccolingual cortices are universal to all expansile lesions.

The flowcharts focus on the most common and important lesions and are not exhaustive with regard to the rarer lesions, particularly if they respond well to the initial treatment—i.e., they are very unlikely to recur.

Multiple Radiopacities

Multiple or widely distributed lesions suggest a systemic rather than a local cause. Therefore, it is necessary to identify such lesions early. Although these lesions are not common, failure to identify them early may have significant implications for the patient's continued well-being. Polyostotic fibrous dysplasia, particularly the McCune-Albright syndrome, will have already been diagnosed early in life. Paget's disease and Gardner's syndrome are very important lesions that present later in life. Before proceeding to these lesions, a unique radiological phenomenon should be introduced: leontiasis ossea.

LEONTIASIS OSSEA

Leontiasis ossea, although infrequently observed, is important because it represents important lesions such as Paget's disease of bone and hyperparathyroidism. Its name precisely reflects an appearance of a lion's maxilla. Classically, both cheeks are very full and the external nose is small. This last effect is achieved by the following: the external nose, which itself remains essentially unchanged, becomes submerged within the outwardly expanded anterior wall of the maxilla. As a result the external nose is now both relatively smaller and flatter due to a more obtuse angle formed between the alae.

The presentation on CT of the internal structure is usually ground glass or an extensive network of serpentine channels within the radiopacity. Its frequently bilateral presentation usually distinguishes it from fibrous dysplasia.

PAGET'S DISEASE OF BONE

Paget's disease of bone (PDB) was originally called "osteitis deformans" by Paget himself, which vividly describes this disease. It is characterized by rapid bone remodeling and the deposition of structurally abnormal bone.[20] Although it classically affects individuals older than 40 year of age, a subset of patients are juveniles. The last is called "early-onset familial Paget's disease of bone" and is primarily genetic.[21] Although its etiology encompasses both genetic and environmental factors, its declining prevalence and its severity at the time of writing (2010) suggests amelioration with regard to those environmental factors.[20] Most cases of PDB occur in communities of Northwestern European descent, particularly from the United Kingdom.[20] Although it has been reported to affect 3% of Britons and White Americans over 50 years of age,[22] it is less frequent in Africans and East Asians[23,24]

The presentation of PDB in East Asians appears to be accompanied with symptoms, whereas that in patients of European origin is largely symptom-free.[23,24]

Although the radiological presentation is broadly similar to fibrous dysplasia, PDB is generally bilateral and first presents over the age of 40 years. It classically presents with a "cotton wool" expansion of the outer table of the skull and a wholly radiopaque vertebral body. When the jaws are affected the lumen of the maxillary antrum is frequently spared.

The radiological presentation of lesions in the alveolus is similar to florid osseous dysplasia, but extends into basal bone. Note that, although classically polyostotic, at least 7 cases of monostotic Paget's affecting the mandible alone have been reported in the literature.[25] Although bone scintigraphy and serum alkaline phosphatase are sensitive screening modalities for PDB, the latter, which is raised in 86% of cases, may not be raised during the more inactive periods of the disease. It is also higher for polyostotic disease than for monostotic disease.[26] Takata et al. have set out an algorithm for the diagnosis of PDB.[26] Bisphosphonates are used to treat PDB.[26]

Although the list of complications is long and covers musculoskeletal, neurological, cardiovascular, metabolic, and neoplastic complications, the complication that may concern the oral and maxillofacial clinician most is sarcomatous change of PDB affecting the jaws.[26] Cheng et al. in their synthesis of the English-language literature reported significantly more cases of osteosarcoma secondary to PDB of the jaws in contrast to osteosarcoma arising from PDB elsewhere in the skeleton and osteosarcoma arising within the jaws.[27] Osteosarcoma arising from PDB of the jaws has a predilection for females and those of sub-Saharan African origin. Although the 5-year survival is poor (21%) it is better than that for osteosarcoma arising from PDB elsewhere in the skeleton, which is 5%. The survival of the former is lower than that for primary osteosarcoma arising within the jaws.[27]

GARDNER'S SYNDROME

The main clinical feature of *Gardner's syndrome* (GS), an autosomal disease, is *familial adenomatous polyposis* (FAP). GS affects only 10% of FAP.[28] Any one or more of these polyps at any time can undergo malignant change. Polyps arise after 20 years of life.[29] Therefore, it is important to diagnose the syndrome as early as possible. Although the taking of a good family history may assist in identification of polyps, the vigilance of the oral and maxillofacial practitioner is crucial because s/he is more likely to observe the hard tissue lesions earlier than the onset of symptoms and signs of FAP. Those hard tissue lesions affecting the jaws are osteomas (Figure 10.11), odontomas, and supernumerary and impacted teeth.[30] They will be most likely to be observed as an incidental finding on a panoramic radiograph. Takeuchi et al. reported the largest series of GS cases.[31] They found 23 cases of GS out of 48 cases of FAP. The average age when diagnosed was 26 years of age. They followed them for an average of 7 years and noted that in one-half of the cases the number and size of the osteomas continued to increase. Although one-third developed colonic cancer, there was no significance between malignant transformation and the extent of the jaw lesions; indeed, the 4 cases displaying widespread lesions were not associated with malignant change.

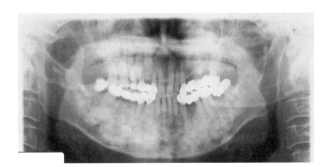

Figure 10.11. Panoramic radiograph displaying multiple osteomas in both the alveolar and basal processes of both jaws of a patient with Gardner's syndrome (polyposis coli). The osteomas have almost completely obliterated the maxillary antrum. See Figure 17.21 for other views of this case. Reprinted with permission from Lee BD, Lee W, Oh SH, Min SK, Kim EC. A case report of Gardner syndrome with hereditary widespread osteomatous jaw lesions. *Oral Surg Oral Med Oral Pathol Oral Radiol Endod* 2009;107: e68–72.

The multiple osteomas affecting the mandible of a middle-aged East Asian woman in Lee et al.'s case have a radiological presentation similar to familial gigantiform cementoma (Figure 10.11).[32] The HCT of this case additionally revealed an osteoma expanding into the orbit displacing the optic nerve and the globe (eyeball) (see Figure 17.21). In addition to similar familial gigantiform cementoma-like lesions in two of her adult children there was a definite family history of abdominal tumors. On endoscopy the patient was found to have multiple intestinal polyps. The sole dental anomaly was an impacted premolar.

Fonseca et al.[33], Madani et al.[34], and Ramaglia et al.[28] reported peripheral (periosteal) osteomas; such osteomas are rarely seen in nonsyndromic cases.

Poorly Defined Radiopacities

Poorly defined lesions are generally suggestive of aggressive disease such as malignancies or infections. This criterion by itself is not entirely decisive with regard to those jaw lesions, which frequently present as radiopacities. Fibrous dysplasia, a fibro-osseous lesion affecting the jaws, also presents with a poorly defined margin, which is central in differentiating it from another fibro-osseous lesion. Nevertheless, there is one other criterion that can assist in the identification of these aggressive lesions, the periosteal reaction. The periosteal reaction is a prominent feature of general radiology but other an expansion of the cortices it is infrequently observed in the jaws, except in regard to chronic infection and some malignant neoplasms, such as the osteosarcoma.[35]

OSTEOSARCOMA

Osteosarcoma is the most common of the sarcomas affecting the jaws; in an American National Cancer Database Report osteosarcoma accounted for 78% of sarcomas affecting the mandible in contrast to chondrosarcoma's 14% and Ewing's sarcoma's 8%.[36] Overall chondrosarcoma has a 75% 5-year survival rate, whereas both osteosarcoma and Ewing's sarcoma is 50%. The 5-year survival rate of osteosarcoma secondary to Paget's disease was very low, about 21%.[27] Guo et al.,[37] comparing Chinese, Japanese, and American databases, observed that the relative frequencies of osteosarcoma were higher in China and Japan in contrast to the United States. There were far fewer cases in people over 50 years of age in the two East Asian countries, which could be ascribed to their lower incidence of Paget's disease. Although both chondrosarcoma and Ewing's sarcoma were higher in the United States, chondrosarcoma first presented younger in the Chinese.[37] Van Es et al. reported that the 10-year survival of osteosarcoma in their Dutch report was 59%.[38]

Unlike its manifestation in the extragnathic skeleton, which mainly occurs largely in the adolescent, osteosarcoma affecting the jaws occurs later in life, usually during the fourth[38-40] and fifth[41-43] decades. In a Nigerian report the mean age was 27 years, 31 and 23 years for the maxilla and mandible, respectively.[44]

Most presented as swellings. Mardinger et al. reported that the maxillary osteosarcoma was larger (13 cm^2) than that affecting the mandible (8 cm^2).[39] Mental paresthesia was observed between 7%[42] and 21%[40] to 36%.[39]

The variety in radiological presentation may merely reflect the ethnic origin of the community reported. Ogunslewe et al. reported that most cases showed a nonspecific radiolucent lesion.[44] Givol et al. reported that overall 78% had poorly defined margins and 29% displayed a locular pattern.[40] Forty-one percent were "mixed," 29% were radiolucent, and 29% were radiopaque.[40] Fernandes et al. reported a "sunray appearance" (Figure 10.12) in 54%[42] and Givol et al. reported 48% of the cases in their synthesis with a "periosteal reaction."[40] In almost every such case, this was observed on an occlusal projection. Almost all the lesions in Givol et al.'s own case series, displaying a "periosteal reaction," affected the posterior mandible.[40]

Givol et al. stated that soft-tissue involvement was reported in 33% and optimally displayed on HCT.[40]

OSTEOMYELITIS

Osteomyelitis of the jaws frequently arises from a dental infection. Kahn et al.'s Figure 3 appears, due to its expansion of the whole bone and diffuse bone pattern (Figure 10.13), similar to a case of fibrous dysplasia.[45] Petrikowski et al. reviewed 10 cases each of osteomyelitis, fibrous dysplasia, and osteosarcoma—three lesions with radiological similarities.[46] They reported that the only two features that most usefully distinguish osteomyelitis are "seques-

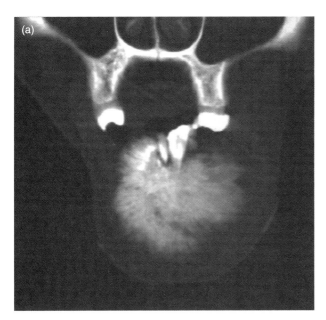
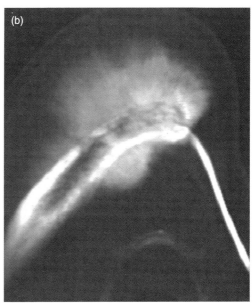

Figure 10.12. Computed tomography (CT) of a case of recurring osteosarcoma. (a) Coronal CT (bone window) exhibiting the sunburst pattern. (b) Axial CT displaying the above affecting almost the whole of the remaining body of the mandible.

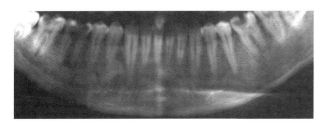

Figure 10.13. Panoramic radiograph displaying a diffuse sclerosing osteomyelitis arising from a carious first molar. This osteomyelitis has affected both the alveolar and basal processes, resulting in the accentuation of the mandibular canal (compare with the normal contralateral side).

tra and laminations of periosteal new bone" (Figure 10.14).[46] This new bone is called an involucrum.

Figure 10.15 displays a case of osteomyelitis, which began in the anterior mandible, that over the years gradually involved almost the entire mandible.

In a small number of cases, particularly the diffuse osteomyelitis[47] may be a manifestation of SAPHO syndrome.[45] This is a localized rheumatic disease with an idiopathic etiology. It presents with synovitis, acne, pustulosis, hypertelorism, and osteitis—hence its acronym, SAPHO. This syndrome is linked with spondyloarthopathies. Diagnosis requires a more general review of the patient for skin lesions and scintography to detect other

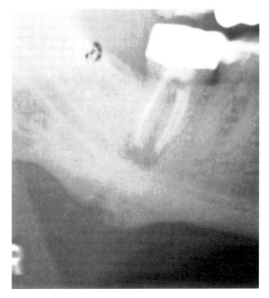

Figure 10.14. Panoramic radiograph a periosteal reaction at the lower border of the mandible apical to the root-filled first molar resulting in an involucrum. This is represented by a suggestion of laminations running approximately parallel to the lower border of the mandible giving rise to an onion-skin appearance. A wide draining tract is obvious running, through this onion-skin structure, from the periapical radiolucency to the most dependant part of the periosteal reaction. Sclerosing osteomyelitis is observed in the bone adjacent to this tooth, particularly inferiorly and distally.

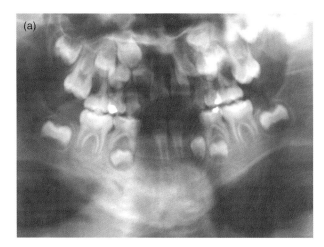

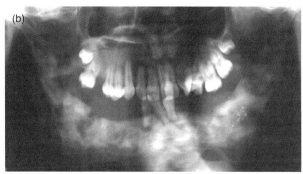

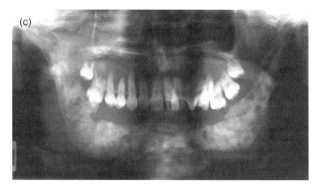

Figure 10.15. A consecutive series of panoramic radiographs displaying the progression of osteomyelitis from the midline (a) to affect the entire mandible except for the condyles (c).

skeletal lesions. If bone resorption is present it may be treated by bisphosphonates.[45]

BISPHOSPHONATE-ASSOCIATED OSTEONECROSIS

In addition to treating osteosarcoma, Paget's disease of bone, and SAPHO syndrome, bisphosphonates are also central in the treatment of osteoporosis, multiple myeloma, and metastatic disease.

Although osteonecrosis of the jaw has been a long-recognized clinical phenomenon, it briefly peaked in incidence as radio-osteonecrosis until bone-saving radiotherapy was developed. Recently, it has become increasingly observed as *bisphosphonate-associated osteonecrosis* (BON). It is now a recognized risk of bisphosphonate therapy, particularly if intravenous and/or of long duration (over 3 years).[48]

It presents clinically as poor wound healing, spontaneous or postsurgical breakdown of soft tissue to expose the bone to the oral environment, and osteomyelitis. This may or may not be accompanied by pain.[48]

Panoramic radiography is of limited value for the assessment of BON. It displayed only a nonspecific osteolysis in all patients.[49] It identified a sequestrum (Figures 10.16, 10.17) in only two-thirds of the cases identified by HCT.[50] A periosteal reaction was frequently found; this was also confirmed by Bedogni et al.[51] Chiandussi et al. found that the bone scan was most sensitive for identifying early-stage osteonecrosis.[52] Furthermore, in such cases *single photon emission computed tomography fused with computed tomography* (SPECT/CT) may enhance the bone scan by distinguishing the osteonecrotic nidus from the adjacent hyperactive viable bone. The reader is reminded that bisphosphonates are part of the treatment for multiple myeloma and metastatic breast and prostrate cancers.[48] Therefore, areas of hyperactivity may represent metastasis.

Both HCT's and MRI's definition of the extent of the osteonecrosis was invaluable for distinguishing between the osteonecrotic and osteomyelitic patterns of BON representing exposed and unexposed bone, respectively.[51] The osteonecrotic pattern gave a low hypointense signal on T1-weighted and T2-weighted and *inversion recovery* (IR) images, suggesting a low water content, which is consistent with the paucity of cells and blood vessels. The osteomyelitic pattern was characterized by a hypointense T1-weighted, a hyperintense T2-weighted and IR images. These suggest an abundant cellular and vascular tissue with osteogenesis.[51]

FIBRO-OSSEOUS LESIONS

My review of the differential diagnoses of those lesions, presenting in the Hong Kong Chinese, which frequently present as radiopacities, indicated that *fibro-osseous lesions* (FOLs) appeared

Chapter 10: Radiopacities **165**

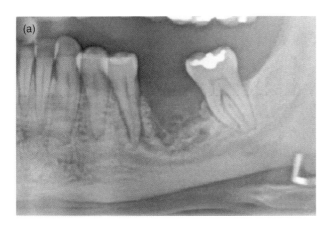

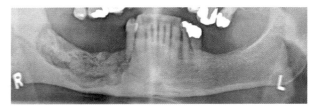

Figure 10.17. Panoramic radiograph displaying bisphosphonate osteonecrosis affecting the entire posterior alveolus of the right mandible. Posteriorly this reaches down to the mandibular canal and reaches the lower border of the mandible anteriorly. Figure courtesy of Dr. Michele Williams, British Columbia Cancer Agency.

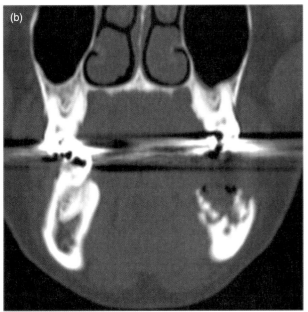

Figure 10.16. Panoramic radiograph and computed tomograph (CT) of a case of bisphosphonate osteonecrosis. (a) Panoramic radiograph displays affected edentulous site. The affected bone is delimited by a broad radiolucent band running parallel and above the mandibular canal. The bone above it is being sequestrated (b). The coronal CT exhibits the sequestrum. Figures courtesy of Dr. Michele Williams, British Columbia Cancer Agency.

terms that appear to be currently in use, this simplified figure is still able to display the "lumping" and "splitting," which appear to attend frequently the development of most classifications and systems of nomenclature.

The late Charles Waldron wrote "In absence of good clinical and radiologic information a pathologist can only state that a given biopsy is consistent with a FOL. With adequate clinical and radiologic information most lesions can be assigned with reasonable certainty into one of several categories."[53] Conversely in the absence of such information Eisenberg and Eisenbud stated "pathologists today will often rightly decline to render a definitive diagnosis. ... Instead, the pathologist will resort to the noncommittal designation of benign fibro-osseous lesions [their italics]. This is the only acceptable approach considering the potential for inappropriate treatment otherwise."[54] Therefore the identification or clarification of the majority of histopathologically proven FOLs affecting the jaws is made upon clinical and radiological features.

FIBROUS DYSPLASIA

Jundt[55] defined *fibrous dysplasia* (FD) as "a genetically based sporadic disease of bone that may affect single or multiple bones ... FD occurring in multiple adjacent craniofacial bones is regarded as monostotic (craniofacial FD). FD may be part of the McCune-Albright syndrome."

FD is an important lesion affecting the maxillofacial region because it can cause severe deformity and asymmetry, and, most devastating of all, blindness.

Jundt's basis for referring to FD as a genetically based sporadic disease of the bone is that "Mutations in the gene (GNAS I) encoding for the α-subunit of a signal transducing G-protein (Gs-α)

frequently. These FOLs are fibrous dysplasia, ossifying fibroma, and osseous dysplasia. As observed in Figures 10.1 and 10.2, FOLs are central in the differential diagnosis of a radiopacity affecting the jaws. Although they display a similar histopathology, a spectrum between cementoid and osteoid, their clinical and radiological presentations and treatment outcomes differ. Figure 10.18 displays the development of the nomenclature and classification of FOLs. Although it includes only those

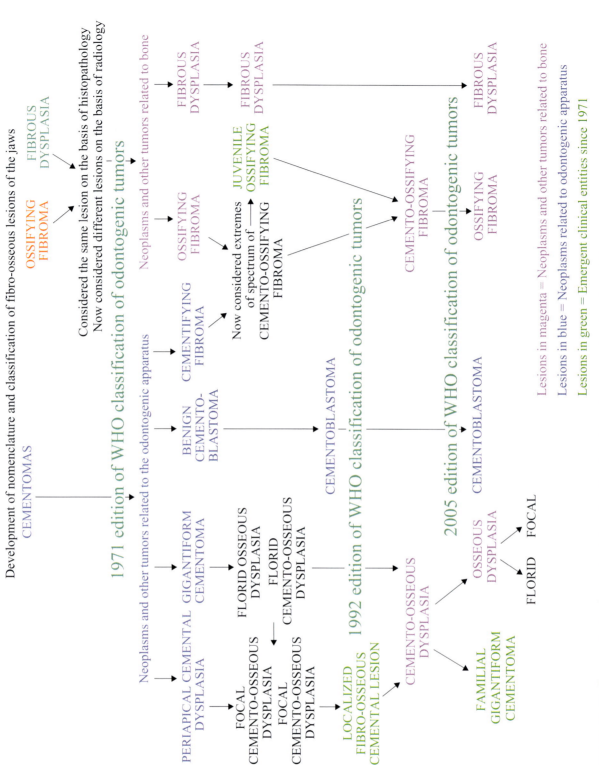

Figure 10.18. Development of the nomenclature and classification of fibro-osseous lesions of the jaws.

Table 10.2. Fibrous dysplasia: systematic review

Feature	Total	Western	East Asian	sub-Saharan	LatinAmer
Male:Female	48%:52%	46%:54%	50%:50%	42%:58%	88%:12%*
Mean number per year per report	1.6	0.9	1.9	3.2	0.1
Mean prior awareness	5.2 years	5.7 years	3.6 years	6.3 years	4.5 years
Mean age	24 years	25 years	24 years	23 years	16 years
Mand:Max	42%:58%	48%:52%	41%:59%	39%:61%	0%:100%*
Mand:Ant:Post	14%:86%*	7%:93%*	100%:0%*	0%:100%*	INA
Max:Ant:Post	19%:81%	25%:75%*	12%:88%*	100%:0%*	0%:100%*
Swelling: Y:N	93%:7%	98%:2%	90%:10%	95%:5%*	100%:0%*
Pain: Y:N	19%:81%	12%:88%	26%:74%	5%:95%*	50%:50%*
Incidental: Y:N	2%:98%	2%:98%	3%:97%	100%:0%*	100%:0%*
Radiolucen: Y:N	5%:95%	7%:93%	4%:96%	8%:92%*	INA
GroundGlas: Y:N	38%:62%	54%:46%	38%:62%	28%:72%	INA
Expansion: Y:N	100%:0%	100%:0%*	100%:0%	100%:0%*	INA
LBMd: Y:N	100%:0%*	100%:0%*	100%:0%*	100%:0%*	INA
Antrum: Y:N	98%:2%*	100%:0%*	100%:0%*	88%:12%*	INA
ToothDispl: Y:N	35%:65%	41%:59%*	28%:73%*	75%:25%*	INA
RootResorp: Y:N	2%:98%	100%:0%*	2%:98%	INA	INA
Reactivat: Y:N	18%:82%	4%:96%	16%:84%*	72%:28%*	18%:82%*

*Advises that the percentages were derived from either one report or from a synthesis of no more that 50 cases. Ant:Post, Anterior:Posterior; GroundGlas, Ground Glass; INA, Information not available; LatinAmer, Latin American; LBMd, downward expansion of the lower border of the mandible; Mand:Max, Mandible:Maxilla; Reactivat, Reactivation; subSaharan, sub-Saharan African; ToothDispl, Tooth displacement; ToothResorp, Tooth resorption; Western, predominantly Caucasian; Y:N, Yes:No.

lead to increased c-AMP production affecting proliferation and differentiation of preosteoblasts."[55]

FD can present either as a local lesion (the monostotic form) or as a systemic lesion (the polyostotic form). When this last form is combined with hormonal changes, it is now the *McCune-Albright syndrome* (MAS), of which precocious puberty is perhaps the most striking feature.

The global distribution of reports included in the systematic review[56] upon which much of the following is derived is set out in Figure 1.44 and their details in Table 10.2.

The mean number of cases of FD per year globally was 1.6. Although this was highest for the sub-Saharan African global group and least for the Western, the difference was not significant.[56]

Although it is generally accepted that FD is a disease affecting children and adolescents and burns out at the end of puberty in most cases, my recent systematic review refutes this contention. Indeed, the majority first present over 20 years old (some even as late as the 8th decade) with an overall mean of 24 years (Table 10.2). In addition, the period of the patient's prior awareness before first presentation is 5.2 years.[56] Taken together, the average patient becomes first aware of his/her disease when 19 years old, precisely when it is traditionally expected to "burn out."[56] Nevertheless, almost all polyostotic cases first present early in infancy.

Therefore, if it is supposed that all FDs arise during childhood or puberty, most remain undetected only to become active or be reactivated later and thus be detected for the first time later in life. This conclusion is feasible if the classical division of FD into monostotic, polyostotic, and MAS forms is considered to reflect the timing of the mutation and, thereby, the initial size of the mass of FD precursor cells.[57,58] The polyostotic form may arise in fetal life, whereas the monostotic form may arise postnatally.[58] In addition to discussing how the different forms of FD arose, the two cited authorities together comprehensively discuss the genetic basis of this lesion. Some of the discussion on both topics was detailed in their earlier publications, which they cite clearly.

The systematic review demonstrated that the distribution between the sexes is almost equal, with females prevailing slightly. The mean overall age at first presentation is 24 years, ranging between 5 to 79. The second decade attracted most cases upon first presentation, 36% percent of all cases of which 67% are males. Males slightly predominate in the third decade but are in the minority in all other decades.[56]

The polyostotic form (with/without endocrinopathies) is easy to diagnosis because many bones are affected and it occurs in childhood, compelling the patient's parents to seek treatment. The McCune-Albright Syndrome (MAS) is just one of a series of lesions associated with precocious puberty, which has been extensively discussed by Fahmy et al.[59] They also provide an algorithm. A diagnosis of precocious puberty requires radiography of the bones of the wrist to determine the developmental age. Café-au-lait spots lead towards a diagnosis of MAS, which should then indicate a bone scan or other radiology to determine the presence of polyostotic FD.[59]

The monostotic form may not be so easy to diagnose. The monostotic form accounts for up to 85% of cases. Monostotic means "one bone," which is the correct term when applied to FD affecting the mandible but is not strictly true when applied to FD affecting the maxilla. FD affecting the maxilla may involve one or more contiguous bones, such as the zygomatic (malar) and palatine bones. Therefore, "craniofacial fibrous dysplasia" has been coined for such cases and used by Jundt in the WHO's 2005 edition.[55] The maxillofacial subset of the craniofacial FD accounted for 13% in a recent systematic review.[56]

Although bone biopsy of FD is generally avoided in medicine, particularly in those cases where pathological fracture may be high, this appears not to be the case of FD affecting the jaws. Only one case of apparent pathological fracture and nonunion has been reported for FD affecting the mandible.[60] Therefore, a biopsy, required to confirm FD histopathologically, should not be contraindicated solely for this reason. But it is a generally accepted principle that the biopsy should be deferred until all necessary imaging has been completed; otherwise, the radiological presentation may be compromised (see Figure 1.2). Biopsies are still necessary for the histopathological diagnosis of FOL, which in conjunction with a poorly defined margin of a radiopaque lesion on conventional radiography is necessary for the diagnosis of FD. Biopsy is also required to refine our understanding of the genetics of FD. It is also required for the identification of markers for future adverse conduct such as reactivation.

In spite of the fact that, at first presentation, the FD lesions in one recent report were so large that many affected all or most of the hemimandible or hemimaxilla, very few were discovered as incidental findings. Furthermore, the clinicians offered solely "fibrous dysplasia" as their provisional diagnosis only on the basis of their clinical and radiological findings.[60]

Only 2% are discovered as incidental findings in the systematic review; the rest presented with symptoms.[56] Ninety-three percent of cases first present with swelling. Swelling was significantly more frequent in the Western global group than in the East Asian global group. Overall, 19% presented with pain.[56] A higher proportion of East Asian cases present with pain in comparison to other global communities. In the jaws the maxilla is affected in 58% of cases; the Western global group approached equal distribution between the jaws. FD displays an overwhelming predilection for the posterior sextants of both jaws.[56]

FD affecting the face and jaws differs radiologically and histologically from that of the rest of the skeleton. FD affecting the jaws is poorly defined, according to Slootweg and Müller's 1 mm criterion (see Chapter 1),[61] whereas that of the extragnathic FD is generally well defined.[62] A possible reason for this difference is that the jaws are derived from membrane and long bones are from cartilage.[63] Support for this contention comes from the WHO's second edition (edited by Schajowicz) of the classification of neoplasms affecting the extragnathic skeleton, which included cartilage as an expected histopathological element of FD affecting the skeleton outwith the jaws.[62] Conversely, the WHO's 2005 edition of the classification of the odontogenic neoplasms did not include cartilage as an expected feature of FD affecting the jaws.[55]

Radiologically overall, 38% present with ground glass as the predominant pattern (Figure 10.19b); it was significantly more frequently observed in Western than in sub-Saharan African communities. Seven percent are predominantly sclerotic and 5% radiolucent (Figure 10.20). Seven percent presented as peau d'orange (Figures 10.21, 10.22).[56]

The lower proportion of ground-glass pattern in this systemic review conflicts with Jundt's definition of FD.[55] His specification of ground glass as the sole pattern in his definition, may be understood as perhaps being derived from the presentation of FD on computed tomography,[64] the modality

Chapter 10: Radiopacities **169**

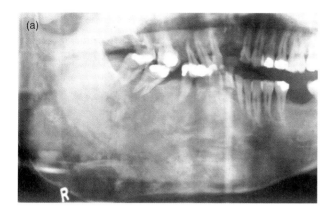

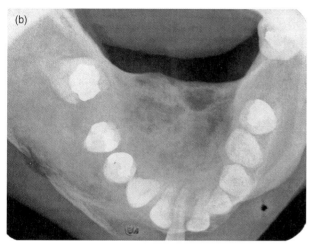

Figure 10.19. Conventional radiographs of bimaxillary fibrous dysplasia. (a) Panoramic radiograph of a case of fibrous dysplasia affecting the right hemimandible and the right hemimaxilla. The right dysplastic lesion of the mandible has crossed the midline of the mandible. The midline has been denoted by the image of the post of the bite block. The dysplastic areas are readily distinguishable from those that remain normal; the wide zone of transition between the dysplastic and normal adjacent bone means that there is poorly-defined margin between them. The fusiform expansion of the mandible is obvious in the vertical dimension. The dysplastic lesions of both jaws have displaced the teeth toward each other resulting in an open bite of the largely unaffected left side. The lamina dura of the teeth within the affected regions are absent, in comparison to those in the unaffected regions. (b) True occlusal radiograph. The teeth still generally follow a catenary curve, although a central incisor has been displaced buccally. The patterns of the dysplastic bone vary throughout from ground glass, to peau d'orange, to radiolucency progressing from the right to left paramedial regions. The buccal dysplastic cortex is very thin in comparison to the normal contralateral side.

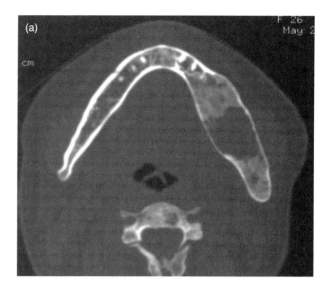

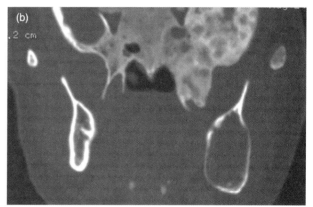

Figure 10.20. Computed tomography of a polyostotic case. (a) Axial CT (bone window) displays fusiform expansion and a radiolucent area in the center of the lesion. This is generally unusual in monostotic cases. (b) Coronal CT (bone window) through the above radiolucency displaying expansion in all directions in comparison with the normal contralateral side. The ipsilateral maxilla also is affected in this polyostotic case. See Figure 17.25, which displays a sagittal section exhibiting involvement of the basiocciputal.

he clearly favored. Nevertheless by reference to Figures 10.19 to 11.23 and 11.30 to 11.36. It can been seen that FD has a wide range of presentations both on conventional radiographs and computed tomography. The spatial resolution of HCT is poorer than that for conventional dental radiography, the spatial resolution of the former is measured in line pairs per centimeter in comparison to the latter's measurement in line pairs per millimeter. The latter is the gold standard for determining the degree of marginal definition, as earlier defined by Slootweg and Müller in Chapter 1.

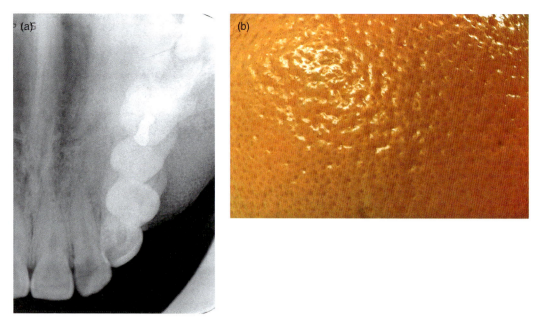

Figure 10.21. (a) Standard occlusal image of the maxilla displaying dysplasia. There is no displacement of teeth. The buccal cortex has been expanded and exhibits peau d'orange (orange peel) appearance, whereas the palatal portion displays a ground-glass appearance. (b) Photograph of orange skin (peau d'orange) displays its characteristic stippled surface. Figure (a) reprinted with permission from MacDonald-Jankowski DS, Li TK. Fibrous dysplasia in a Hong Kong community; the clinical and radiological presentations and the outcomes of treatment. *Dentomaxillofacial Radiology* 2009;38:63–72.

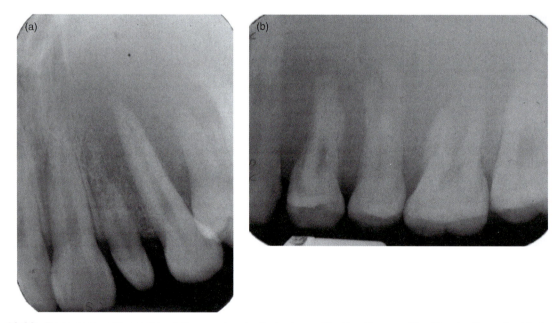

Figure 10.22. Periapical radiographs of fibrous dysplastic lesions affecting the maxilla. The teeth embedded within the dysplastic bone have no lamina dura although a periodontal space is apparent. Figure (b) exhibits a ground-glass pattern, whereas (a) also displays a peau d'orange pattern. In Figure 10.22b, the roots within the dysplastic bone display an abnormal shape and root resorption.

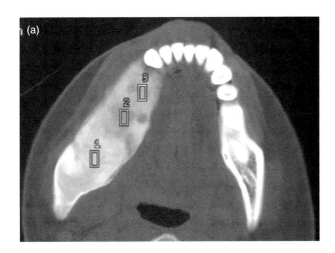

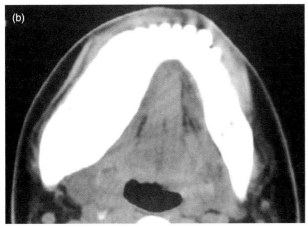

Figure 10.23. Axial bone window (a) and soft-tissue window (b) computed tomographs of the same case displaying fusiform shape. **Note:** The area covered by the image of the bone in a soft-tissue window is larger than that covered by the image of the same bone in the bone window.

The classical fusiform (spindle) shape of FD (Figure 10.23), is even observed if the lateral wall of the antrum is affected (see Figure 11.32c).[64] Only one report in a recent systematic review on FD reported three of its cases as multilocular radiolucencies.[56] These cases appeared in a report 4 decades ago and can be discounted because they may represent cherubism (see Chapter 9), which had at that time been considered as a manifestation of FD.

All cases reported buccolingual expansion and displacement and/or thinning of the lower border of the mandible. The maxillary antrum is involved in nearly every case of adjacent FD (see Figures 11.30–11.36).[56]

The displacement of the mandibular canal differed between reports. It was displaced downward (see Figures 1.7, 10.19) in a Canadian report,[46] either upward or downward in a Hong Kong Chinese report.[60] This may reflect differences between different global communities. Petrikowski et al.[46] suggested that the loss of the lamina dura within the dysplastic lesion can be used to confirm a radiological diagnosis of FD (see Figure 10.22).

Teeth were displaced in 35% of cases (see Figures 1.7 and 10.19), significantly more in Western than in East Asian reports. Root resorption (see Figure 10.22b) was observed in only 2 cases in an East Asian report.[56] In addition, 1 of these cases displayed abnormally shaped roots (see Figure 10.22b). This may have been induced by the adjacent dysplastic process during the development of their root.[60]

As already stated, in addition to an FOL histopathology, a radiologically poorly defined margin is an essential criterion for a firm diagnosis of FD. This feature is really reliably displayed by conventional radiography by virtue of its superior spatial resolution.[64] Nevertheless, CT (both HCT and CBCT) can more readily display the full extent of the lesion, particularly within the more anatomically complex maxilla (see Figures 4.5, 11.30–11.36).[64]

The cortex when displayed on HCT is generally intact, except when adjacent to the teeth in the maxilla. The margins were generally poorly defined, but well defined on at least some sections of each maxillary case.[64]

All cases displayed expansion, which was fusiform in the mandible (see Figure 10.23), and an enlargement of the normal contour in the maxilla (see Figure 11.30a). Although the maxillary antrum when affected is completely obturated, Figure 11.32 displays an FD lesion at an interim stage in the obliteration of the maxillary antrum. It exhibits a rounded dome shape, which while suggestive of a benign neoplasm, is actually a fusiform expansion of the lateral wall. All maxillary cases extended back to the pterygoid process but did not displace it.[64]

The bone windows generally displayed a ground-glass pattern; one also displayed cystlike radiolucencies (see Figures 11.30, 11.34.). The soft-tissue window, which depicts mineralized tissue as white, showed that 5 cases were completely mineralized. Compare the bone windows of Figures 4.5a and 10.23a with their corresponding soft-tissue windows, Figures 4.5b and 10.23b.[64]

Clinical implications

Surgery is generally indicated if there is a threat to vision, which occurs particularly when the FD reduces the diameter of the optic canal. Steroids are given immediately to safeguard vision; surgery then follows.[65] Although blindness is a real risk for FD primarily involving the skull base (optic canal), there appears to be no report of blindness caused directly by FD arising from the jaws. The fact that not one case of proptosis was reported within a total of over 336 maxillary cases of FD in a systematic review[56] indicates that this is not a frequent finding of cases of FD arising primarily in the jaws. This is particularly surprising considering the substantial increase in the vertical dimension of the maxilla observed in one HCT series.[64] Nevertheless, occasionally proptosis does happen in such cases (see Figure 11.30).[64]

Although surgery of FD during its active growth period may cause increased growth, in the past it was generally the view that it would best to await "burnout" of the lesion's growth, which should coincide with the cessation of the individual's growth, namely early adulthood. As already noted, the view that burnout occurs in late adolescence, has been refuted by the systematic review[56]; the majority of cases in the systematic review first presented older than the second decade.

It is now clear that not every known case of FD presenting during childhood and adolescence burn out; some are activated or reactivated in adult life by a precipitating factor such as pregnancy. A woman was blinded in one eye by the activation of FD during her pregnancy.[66] Reactivation is not confined just to females with FD; males have also occasionally exhibited reactivation.[61] Furthermore, Jacobsson et al. reported recurrent episodes of pain and swelling.[67]

I have used the term "reactivation" rather than "recurrence" because the former more accurately describes the FD's response to life events, which will be discussed later, rather than to a failure to ablate completely the lesion in the manner of a neoplasm, which is generally implied by the term "recurrence." Although it is clear from the preceding paragraph that the majority of FDs do not comfortably rest within the definition of a harmatoma, neither are they neoplasms. Therefore, it is for this reason that FDs are placed at Point 2 in the "Scale of Severity of Outcomes" (see Table 1.1).

Futhermore, waiting until burnout in late adolescence is infrequently a real option because the psychological injury provoked by such deformity in early adolescence precipitates the need for its surgical reduction. This is usually achieved conservatively by "shaving."

The readers should be aware that there is a misunderstanding by some head and neck surgeons serving some communities, that FD and OF are part of the same disease spectrum[68] and that FD is a true neoplasm. As a result, FD of the face and jaws in these communities have been routinely resected.[69]

An autopsy of a woman whose first diagnosis of FD had been made 60 years earlier revealed that the dysplastic bone was similar to that seen in active cases with no evidence of involution to normal lamellar bone.[70] Therefore, as Posnick suggests, FD should be considered as a lifelong disease and merit lifelong follow-up.[71] Although he also suggests that this follow-up should be supplemented by HCT,[72] he did not disclose the precise clinical indications for this. Nevertheless, There have been very few case series that have been followed up for a long period. The longest and largest is that of the Hong Kong Chinese, in which 17 patients were followed up for a mean of 9 years.[60]

Despite the paucity in follow-up, 18% of cases were reactivated.[60] This happens significantly more in the sub-Saharan African global group than in the Western global group. The East Asian global group with an intermediate reactivation rate differs significantly from both of them. It should be noted that the reports and overall numbers of cases in this synthesis of the global groups, other than the Western global group, are small.

The systematic review demonstrated that polyostotic cases occurred in 39% of included reports and constituted only 6% of all cases contained within these reports. It was most likely that these polyostotic cases were already known to the patient due to earlier diagnosis of the perhaps more obvious extragnathic lesions, such as those affecting the limbs. Furthermore, menstruation in an infant girl with MAS would be obvious to her parents.

FD has a reputation to undergo sarcomatous change. This change was most frequently associated with radiotherapy, which was discontinued as a treatment for FD over 4 decades ago. Sarcomatous change is more likely to occur spontaneously with MAS 10 times more frequently than for the monostotic form.[73] Vigilance by oral and maxillofacial clinicians is necessary because the jaws are the most frequent site for this transformation.[74] Nevertheless, only one case has occurred within a sequential case series during follow-up.[75] Regardless, of its frequency a lucency within dysplastic bone with poorly defined borders should indicate further investigation. This should be particularly more ominous if the dysplastic cortex is destroyed with/without spiculated periosteal reaction, and there is a widening of the entire periodontal ligament space. HCT of a sequential series of cases indicated that although the cortex of the dysplastic areas of the jaws varies, it is intact.[64] Breaks in the cortex could also result from biopsies (see Figure 1.2) and surgical shaving. Although the lamina dura of the periodontal ligament is replaced by dysplastic bone the space is not only still patent, but subjectively it appears to be narrower than normal.[46] Therefore any widening should be viewed as suspicious.

Although an association between *aneurysmal bone cysts* (ABC) and FD, particularly affecting the base of the skull, is well known, ABCs occurring within FD of the jaws is infrequent. Indeed the best published cases of such affecting a mandible[76] and affecting the maxilla[77] have been reported in medical sources.

OSSIFYING FIBROMA (ICD-O 9262/0)

Slootweg and El Mofty defined the ossifying fibroma as "a well-demarcated lesion composed of fibrocellular tissue and mineralised material of varying appearances. Juvenile trabecular and juvenile psammomatoid ossifying fibroma are two histologic variants of ossifying fibroma."[78]

The term *ossifying fibroma* (OF) has been adopted by the 2005 edition of the WHO[78] in preference to its second edition's "cemento-ossifying fibroma."[79] The latter was a histopathological descriptive term denoting the varying mixture of osseous and cementoid elements. As both elements are now recognized as variants of abnormal bone, such a term became redundant. Nevertheless, the form affecting extragnathic skeleton, which, prior to 1993 had been called ossifying fibroma, is now known as *osteofibrous dysplasia*.[80] Schajowicz recognized the continued use of the term ossifying fibroma but confined it to the jaws.[80]

The OF is a well-defined benign neoplasm that has a capsule (Figure 10.24). It is readily enucleated and does not recur in the majority of cases. Those recurring cases generally represent the juvenile OF form, which affects individuals in the first 2 decades. Cases of this lesion had been reviewed by Slootweg et al. and divided into the psammomatoid and

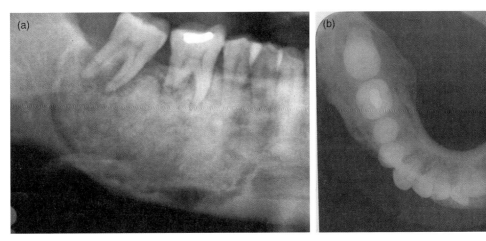

Figure 10.24. Conventional radiography of ossifying fibroma (OF) affecting the posterior mandible. (a) Panoramic radiograph displaying an OF with a well-defined periphery. The central radiopacity is separated from the normal adjacent bone by a radiolucent space representing the capsule. The central radiopacity displays both peau d'orange and cotton wool sclerosis. The mandibular canal has been displaced. The lower border of the mandible has also not only been displaced but is also eroded. The roots of the first molar tooth have been resorbed and the roots of the second molar appear to be displaced distally. (b) True occlusal radiograph displaying buccolingual expansion. Reprinted with permission from MacDonald-Jankowski DS, Li TK. Ossifying fibroma in a Hong Kong community; the clinical and radiological presentations and the outcomes of treatment and follow-up. *Dentomaxillofacial Radiology* 2009;38:514–523.

Table 10.3. Ossifying Fibroma: systematic review

Feature	Total	Western	East Asian	subSaharan	LatinAmer
Male:Female	29%:71%	31%:69%	26%:74%	33%:67%*	25%:75%*
Mean number per year per report	1.7	1.1	1.6	3.0	0.5
Mean prior awareness	1.7	0.9	2.8	1.4	INA
Mean age	31 years	32 years	35 years	19 years	33 years*
Mand:Max	75%:25%	78%:22%	83%:17%	48%:52%	65%:35%*
Mand:Ant:Post	37%:63%	50%:50%	17%:83%	33%:67%*	18%:82%*
Max:Ant:Post	32%:68%	35%:65%*	25%:75%*	50%:50%*	17%:83%*
Swelling: Y:N	66%:34%	42%:58%	79%:21%	100%:0%*	INA
Pain: Y:N	16%:84%	13%:87%	20%:80%	0%:100%*	INA
Incidental: Y:N	31%:69%	43%:57%	23%:77%	0%:100%*	INA
Radiolucent	26%:74%	36%:64%	21%:79%	10%:90%*	INA
Uni:Multiloc	80%:20%	76%:24%	90%:10%*	INA	INA
Cortex: Y:N	53%:47%	53%:47%*	52%:48%	INA	INA
Expansion: Y:N	84%:16%	72%:28%*	86%:14%	100%:0%*	INA
LBMd: Y:N	50%:50%*	25%:75%*	45%:55%*	100%:0%*	INA
Antrum: Y:N	90%:10%*	100%:0%*	100%:0%*	67%:33%*	INA
ToothDisplace	27%:73%	28%:72%	25%:75%*	INA	INA
RootResorption	20%:80%	25%:75%	4%:96%*	INA	INA
Recurrent: Y:N	12%:88%	16%:84%	7%:93%*	10%:90%*	INA

*Advises that the percentages were derived from either one report or from a synthesis of no more that 50 cases. Ant:Post, Anterior:Posterior; INA, Information not available; LatinAmer, Latin American; LBMd, downward expansion of the lower border of the mandible; Mand:Max, Mandible:Maxilla; subSaharan, sub-Saharan African; ToothDispl, Tooth displacement; ToothResorp, Tooth resorption; Uni:Multiloc, Unilocular:Multilocular; Western, predominantly Caucasian; Y:N, Yes:No.

the WHO (now trabecular) types.[81] The former affected young adults and was considered an aggressive variant of OF whereas the latter was confined to children under 15 years of age. Slootweg now includes both the above-mentioned juvenile types of JOF within the OF as a subtype,[78] affirming Brannon and Fowler's[82] earlier contention that JOF was not a separate entity. They found both its psammomatoid and WHO (now trabecular) histopathology, as defined by Slootweg et al.,[81] within the different parts of the same lesion at different times. They added that, "because the initial treatment for *all* (their italics and bold text) OF is assured complete surgical excision and because follow-up is recommended for all, the necessity of the diagnosis of 'JAOF' (my comment: Juvenile Aggressive OF, a synonym for JOF) may be unwarranted."[82] They determined that all OFs need to be enucleated completely to prevent recurrence.[82] Nevertheless, OFs do recur after careful surgery. A need for long-term follow-up is evident in Meister et al.'s report; their 4 OFs followed up for 18 years all recurred.[83] There is so far no radiological marker to determine which lesion is likely to recur.

A hitherto under-considered aspect of OF is a possible association with a familial hyperparathyroidism.[80] This is hyperparathryoidism–jaw tumor syndrome and has an autosomal dominant transmission.[84] Unlike primary hyperparathyroidism, which affects the older adult, this recently identified syndrome presents in adolescents. A secreting carcinoma is the cause in 10–15%. Furthermore, its course is more aggressive, causing more severe hypercalcemia, which may actually present with a hypercalcemic crisis.[84]

The global distribution of reports included in the systematic review[85] upon which much of the following is derived is set out in Figure 1.45 and their details in Table 10.3.

The mean number of cases of OF per year globally was 1.7,[85] similar to the FD.[56] Again as for the FD, although this was highest for the sub-Saharan African global group and least for the Western, the difference was not significant.[85]

OF has a predilection for females (71%) and is the similar for all global groups. Their mean age at first presentation is 31 years of age. The East Asian global group with the oldest mean age is significantly older than the sub-Saharan Africans with the youngest mean age.[85] The peak decade for first presentation is the third and fourth decade equally; they account for nearly one-half of all cases. Sub-Saharan Africans present in the second decade, whereas the other three global groups present in the fourth decade.[85] Males account for nearly one-half of cases in the second decade, but only for 15 to 20% in the subsequent 3 decades. The period of prior awareness for OF is 1.7 years.[85]

Thirty-one percent are detected as incidental findings, whereas 66% first present with swelling and a 16% with pain. Swelling presents significantly more frequently in the East Asian than in the Western global group and vice versa for those discovered as incidental findings. The mandible is affected in 75% of cases except for sub-Saharan Africans, among whom both jaws are affected equally. This was particularly significant in comparison to the 83% : 17% mandible : maxilla ratio in East Asians. Mandibular cases of OF are equally distributed between the anterior and posterior sextants only in the Western global group. The other global groups display a predilection for the posterior sextant, particularly in East Asians with an anterior : posterior sextant ratio of 17% : 83%. Overall the anterior sextant of the maxilla is affected in only 32%.[85]

Although the internal structure of OF on conventional radiography is similar to FD, reflecting their similar histopathology, the OF has a capsule. This capsule is represented by a well-defined radiolucent line sharply separating the lesion from the adjacent normal bone.[86] This feature can be easily and cheaply appreciated on a single conventional image, such as a panoramic radiograph (see Figure 10.24a). This image can be, if necessary, supplemented by intraoral images.

The predominant radiological pattern (radiolucency, radiopacity within a radiolucency, or completely radiopaque) varied significantly between global groups. It also varied between reports within these groups, indicating that the pattern will vary with the community reported, which may be in turn be influenced by the age at first presentation. Fifty-eight percent present with the classical presentation of a radiopacity within a radiolucency (see Figure 10.24a). Twenty-six percent are radiolucent (Figure 10.25), and 16% are completely opaque. Sub-Saharan Africans display significantly more radiolucent lesions. This is likely to reflect their younger age at first presentation. The OFs in an East Asian report are equally divided between

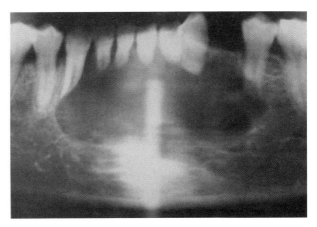

Figure 10.25. Panoramic radiograph of an ossifying fibroma (OF) presenting as its initial or early radiolucent stage appear as radiolucencies. The presentation in this stage can be as multilocular radiolucencies. This case also displays, in addition to tooth displacement, root resorption, which is strongly suggestive of a solid ameloblastoma. **Note:** Generally, lesions are infrequently as well-displayed and accurately displayed in the midline of a panoramic radiograph as in this case. This investigation should normally be supplemented by pulp-testing and intraoral radiographs. Reprinted with permission from MacDonald-Jankowski DS. Fibro-osseous lesions of the face and jaws. *Clinical Radiology* 2004;59:11–25.

those that have a round or oval shape.[87] The oval-shaped OFs are significantly larger. Half of them present in females of 45 years old or older.[87]

Fifty percent present with a cortex or marginal sclerosis.[85]

Eighty-four percent present with buccolingual expansion (Figure 10.26). Although the OF can display substantial buccolingual expansion, this is not always so because the buccolingual expansion may not always reflect the mesiodistal expansion of the neoplasm (see Figure 10.24).[85]

The lower border of the mandible presents with erosion and/or displacement (see Figures 1.8, 10.19, 10.24a) in one-half of mandibular cases.[85] The maxillary antrum is involved by 90% of cases of OF subjacent to it.[85]

Teeth are displaced in 27% of cases (see Figures 1.8, 10.24a) and roots are resorbed in 20% of cases (see Figure 10.24a, 10.25).[85] The sole East Asian report in the systematic review displayed significantly fewer cases of root resorption in comparison to one of two Western reports.[85] There was no overall significant difference between the East Asian and Western global groups.[85]

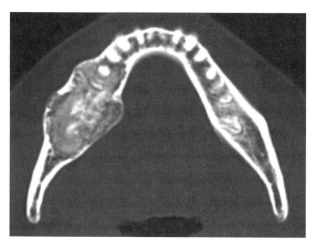

Figure 10.26. Axial computed tomography (bone window) of a ossifying fibroma displaying buccolingual expansion. The periphery of the lesion is well defined. The cortex is eroded at several points.

Because the signal intensity on T1-weighted and T2-weighted MRI images is dependent upon a number of factors, such as the amount of bone trabeculae and degree of cellularity,[88] FD and OF both show an intermediate signal on T1-weighted and a hypointense signal on T2-weighted images.[35] The hypointense signal intensity on T2-weighted images is caused by numerous bony trabeculae.[89] In the early stage of FD, there may be areas of T2-weighted hyperintensity.[35] This may correlate with the bone resorption phase of early FD. Although intravenous contrast (Gadolinium) produces a moderately enhanced signal for OF, it is often marked for FD. Although both FD and OF can be mistaken for meningioma on MRI,[89] MRI offers greater specificity where there is neurovascular and ocular involvement.[65]

Clinical implications

Twelve percent recur after treatment, which is generally enucleation.[85] There was no difference between global groups. There is some circumstantial evidence that the onset of menopause may initiate, reactivate, or accelerate growth. Those cases first presenting during menopause were significantly larger than those first presenting between attainment of the peak bone mass and menopause.[87]

OSSEOUS DYSPLASIAS

As can be seen by reference to Figure 10.18, of all lesions affecting the face and jaws this group of lesions have perhaps seen the most changes in their nomenclature and classification, certainly since the WHO's first edition.[90] This change is still taking place. This FOL has now been divided into two broad categories, the focal and the florid forms. Although they are not neoplasms they have exceptionally been included in the recent 2005 edition of the WHO classification of odontogenic neoplasms,[91] which excluded all the other nonneoplasms, with the other exceptions of FD and simple bone cyst. Although they are generally not considered to be odontogenic lesions, they are undoubtedly related to the presence of teeth. Almost all appear above the mandibular canal and thus are confined to the alveolar process. This suggests, at least, some odontogenic influence upon their genesis. For the same reasons given for OF, the WHO's 2005 edition[91] preferred to use "osseous dysplasia (OD)" rather than "cemento-osseous dysplasia" used in the second edition.[92] The term *osseous dysplasia* is adopted for this text.

Florid osseous dysplasia

The global distribution of reports of *florid osseous dysplasia* (FOD) included in the systematic review[93] upon which much of the following is derived is set out in Figure 1.46 and their details in Table 10.4. This lesion was initially associated with middle- to old-aged females of sub-Saharan African origin, but also more recently in those of East Asian origin.

The mean number of cases of FOD per year globally is 1.2.[93] The mean number of cases is highest for the Western global group but surprisingly least for the sub-Saharan global group. This may reflect the greater ethic diversity within the former, particularly the United States of America where those of non-European origin account for nearly one-third of the population, and the lower life expectancy in the latter. In the sub-Saharan African global group, potential FOD victims are more likely to die earlier in life of other causes before they could acquire the FOD lesions.[94]

FOD has been known under a variety of names.[93] It overwhelmingly affects females (97%). The mean age at first presentation is 49 years old (Table 10.4). The age range is 21 to 83 years old. Half are discovered as incidental findings; 48% first present with pain, 31% with swelling, and 30% with a discharge or a fistula. Comparison of two case series derived from the same community showed that the series of cases observed as incidental findings on radiographs were significantly

Table 10.4. Florid Osseous Dysplasia: systematic review

Feature	Total	Western	East Asian	subSaharan	LatinAmer
Male:Female	3%:97%	2%:98%	%:96%*	0%:100%*	INA
Mean number per year per report	1.2	2.2	0.6	0.2	INA
Mean age	49 years	48 years	51 years	52 years	INA
Mand: Y:N	100%:0%	100%:0%	100%:0%		INA
Mand:Ant: Y:N	64%:38%	78%:22%*	54%:45%*	20%:80%*	INA
Mand:Post: Y:N	99%:1%	100%:0%	96%:4%*	100%:0%*	INA
Max: Y:N	68%:32%	68%:32%	71%:29%*	60%:40%*	INA
Max:Ant: Y:N	53%:47%	70%:30%*	40%:60%*	0%:100%*	INA
Max:Post: Y:N	100%:0%	100%:0%*	100%:0%*	100%:0%*	INA
Swelling: Y:N	31%:69%	35%:65%	24%:76%*	INA	INA
Pain: Y:N	48%:52%	35%:65%	69%:31%*	INA	INA
Incidental: Y:N	49%:51%	54%:46%	40%:60%*	INA	INA

*Advises that the percentages were derived from either one report or from a synthesis of no more that 50 cases. Ant:Post, Anterior:Posterior: INA, Information not available: LatinAmer, Latin American: Mand:Max, Mandible:Maxilla: subSaharan, sub-Saharan African: Western, predominantly Caucasian; Y:N, Yes:No.

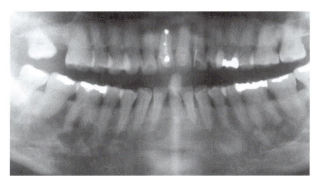

Figure 10.27. This panoramic radiography displays multiple well-defined radiolucencies, many containing central opacities. All lesions are confined to the alveolus. This a florid osseous dysplasia and is most frequently found in middle- to old-aged females of sub-Saharan African or East Asian origin. Reprinted with permission from MacDonald-Jankowski DS. Fibro-osseous lesions of the face and jaws. Clinical Radiology 2004;59:11–25.

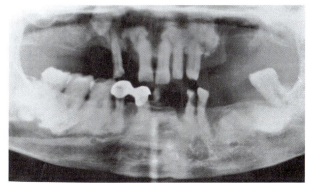

Figure 10.28. This panoramic radiography displays multiple well-defined radiopacities affecting most sextants. They are largely confined to the alveolus. This is florid osseous dysplasia and is most frequently found in middle- to old-aged females of sub-Saharan African or East Asian origin.

older than the series of cases that presented with symptoms.[93] More than one sextant needs to be affected to fulfill a diagnosis of FOD. The presentation on a panoramic radiograph or on a full-mouth survey (using intraoral film or digital detectors) is usually bilateral. The mandible is affected in 100% of cases and the maxilla in 67%.[93]

The radiological presentation of lesions of either form of OD ranges from a radiolucency, to one with one or more small central radiopacities (Figures 10.27 and 11.39), on to a substantial radiopacity with a radiolucent periphery, to complete radiopacity that abuts directly onto adjacent normal bone (Figures 1.9, 3.1, 10.28) According to Kawai et al. they are not likely to change from one pattern to another, although a patient can acquire more lesions with time.[95]

Diagnosis is readily achieved by conventional radiography; as detailed in the previous paragraph, there is little need for computed tomography in

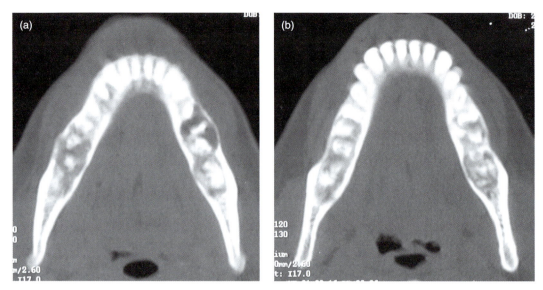

Figure 10.29. These are axial bone windows sections of computed tomography. The radiolucencies contain central radiopacities. Although extensive mesiodistally, they cause very little buccolingual expansion. The cortex displays erosion in places.

the uncomplicated case. Nevertheless, computed tomography has revealed some more detail not appreciated by conventional radiography (Figure 10.29), such as the central positioning of the osseous dysplastic tissue within the lesion. Radiolucent FOD lesions were more likely to display buccolingual expansion than those that contained high-density masses (osseous dysplastic tissue).[96] These high-density masses had CT numbers of 772 to 1587 HU, which is equivalent to cementum or cortical bone.[96] HCT invaluably assisted in the investigation of a recurrent OF, which was subsequently diagnosed as an OD.[97] This lesion eroded the buccal cortex, but caused no expansion. HCT imaging displayed another FOD lesion that both expanded and breached the cortex.[98] The HCT also permitted precise localization of the mandibular canal to the lesions.[98]

Focal osseous dysplasia

The global distribution of reports on *focal osseous dysplasia* (FocOD) included in the systematic review[94] upon which much of the following is derived is set out in Figure 1.47 and their details in Table 10.5.

Table 10.5. Focal Osseous Dysplasia: systematic review

Feature	Total	Western	East Asian	subSaharan	LatinAmer
Male:Female	12%:88%	12%:88%	14%:86%	6%:94%	INA
Mean number per year per report	4.9	20	1.1	INA	INA
Mean age	44 years	39 years	47 years	INA	INA
Mand:Max	85%:15%	91%:9%	85%:15%	INA	INA
Mand:Ant:Post	20%:80%	27%:73%*	1%:99%	INA	INA
Max:Ant:Post	26%:74%	32%:68%	20%:80%*	INA	INA
Swelling: Y:N	25%:75%	24%:76%	30%:70%	INA	INA
Pain: Y:N	28%:72%	24%:76%	44%:56%	INA	INA
Incidental: Y:N	64%:36%	74%:26%	14%:86%	INA	INA
Numb: Y:N	17%:83%*	INA	17%:83%*	INA	INA

*Advises that the percentages were derived from either one report or from a synthesis of no more that 50 cases. Ant:Post, Anterior:Posterior; INA, Information not available; LatinAmer, Latin American; Mand:Max, Mandible:Maxilla; subSaharan, sub-Saharan African; Western, predominantly Caucasian; Y:N, Yes:No.

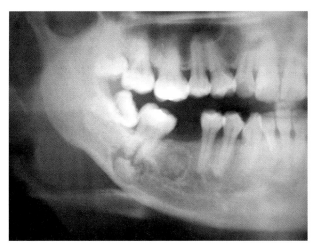

Figure 10.30. This panoramic radiograph displays two osseous dysplastic lesions only within one sextant. In the absence of symptoms, they can be safely left to occasional radiographic review as they are most likely to represent focal osseous dysplasia. One is wholly confined to the alveolus, whereas as the more distal lesion although classically sited at the apex of an erupted tooth is wholly within the basal bone of the mandible. Although, this is an exception to the rule that osseous dysplasia affects the alveolus only, the hypoplastic mandible at that site was insufficiently deep vertically to allow accommodation of the distal osseous dysplastic lesion within the alveolus. Figure courtesy of Dr. Ian Matthew, Faculty of Dentistry, University of British Columbia.

The mean number of cases of FocOD per year globally was 4.9.[94] As already observed for FOD, this was highest for the Western global group and least for the sub-Saharan African global group. The explanation is the same as that for FOD. The FocODs younger age of first presentation in comparison to that of the FOD is not reflected in better reporting in the sub-Saharan African and Latin American global groups. The reason simply may be that it has only recently become recognized as a separate clinical entity.

FocOD is confined to a single sextant (Figure 10.30).[91] It may present as a single lesion or as a group of juxtapositioned lesions.[91] If it affects more that one sextant, certainly if it is bilateral or affects both jaws, it should be considered to be a case of FOD.

Prior to its recognition as a discrete entity its frequent solitary clinical presentation led to its inclusion among OFs. This was the experience both of Summerlin and Tomich[99] and of Su et al.[100,101] Unfortunately, neither disclosed how many had been misdiagnosed. The first clear differentiation of the OD as a separate broad entity was established by both the WHO's second edition[92] and the Waldron et al. report.[102] Unlike OFs, OD (including both FODs and FocODs) would not "shell-out," and therefore the specimen was delivered for its histopathological examination in fragments. These centrally important criteria were reiterated by Melrose.[103] Therefore, the definitive feature that distinguishes between OF and OD is the presentation of the gross surgical specimen, which is histopathologically a FOL.

Periapical cemental dysplasia, first appeared as an independant entity in the WHO's first edition,[90] became indistinguishable from FOD in the 2nd edition only to return to its 1st edition definition that confined it to the anterior sextant of the mandible in the WHO's 2005 edition.[91] This lesion appears initially as periapical radiolucencies associated with noncaries lower incisors (Figure 9.12). With the exception of its histopathological outdated name and its anterior location there appears to be nothing to else to distinguish it from FocOD. It is not an infrequent experience for the dental specialist to observe root-treated lower incisors (Figure 1.25), whose apices are now associated with mature OD lesions. Avoidance of unnecessary treatment can be achieved by pulp-vitality testing of all periapical lesions, especially radiolucencies.

Although it is generally accepted that symptomless FocODs need no treatment, their clinical importance has been emphasized by their presence in edentulous sites required for osseointegrated implants.[94] As an essential prelude to its management in this regard, a better understanding of the frequency and presentation of the FocOD in the global literature is required.

FocOD overwhelmingly affects females (88%)[94] (Table 10.5). The mean age is 44 years of age. Sixty-four percent present as incidental findings. Another quarter present with swelling, 28% with pain and 17% with numbness. The last was reported only in East Asians. Eighty-five percent affect the mandible. The posterior sextants of the mandible and the maxilla are affected in 80% and 74%, respectively.[94]

Just over half (53%) of cases were well defined. Of these, 40% display a sclerotic periphery; the other 60% of cases are nonsclerotic. Forty-nine percent of the cases appear in dentate areas of the jaws (teeth were still present) and the other 51% in edentulous areas (teeth had been extracted). In the dentate areas, no tooth displacement or root resorption are observed.[94]

The predominant radiological patterns are radiolucency (31%), a central radiopacity within a radiolucency (separated from the adjacent bone by a radiolucent space) (37%), and a complete radiopacity (32%). East Asian reports display significantly more complete radiopacities and fewer radiolucencies than those of Western communities.[94] This may reflect the older mean age of the East Asian global group at first presentation (47 years) in comparison to that of the Western global group (39 years).[94]

Familial gigantiform cementoma

The familial gigantiform cementoma have similar radiological and histopathological presentations, but they differ in their clinical presentations and behavior from the already-discussed "conventional" ODs. Although they clearly fulfill the histopathological criteria for FOLs, they are not ODs. Young et al., reported this phenomenon in 55 members over five generations of the same kindred. They coined the term *familial gigantiform cementoma* (FGC) to differentiate them from the FODs.[104] The application of this term for this lesion, which was so different in presentation and prognosis from FOD, was affirmed by Waldron.[105] There is little doubt that these lesions are neoplastic, and although no oncogenes have been reported, their mode of inheritance is autosomal dominant.[104] In addition to a family history they largely affect Caucasian kindreds[105,106] and the young of both genders equally. In one recent report the youngest of 6 affected members of 1 kindred were a 13-year-old girl and a 17-year-old boy.[106] Of the 5 individuals for whom blood work was available, it was the boy who had a raised serum alkaline phosphatase. There have been cases that present as FGC, but have no family history. Such cases can be provisionally termed *spontaneous gigantiform cementoma*. So far, it has reportedly affected at least 2 males of East Asian origin,[107,108] and perhaps 2 American females.[109] All displayed very aggressive behavior and achieved substantial dimensions. The question of whether these lesions constitute a separate clinical entity or are a manifestation of the rarely occurring multiple OFs or even hyperparathyroidism–jaw tumor syndrome,[84] warrants further and fuller reported cases. Already such cases have been linked to hereditary lesions such as Gardner's syndrome[32] and neurofibromatosis type 1.[110]

When patients present with extensive bilateral, almost symmetrical, involvement by multiple radiopacities of both the alveolar and basal processes of both jaws, Gardner's syndrome should also be considered (see Figure 10.11). Lee et al. reported such a case in a middle-aged woman.[32]

FGC was reported in 8 Caucasian females of the 55 patients diagnosed with neurofibromatosis type 1 (NF1), an autosomal dominant disease, with some malignant transformation potential.[110] All these patients had been evaluated with a panoramic radiograph. Missing and unerupted teeth, overgrowth of the alveolus, and dilated mandibular canals and mandibular foramina are well recognized dental manifestation of NF1. These features occurred in one-third of all adult females with NF1; males and children who accounted for half of the 55 cases were not affected. The radiolucent-staged lesions appeared as periapical radiolucencies suggestive of inflammatory disease. The vast majority of adjacent teeth in this report responded positively to pulp-vitality testing. The reader should be cautious because not one such case in Visnapuu et al.'s report has been confirmed by histopathological examination.[110]

Moshref et al. report 4 males of an Iranian kindred who presented with FGC. Three individuals also had multiple long bone fractures.[111]

Treatment implications

Either form of OD is best kept under review. Although surgery should be avoided unless the lesions produce symptoms, this option may no longer be tenable if the edentulous site is required for an osseointegrated implant.[94] In such a case in absence of published evidence currently, it is advised to surgically ablate the lesion by lateral trepanation and curettage and then allow it to heal first. This should minimize a real risk of failure of the implant by insertion into abnormal tissue. Only FGC or spodratic gigantiform cementoma require routine surgical ablation because of their aggressive behavior.[107,108,109]

OSTEOMA

The osteoma is a rare benign osteogenic neoplasm. It features compact and/or cancellous bone. It can arise from either the periosteal or endosteal bone surfaces to be a peripheral or central osteoma, respectively. Although multiple osteomas of the jaws are a hallmark of Gardner's syndrome (familial adenomatous polyposis) (see Figure 10.11), nonsyndromic cases are typically solitary (Figure

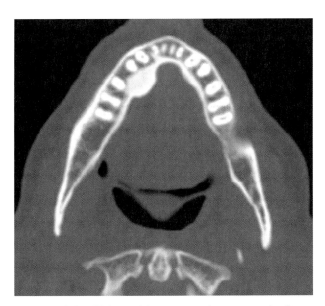

Figure 10.31. This axial computed tomograph (bone window) reveals a dense radiopacity on the lingual cortex of the mandible. This is an osteoma. Figure courtesy of Dr. Montgomery Martin, British Columbia Cancer Agency.

10.31). Kaplan et al. synthesized the literature and found 91 solitary cases of which all but 6 were peripheral.[112] They state that while peripheral osteomas pose little diagnostic challenges, the central osteoma has to be distinguished from ossifying fibroma, condensing osteitis, idiopathic osteosclerosis, osteoblastoma, cementoblastoma, and complex odontoma.[112] Central osteomas are also more likely to recur.[113] The largest case series of osteomas affecting the face and jaws was reported in a Spanish community. The 132 osteomas arising in 106 patients affected the mandible, the maxilla, and the paranasal sinuses, 54%, 16% and 21%, respectively. In contradiction of Kaplan et al.'s synthesis,[112] peripheral osteomas Larrea-Oyarbide et al.'s accounted for 59% of those affecting the jaws.[113] Another report of 14 consecutive peripheral osteomas affecting an Israeli community found that 12 arose from the jaws.[114]

OSTEOBLASTOMA

The osteoblastoma is a slow-growing benign neoplasm of bone, accounting for 1% of all primary bone neoplasms. Jones et al. compared 24 of their own cases to that of their synthesis of 77 cases of osteoblastomas and osteoid osteomas of the jaws reported elsewhere in the literature.[115]

The osteoblastoma presented five times more frequently in the mandible. Jones et al. found overall that the osteoblastoma predominantly affects the left posterior mandible.[115] Although 60% of their own 24 cases at the time of discovery were symptom-free, the rest were associated with pain and tenderness; this was contrary to the results of their synthesis. Radiologically they are generally well defined and equally distributed between radiolucent, radiopaque, and mixed lesions. Although the osteoblastoma and the histopathologically similar osteoid osteoma are held to be different lesions extragnathically, Jones et al. discussed the latter as an earlier manifestation of the former in the jaws.[115]

The 9 cases followed up did not recur.[115]

CEMENTOBLASTOMA (ICD-O 9273/0)

van der Waal defined the cementoblastoma as "characterized by the formation of cementum-like tissue in connection with the root of a tooth."[116] Furthermore, according to him: "Radiologically, the tumour is well-defined and is mainly of a radiopaque or mixed-density, surrounded by a thin radiolucent zone. Root resorption, loss of root outline and obliteration of the periodontal ligament space are common findings"[116] (Figure 10.32).

The cementoblastoma is a relatively rare lesion. The largest synthesis of 70 cases is derived from single case reports or small case series.[117] Brannon reported 44 cases from the U.S. Army pathology files[118] and Dominguez et al. reported 25 Latin American cases.[119]

Although the cementoblastoma is a benign odontogenic neoplasm, Figure 10.33 demonstrates that the cementoblastoma can grow quickly. It presents in about one-half of cases in children and adolescents as pain on biting. The largest case series is that of Brannon et al.[118] The cementoblastoma affects males and females equally and affects the mandible thrice as frequently as the maxilla.[117–119] The mandibular first molar is most frequently affected.[117–119] The mean ages for 2 syntheses and 2 case series are 23,[117] 22,[118] 21,[118] and 26[119], years, respectively. Brannon et al. reported that the deciduous dentition is rarely affected.[118] Swelling in both syntheses was nearly 60%,[117,118] whereas it was 86% in the military series.[118] Awareness of lesions prior to presentation was 1 year.[118]

The mean size of the lesion at presentation is 2.1 cm.[118] In a synthesis[117] and the Spanish-language

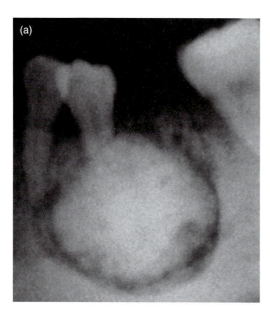 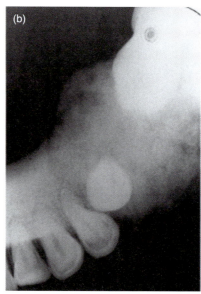

Figure 10.32. The panoramic radiograph (a) and true occlusal radiograph (b) are of a cementoblastoma. This presents on the panoramic radiograph as a well-defined radiopacity separated from the adjacent bone by a radiolucent margin, which is continuous with the periodontal ligament space of the affected tooth. The lesion was fused directly to the root, which it has partially resorbed. The lesion is substantial in size and is close to the alveolar crest. The lesion has also expanded to encompass a retained root, which displays resorption by the capsule. The mandibular canal is displaced inferiorly but around the inferior aspect of the lesion. Reprinted with permission from MacDonald-Jankowski DS, Wu PC. Cementoblastoma in the Hong Kong Chinese: a report of 4 cases. *Oral Surgery, Oral Medicine and Oral Pathology* 1992;73:760–764.

report,[119] larger cementoblastomas were found outside the posterior mandibular sextant. In the former, 3 were located in the maxilla, almost affecting the entire hemimaxilla.[117] In the Spanish-language report, 2 extensively crossed the midline, 1 in each jaw, occupying the entire anterior sextant.[119]

According to van der Waal, the diagnosis of cementoblastoma cannot be made on the histopathology alone. Although its histopathology can be indistinguishable from osteoblastoma or osteoid osteoma or well-defined osteogenic sarcoma, this neoplasm is easy to distinguish by its radiological presentation, which is almost pathognomonic.[116]

Radiological imaging displays a well-defined (94%) circumscribed radiopacity (67%) or mixed density (radiopacity within a radiolucency) (28%).[118] The attached tooth root displays not only root resorption but is also fused to the cementoblastoma. This is represented by loss of root outline (see Figures 10.32, 10.33). Despite this radicular disruption, Brannon et al. reported that the affected tooth was vital in 19 out of 21 cases.[118] Complete enucleation, including extraction of the affected tooth is required to avoid recurrence. Brannon et al.'s synthesis of the literature revealed that 37 percent of cases followed up recurred within a mean of 5.5 years.[118] Brannon et al. noted that the cases that recurred displayed more expansion and cortical perforation.[118]

Dominquez et al.'s report additionally correlated the radiology and histopathology. The 2 radiolucent cases (these appear to be the first such lesions to be reported) were represented by cementoid trabeculae and nodules; the 9 mixed cases displayed remodeled trabeculae, which appeared pagetoid. The 6 dense cases were represented by radial trabeculae.[119]

In almost every case, the apices of permanent teeth are affected. Occasionally, the cementoblastomas have affected deciduous teeth. In one case the cementoblastoma primarily affected the coronal half the root (Figure 10.34).[117] Growth can be rapid (compare see Figure 10.33a,b). Furthermore, the prior extraction of the tooth leaving the neoplasm behind deprives the subsequent clinician of an easy diagnosis. This happened in one case in a case series.[117] The definitive diagnosis was exceptionally entirely furnished by the histopathologist, who

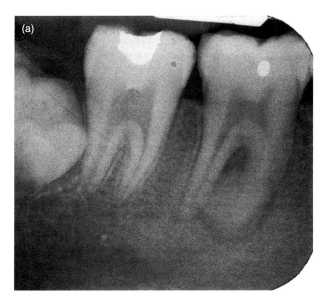 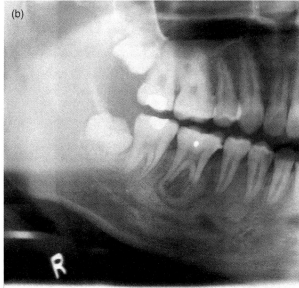

Figure 10.33. The periapical image (a) displays the cementoblastoma as a small round radiopacity fused to the mesial root-affected first molar mandibular tooth, which also displays root resorption. The panoramic image (b) is of the same lesion a year later. Meanwhile the lesion has now grown both mesially and distally. This growth in the latter direction has reached the distal root of the same tooth. Its periodontal ligament space on its mesial aspect appears wider. The panoramic radiograph also displays a well-defined round radiopacity a the apex of a carious first premolar. This lesion has no radiolucent space. The reviewer may observe the hint of such a space, this is the Mach band effect. This "space" is traversed by trabeculae. This radiopacity is termed *condensing* (or *sclerosing*) *osteitis* and represents a sclerotic reaction to infection. The whole area of bone containing both lesions is more sclerotic than the rest. The mandibular canal, which is normally rendered visible as a radiolucent line defined by its cortex, is now obvious because it is contrasted against the sclerotic bone. This zone is diffuse osteosclerosis suggestive of sclerosing osteomyelitis and classically affects both the alveolar and basal processes of the mandible. In this case it may have arisen from the periapical infection of the first premolar.

observed the retained root encased by the neoplasm. It displayed the typical root resorption and fusion with the neoplasm.

ODONTOMA (ICD-O 9281/0 AND 9282/0)

The odontoma is perhaps the most common odontogenic neoplasm, certainly in reports arising from North America and Europe. It is recognized as one of two basic types; the complex and compound types, each with its own ICD-O code, 9282/0[120] and 9281/0,[121] respectively. Praetorius and Piatelli defined the complex type as "a tumour like malformation (harmatoma) in which enamel and dentin, and sometimes cementum, is present[120]", and the compound type as "a tumour like malformation (harmatoma) with varying numbers of tooth-like elements (odontoids).[121]"

The compound type, first presents frequently in children as noneruption of permanent teeth. It classically presents on a radiograph as a "bag of teeth" (Figure 10.35) about the size of a normal tooth in that site, whereas the complex type frequently presents as a well-defined radiopacity (see Figure 1.14 a,b), which can become very large (see Figures 11.37 and 11.38). This radiopacity is frequently demarcated from the adjacent normal bone by a radiolucent space.

The details of the global distribution of reports of complex and compound odontomas are set out in Tables 10.6 and 10.7, respectively.

The mean number of cases of odontomas per year globally was 1.7 and 2.5 for complex and compound odontomas, respectively. The mean number of cases odontomas per year is least for the sub-Saharan African global group. This may reflect a tendency not to refer these lesions for histopathology rather than a true lower prevalence in this global group.

The mean ages of first presentation of both forms is similar; the complex odontoma first presents consistently a few years later in life. The

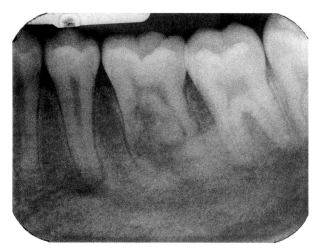

Figure 10.34. This periapical radiograph displays a unique cementoblastoma arising from closer to the furcation rather than classically from the apex. The mandibular canal is rendered obvious by increased density. Reprinted with permission from MacDonald-Jankowski DS, Wu PC. Cementoblastoma in the Hong Kong Chinese: a report of 4 cases. *Oral Surgery, Oral Medicine and Oral Pathology* 1992;73:760–764.

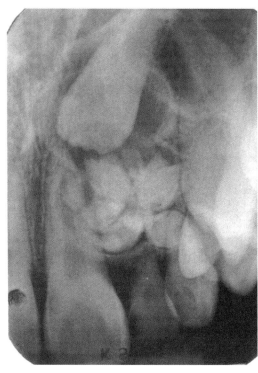

Figure 10.35. This periapical radiograph displays a compound odontoma preventing the eruption of a permanent incisor tooth. The presence of denticles on the radiograph are pathonemonic for a compound odontoma. Reprinted with permission from MacDonald-Jankowski DS. Florid osseous dysplasia in the Hong Kong Chinese. *Dentomaxillofacial Radiology* 1996;25:39–41.

male:female ratio is almost equal in all respects. The complex form has a greater predilection for the mandible, whereas the compound form shows a predilection for the maxilla. The complex form also displays a clear predisposition (77%) for the posterior sextant of the mandible, whereas the compound form exhibits some predilection for the anterior sextant. Although both forms affect the anterior sextant of the maxilla, that of the compound form occurs there most frequently (78%).

Although almost all reports pool the findings of each type together, this is hard to understand because both differ in their radiology, disposition between jaws and sextants for each jaw, clinical presentation, and propensity to recur. Therefore, specific information for each with regard to presenting clinical and radiological features is almost wholly absent. Chen et al.[122] synthesized 43 cases, which clearly illustrated the differences in clinical presentation. They found that just over a half of their complex odontomas first presented as swellings, but few with noneruption of teeth, whereas just under half of the compound odontomas presented with noneruption of teeth and few with swellings.

Praetorius and Piatelli state that the complex type can recur, whereas the compound type never recurs following treatment.[120] This suggests that the complex type has at least one neoplastic feature. There is further circumstantial evidence that the complex type is a neoplasm. In one report, measurements of the complex forms from periapical radiographs displayed an increase in size with advancing age.[123] Furthermore, a synthesis of case reports of large odontomas, in the same report, found that 80% are of the complex type.[123]

CALCIFYING EPITHELIAL ODONTOGENIC TUMOR (ICD-O 9340/0)

Takata and Slootweg defined the *calcifying epithelial odontogenic tumor* (CEOT) as "a locally invasive epithelial odontogenic neoplasm, characterised by the presence of amyloid material that may become calcified."[124]

Table 10.6. Complex Odontoma: systematic review

Feature	Total	Western	East Asian	subSaharan	LatinAmer
Male:Female	52%:48%	54%:46%	51%:49%	54%:46%*	42%:58%
Mean number per year per report	1.7	1.9	1.6	0.5	1.4
Mean age	23 years	22 years	24 years	INA	21 years
Mand:Max	53%:47%	54%:46%	55%:45%	80%:20%*	45%:55%
Mand:Ant:Post	23%:77%	22%:78%	21%:79%	25%:75%*	29%:71%
Max:Ant:Post	60%:40%	64%:36%	48%:53%	0%:100%*	71%:29%

*Advises that the percentages were derived from either one report or from a synthesis of no more that 50 cases. Ant:Post, Anterior:Posterior; INA, Information not available; LatinAmer, Latin American; Mand:Max, Mandible:Maxilla; subSaharan, sub-Saharan African; Western, predominantly Caucasian; Y:N, Yes:No.

Table 10.7. Compound Odontoma: systematic review

Feature	Total	Western	East Asian	subSaharan	LatinAmer
Male:Female	49%:51%	48%:52%	53%:47%	43%:57%*	52%:48%
Mean number per year per report	2.5	3.2	2.2	0.3	1.7
Mean age	20 years	19 years	20 years	INA	INA
Mand:Max	42%:58%	35%:65%	47%:53%	83%:17%*	33%:67%
Mand:Ant:Post	56%:44%	60%:40%	58%:42%	0%:100%*	36%:64%
Max:Ant:Post	78%:22%	76%:24%	80%:20%	0%:100%*	85%:15%

*Advises that the percentages were derived from either one report or from a synthesis of no more that 50 cases. Ant:Post, Anterior:Posterior; INA, Information not available; LatinAmer, Latin American; Mand:Max, Mandible:Maxilla; subSaharan, sub-Saharan African; Western, predominantly Caucasian; Y:N, Yes:No.

Philipsen and Reichart reported on 181 cases synthesized from the literature in 2000. The relative period prevalence of the CEOT is 1%.[125] The CEOT is equally distributed between the sexes. Sixty-seven percent are found in the mandible. For those cases that occur in either a posterior or an anterior sextant, the posterior prevails: 84% and 76% in the mandible and maxilla, respectively. The mean age for the Western global group is 39 (females) and 40 (males) years, whereas that for the East Asian global group is 37 (females) and 37 (males) years.

Kaplan et al. synthesized the literature, with particular emphasis on the CEOT's manifestations on conventional radiology.[126] They observed that swelling or expansion was "the most prevalent clinical manifestation" (72%). Twenty-two percent are symptom-free. The lesions had a mean size of 3.5 (0.5 to 10) cm.[126]

Seventy-eight percent are well defined, of which 28% are corticated.[126] All smaller lesions are well defined, whereas two-thirds of larger lesions are well defined but not corticated. Twenty-eight percent of these larger lesions are poorly defined. The most frequent radiographic presentation is a mixed radiolucent-radiopaque pattern (65%). The radiolucent pattern accounts for 32%. Most of the small lesions are radiolucent (86%), whereas most of the large are mixed (74%).[126] "The coronal clustering (radiopacities close to the crown of the impacted teeth)" is found only in 12% of cases, whereas the "driven snow" pattern is observed only in one case.[126] Although 58% are unilocular and 27% multilocular, those over 3 cm are more likely to be multilocular. Nevertheless, those associations with the maxilla are mainly unilocular. Sixty percent are associated with unerupted teeth, of which 62% were molars.[126]

Philipsen and Reichart report that more than half of these cases are associated with a mandibular molar. In order to distinguish those pericoronal radiolucent CEOTs from dentigerous cysts they suggest that whereas the latter is more frequently associated with the third molar, the former is more frequently associated with the first and second molars.[125]

Tooth displacement is observed frequently (41%), whereas root resorption is infrequent (4%).[126] Kaplan suggested that this infrequency for root resorption may distinguish it from the solid ameloblastoma,[126] which was reported by Struthers and Shear to exhibit root resorption in 81% of cases.[127] Displacement of anatomical structures, such as the mandibular canal, were also infrequent.[126] A feature that was considered a characteristic of this lesion originally by Pindborg was the penetration of the apex of the inferiorly displaced molar through the inferior cortex. Although Kaplan et al. observed mention of this feature in only 7% of cases in their synthesis, they did suggest that its presence should prompt consideration of a CEOT.[126]

Philipsen and Reichart advise that due to the relative indolent behavior of mandibular cases, these cases can be effectively treated by enucleation.[125] Maxillary cases on the other hand tend to grow more rapidly and do not remain well confined. These should be treated more radically. Although recurrences are rare, a 5-year follow-up period must be considered the absolute minimum. The overall recurrence rate is 14%, but that of the relatively identified clear cell variant is higher. It has a 22% recurrence rate.[125]

Anavi et al. synthesized the literature on the clear cell variant of CEOT.[128] Nineteen are identified, of which 12 are central; the other 7 are peripheral. The central cases are evenly distributed between males and females; their mean age at first presentation is 34 years. Three-quarters of the central cases are located in the mandible. Most affect the posterior sextant. All but one are well defined; two-thirds are corticated. Half are radiolucent and half mixed. All but 1 are unilocular. All affect the alveolus, one-quarter occur in edentulous areas. 2 more are associated with root resorption, 2 are associated with tooth displacement without root resorption and 2 are associated with the crowns of unerupted teeth. Two-thirds of cases perforate the cortex.

ADENOMATOID ODONTOGENIC TUMOR (ICD-O 9300/0)

Philipsen and Nikai defined the *adenomatoid odontogenic tumor* (AOT) as "composed of odontogenic epithelium in a variety of histoarchitectural patterns, embedded in a mature connective tissue stroma and characterized by slow but progressive growth."[129]

Philipsen et al. reported a synthesis of 1082 cases.[130] Their relative period prevalence varies widely 0.6% to 38.5%. Gender predilection varies markedly with the radiological variant. Those associated with unerupted teeth (67% were canines) are the follicular (or pericoronal) variant. This variant accounts for 71% of all AOTs and is equally distributed between males and females, whereas the extrafollicular (or extracoronal) variant (27% of the total) has a 67% predilection for females. The rare peripheral (extraosseous or gingival) variant(2%) shows a 86% predilection for females.

The mean age of the follicular AOT (17 years old) is significantly younger than the extrafollicular AOT (26 years old) at first presentation.[131] This early mean age may account for the infrequent appearance of follicular AOTs associated with third molars, which have largely not developed by then. Those few follicular cases presenting with third molars have an older mean age (20.3 years).[130]

Just over half present as painless swellings.[131,132] The mean size is 2.9 (1 to 7) cm.[132] The period of awareness prior to presentation varied from 0.2 to 2 years[132] in Leon et al.'s Latin-American report and from 0.02 to 4 years[131] in Swasdison et al.'s Thai report.

All of the cases of Swasdison et al.[131] and Leon et al.[132] presented as well-defined unilocular radiolucencies. All of Swasdison et al.'s cases were associated with at least one tooth, whereas 80% of Leon et al.'s associated with an unerupted anterior tooth. Both studies reported well-defined margins, which Leon et al. described as "sclerotic." Seventy-seven percent of Leon et al.'s cases identified for variant are follicular (the majority in the anterior maxilla) and 10% are extrafollicular. One-third presented with flecks of calcification.[132] These are best appreciated on intraoral radiographs rather than on panoramic radiographs.[133] The identification of these flecks of calcification facilitates differentiation from those lesions, which typically present as radiolucencies.[133]

Leon et al. also remarked that most of their cases associated with an unerupted tooth mimicked the dentigerous cyst.[132] Swasdison et al. reported that the initial clinical diagnosis of 26% was as dentigerous cysts, 16% as calcifying cystic odontogenic tumors, 4% as ameloblastomas, but only 16% as AOTs.[131] This range of clinical diagnoses reflects the AOT's variable radiology.

None of Swasdison et al.'s cases recurred after enucleation.[131]

CALCIFYING CYSTIC ODONTOGENIC TUMOR (ICD-O 9301/0)

Praetorius and Ledesma-Montes defined the *calcifying cystic odontogenic tumor* (CCOT) as "benign cystic neoplasm of odontogenic origin, characterised by an ameloblastomalike epithelium with ghost cells that may calcify."[134]

This lesion was further refined by the recent international collaborative study,[135] into 4 types. Type 1 (simple cyst CCOT) was the most frequent (70%), followed by Type 2 (odontoma-associated CCOT) (24%), Type 3 (ameloblastomatous proliferating CCOT), and Type 4 (CCOT associated with any odontogenic tumor, other than an odontoma).[135]

Type 1 has a slight predilection for males (56%) and the mandible (55%), whereas Type 2 has a clear predilection for the mandible (73%). Both types display a predilection for the anterior sextants of both jaws. The mean ages for Type 1 and Type 2 are 30.1 (7–76) years and 16.3 (7–34) years, respectively. The awareness of each type prior to presentation is 4.3 (0.5–25) years and 1.6 (0.5–5) years, respectively. Type 1 CCOTs are larger at first presentation than Type 2. Unerupted teeth are association with both types, 50% and 41%, respectively. Both types present with swelling in 92% of cases. Type 1 most frequently appears radiologically as well-defined (92%) unilocular (61%) radiolucencies (78%), whereas Type 2 appears as well-defined (100%) unilocular (92%) radiolucencies containing opacities (77%). The differential diagnosis includes dentigerous cyst, ameloblastoma, and then CCOT for Type 1, whereas Type 2 is identified as an odontoma in nearly two-thirds of cases. Six cases recurred, of which at least 2 were Type 1.[135]

Because Ledesma-Montes et al.'s study did not include any East Asian case series, Iida et al.'s case series of 11 Japanese cases[136] and Li and Yu's case series of 21 Chinese cases[137] may provide a more complete global coverage of the CCOT. Iida et al.'s case series[136] (predilection for males [7] and the mandible [7]; mean age 26.0 [14–82 years]) included 7 of Type 1, 3 of Type 2, and 1 of Type 3. All were unilocular except the Type 3. All were radiolucencies except for 5 Type 1 CCOTs, which contained localized masses or disseminated flecks. Root resorption occurred in 4 out of 5 cases that considered this feature; 3 were Type 1 and 1 was Type 2. Teeth are displaced in all 5 cases (4 Type 1 and 1 Type 2). Seven cases are associated with impacted teeth; 5 are enveloped by the CCOT (3 Type 1, 1 Type 2, and 1 Type 3) and the remaining 2 are adjacent to Type 2 CCOTs. Li and Yu's 16 "cystic" cases[137] (11 Type 1 and 5 Type 2) are well-defined unilocular radiolucencies; some have tiny flecks scattered throughout. The mean age is 31 (12–72) years. There are predilections for males (9) and the maxilla (11). Two are discovered incidentally and 14 by hard swellings. Five have root resorption. None recur after follow-up.

AMELOBLASTIC FIBRO-ODONTOMA (ICD-O 9290/0)

Takeda and Tomich defined the *ameloblastic fibro-odontoma* (AFO) as "a tumour, which has the histologic features of ameloblastic fibroma (AF) in conjunction with the presence of dentin and enamel."[138] AFO is less common than AF and presents between 8 and 12 years of age. Because AFOs are more frequently symptom-free, they are found incidentally during an investigation for noneruption. They present as well-defined radiolucencies with varying degrees of opacification. They may be either unilocular or multilocular. They are often associated with unerupted teeth. They rarely recur after surgery.[138] Figure 11.40 displays a large case of AFO affecting the maxillary antrum.

SCLEROSING OR CONDENSING OSTEITIS AND DENSE BONE ISLANDS OR IDIOPATHIC OSTEOSCLEROSIS

Sclerosing osteitis (SO: also known as "condensing osteitis" [see Figure 10.33b] and "inflammatory sclerosis") and *dense bone islands* (DBI; also commonly known as "idiopathic osteosclerosis"; see Figure 1.10) are radiopacities within the bone, representing thickening of the trabeculae.[139] DBI, as its synonym the idiopathic osteosclerosis suggests, has no known cause. Therefore, while SO is likely to be accompanied by symptoms associated with a necrotic pulp or its sequelae, the DBI would be completely symptom-free and observed as an incidental finding in the examination for another clinical reason.

Although well defined, they do not have a radiolucent margin. Nevertheless, this may not be so easy to determine because the Mach band effect may create the illusion of one and compel consideration of ossifying fibroma and focal osseous dysplasia in the differential diagnosis. To offset this phenomenon, trabeculae are seen crossing the Mach band effect–created radiolucent space.

Both SO and DBI should be distinguished from retained deciduous molar roots and from torus mandibularis. Although the former infrequently have a visible periodontal ligament space, they are typically root-shaped and are found mesial and or distal to the second mandibular premolar. The image of the torus mandibularis on panoramic radiographs is usually displaced more coronally than it appears on clinical examination due to the upward angulation of the central ray.

The global distribution of reports included in the systematic review[140] upon which much of the following is derived is set out in Figure 1.48 and their details in Table 10.8.

The DBI is a frequently encountered clinical phenomenon, particularly on panoramic radiographs. It was observed more frequently in an East Asian community (Hong Kong) in comparison to two Western communities within the United Kingdom.[140] They display a slight predilection for females (58%). Their mean age at first presentation is 31 years of age (Table 10.8.). They predominantly affect the mandible (94%); this percentage is lower among sub-Saharan Africans.[140]

It is important to distinguish between SO and DBI because they represent different disease processes. For most cases this is not difficult. In a dentate sextant, if the lesion is not in direct association with a tooth, DBI is most likely. If associated with a tooth and the tooth is carious and/or heavily restored, SO is most likely, whereas noncarious and/or unrestored is virtually pathognominic of the DBI. It is assumed that pulp-vitality testing has been already performed. A nonvital tooth is indicative of a possible SO, whereas a vital tooth is more certainly a DBI. Difficulty occurs when these lesions arise in the edentulous alveolus. In such a situation, distinguishing between them requires review not just of the radiograph at the time of extraction, but also those prior to that. Generally, it is expected that the SO would regress once the inflammatory cause has been removed by either tooth extraction or by endodontic treatment.

Although DBI has been shown to be labile,[141] environmental factors may influence its prevalence and size. The higher prevalence of DBIs in the Japanese as in the Hong Kong Chinese may be linked to higher fluoridation of the water supply at the relevant time in the patients' lives. Both communities have already a high dietary fluoride intake.[140] The Hong Kong Chinese displayed not only a significant reduction in prevalence between two separate but similar consecutive series of nearly 1000 patients 10 years apart, but they also displayed a significant reduction in size. After the reduction of concentration of fluoride in the water supply, only those in the fifth decade and above displayed very little if no reduction in size. The most likely reason for this phenomenon is that the individuals were already past the age of their peak bone mass (about 30 years of age).[140]

Table 10.8. Dense Bone Island or Idiopathic Osteosclerosis: systematic review

Feature	Total	Western	East Asian	subSaharan	LatinAmer
Male:Female	42%:58%	48%:52%	40%:60%	36%:62%	INA
Mean age	31 years	29 year	32 years	INA	INA
Mand:Max	94%:%	92%:%	98%:2%	74%:26%*	INA
Mand:Ant:Post	14%:86%	14%:86%	10%:90%	59%:41%*	INA
Max:Ant:Post	0%:100%*	0%:100%*	0%:100%*	33%:67%*	INA

*Advises that the percentages were derived from either one report or from a synthesis of no more that 50 cases. Ant:Post, Anterior:Posterior; INA, Information not available; LatinAmer, Latin American; Mand:Max, Mandible:Maxilla; subSaharan, sub-Saharan African; Western, predominantly Caucasian; Y:N, Yes:No.

The mean age of first presentation is 31 years old. They peak in the third decade (38%). Fifty-eight percent are female overall. Ninety-five percent affect the mandible. The mandible and maxilla affect the posterior sextants in 94% and 100%, respectively. DBI are generally not associated with either root resorption or tooth displacement.[140]

Furthermore, in following up the findings in the Hong Kong Chinese,[140] an increased number of DBI would be expected in those patients who grew up in regions of the world where natural water fluoridation concentration is very high; such regions are found in India, Thailand, China, and Ethiopia.

References

1. MacDonald-Jankowski DS. Calcification of the stylohyoid complex in Londoners and Hong Kong Chinese. *Dentomaxillofac Radiol* 2001;30:35–39.
2. Okabe S, Morimoto Y, Ansai T, Yamada K, Tanaka T, Awano S, Kito S, Takata Y, Takehara T, Ohba T. Clinical significance and variation of the advanced calcified stylohyoid complex detected by panoramic radiographs among 80-year-old subjects. *Dentomaxillofac Radiol* 2006;35:191–199.
3. Rizzatti-Barbosa CM, Ribeiro MC, Silva-Concilio LR, Di Hipolito O, Ambrosano GM. Is an elongated stylohyoid process prevalent in the elderly? A radiographic study in a Brazilian population. *Gerodontology* 2005;22:112–115.
4. Sperber GH. *Craniofacial Development*. BC Decker Inc., Hamilton, Canada 2001: pp 55–59.
5. MacDonald-Jankowski DS. The synchondrosis between the greater horn and the body of the hyoid bone: a radiological assessment. *Dentomaxillofac Radiol* 1990;19:171–172.
6. Hernández JL, Velasco J. Neurological picture. Elongated styloid process (Eagle's syndrome) as a cause of atypical craniocervical pain. *J Neurol Neurosurg Psychiatry* 2008;79:43.
7. Ahmad M, Madden R, Perez L. Triticeous cartilage: prevalence on panoramic radiographs and diagnostic criteria. *Oral Surg Oral Med Oral Pathol Oral Radiol Endod* 2005;99:225–230.
8. Friedlander AH, Lande A. Panoramic radiographic identification of carotid arterial plaques. *Oral Surg Oral Med Oral Pathol* 1981;52:102–104.
9. Tanaka T, Morimoto Y, Ansai T, Okabe S, Yamada K, Taguchi A, Awano S, Kito S, Takata Y, Takehara T, Ohba T. Can the presence of carotid artery calcification on panoramic radiographs predict the risk of vascular diseases among 80-year-olds? *Oral Surg Oral Med Oral Pathol Oral Radiol Endod* 2006;101:777–783.
10. Mupparapu M, Kim IH. Calcified carotid artery atheroma and stroke: a systematic review. *J Am Dent Assoc* 2007(Apr);138(4):483–492. Friedlander's comment and Mupparappu's response; *J Am Dent Assoc* 2007;138:491–492.
11. Madden RP, Hodges JS, Salmen CW, Rindal DB, Tunio J, Michalowicz BS, Ahmad M. Utility of panoramic radiographs in detecting cervical calcified carotid atheroma. *Oral Surg Oral Med Oral Pathol Oral Radiol Endod* 2007;103:543–548. Comment in *J Evid Based Dent Pract* 2007;7:172–173. *Oral Surg Oral Med Oral Pathol Oral Radiol Endod* 2007;104:451–452; author reply 452–454.
12. Damaskos S, Griniatsos J, Tsekouras N, Georgopoulos S, Klonaris C, Bastounis E, Tsiklakis K. Reliability of panoramic radiograph for carotid atheroma detection: a study in patients who fulfill the criteria for carotid endarterectomy. *Oral Surg Oral Med Oral Pathol Oral Radiol Endod* 2008;106:736–742.
13. Friedlander AH, Cohen SN. Panoramic radiographic atheromas portend adverse vascular events. *Oral Surg Oral Med Oral Pathol Oral Radiol Endod* 2007;103:830–835.
14. Farman AG. Utility of panoramic radiographs in detecting cervical calcified carotid atheroma by Richard P. Madden et al. *Oral Surg Oral Med Oral Pathol Oral Radiol Endod* 2007;103:549.
15. Rio AC, Franchi-Teixeira AR, Nicola EM. Relationship between the presence of tonsilloliths and halitosis in patients with chronic caseous tonsillitis. *Br Dent J* 2008;204:E4.
16. Cantarella G, Pagani D, Biondetti P. An unusual cause of mechanical dysphagia: an agglomerate of calculi in a tonsillar residue. *Dysphagia* 2006;21:133–136.
17. Suarez-Cunqueiro MM, Dueker J, Seoane-Leston J, Schmelzeisen R. Tonsilloliths associated with sialolithiasis in the submandibular gland. *J Oral Maxillofac Surg* 2008;66:370–373.
18. el-Sherif I, Shembesh FM. A tonsillolith seen on MRI. *Comput Med Imaging Graph* 1997;21:205–208.
19. Chiu HL, Lin SH, Chen CH, Wang WC, Chen JY, Chen YK, Lin LM. Analysis of photostimulable phosphor plate image artifacts in an oral and maxillofacial radiology department. *Oral Surg Oral Med Oral Pathol Oral Radiol Endod* 2008;106:749–756.
20. Cooper C, Harvey NC, Dennison EM, van Staa TP. Update on the epidemiology of Paget's disease of bone. *J Bone Miner Res* 2006;21:P3–8.
21. Ralston SH. Juvenile Paget's disease, familial expansile osteolysis and other genetic osteolytic disorders. *Best Pract Res Clin Rheumatol* 2008;22:101–111.
22. Ankrom MA, Shapiro JR. Paget's disease of bone (osteitis deformans). *J Am Geriatr Soc* 1998;46:1025–1033.
23. Hashimoto J, Ohno I, Nakatsuka K, Yoshimura N, Takata S, Zamma M, Yabe H, Abe S, Terada M, Yoh K, Fukunaga M, Cooper C, Morii H, Yoshikawa H;

Japanese Committee on Clinical Guidelines of Diagnosis and Treatment of Paget's Disease of Bone of the Japan Osteoporosis Society. Prevalence and clinical features of Paget's disease of bone in Japan. *J Bone Miner Metab* 2006;24:186–190.
24. Wang WC, Cheng YS, Chen CH, Lin YJ, Chen YK, Lin LM. Paget's disease of bone in a Chinese patient: a case report and review of the literature. *Oral Surg Oral Med Oral Pathol Oral Radiol Endod* 2005;99:727–733.
25. MacDonald-Jankowski DS. Gigantiform cementoma occurring in two populations, London and Hong Kong. *Clin Radiol* 1992;45:316–318.
26. Takata S, Hashimoto J, Nakatsuka K, Yoshimura N, Yoh K, Ohno I, Yabe H, Abe S, Fukunaga M, Terada M, Zamma M, Ralston SH, Morii H, Yoshikawa H. Guidelines for diagnosis and management of Paget's disease of bone in Japan. *J Bone Miner Metab* 2006; 24:359–367.
27. Cheng YS, Wright JM, Walstad WR, Finn MD. Osteosarcoma arising in Paget's disease of the mandible. *Oral Oncol* 2002;38:785–792.
28. Ramaglia L, Morgese F, Filippella M, Colao A. Oral and maxillofacial manifestations of Gardner's syndrome associated with growth hormone deficiency: case report and literature review. *Oral Surg Oral Med Oral Pathol Oral Radiol Endod* 2007;103:e30–34.
29. Jaiswal AS, Balusu R, Narayan S. Involvement of adenomatous polyposis coli in colorectal tumorigenesis. *Front Biosci* 2005;10:1118–1134.
30. Wijn MA, Keller JJ, Giardiello FM, Brand HS. Oral and maxillofacial manifestations of familial adenomatous polyposis. *Oral Dis* 2007;13:360–365.
31. Takeuchi T, Takenoshita Y, Kubo K, Iida M. Natural course of jaw lesions in patients with familial adenomatosis coli (Gardner's syndrome). *Int J Oral Maxillofac Surg* 1993;22:226–230.
32. Lee BD, Lee W, Oh SH, Min SK, Kim EC. A case report of Gardner syndrome with hereditary widespread osteomatous jaw lesions. *Oral Surg Oral Med Oral Pathol Oral Radiol Endod* 2009;107:e68–72.
33. Fonseca LC, Kodama NK, Nunes FC, Maciel PH, Fonseca FA, Roitberg M, de Oliveira JX, Cavalcanti MG. Radiographic assessment of Gardner's syndrome. *Dentomaxillofac Radiol* 2007;36:121–124.
34. Madani M, Madani F. Gardner's syndrome presenting with dental complaints. *Arch Iran Med* 2007;10: 535–539.
35. Wenig BM, Mafee MF, Ghosh L. Fibro-osseous, osseous, and cartilaginous lesions of the orbit and paraorbital region. Correlative clinicopathologic and radiographic features, including the diagnostic role of CT and MR imaging. *Radiol Clin North Am* 1998;36: 1241–1259,
36. Damron TA, Ward WG, Stewart A. Osteosarcoma, chondrosarcoma, and Ewing's sarcoma: National Cancer Data Base Report. *Clin Orthop Relat Res* 2007; 459:40–47.
37. Guo W, Xu W, Huvos AG, Healey JH, Feng C. Comparative frequency of bone sarcomas among different racial groups. *Chin Med J (Engl)* 1999;112: 1101–1104.
38. van Es RJ, Keus RB, van der Waal I, Koole R, Vermey A. Osteosarcoma of the jaw bones. Long-term follow up of 48 cases. *Int J Oral Maxillofac Surg* 1997;26: 191–197.
39. Mardinger O, Givol N, Talmi YP, Taicher S. Osteosarcoma of the jaw. The Chaim Sheba Medical Center experience. *Oral Surg Oral Med Oral Pathol Oral Radiol Endod* 2001;91:445–451.
40. Givol N, Buchner A, Taicher S, Kaffe I. Radiological features of osteogenic sarcoma of the jaws. A comparative study of different radiographic modalities. *Dentomaxillofac Radiol* 1998;27:313–320.
41. Bennett JH, Thomas G, Evans AW, Speight PM. Osteosarcoma of the jaws: a 30-year retrospective review. *Oral Surg Oral Med Oral Pathol Oral Radiol Endod* 2000;90:323–332.
42. Fernandes R, Nikitakis NG, Pazoki A, Ord RA. Osteogenic sarcoma of the jaw: a 10-year experience. *J Oral Maxillofac Surg* 2007;65:1286–1291.
43. Yamaguchi S, Nagasawa H, Suzuki T, Fujii E, Iwaki H, Takagi M, Amagasa T. Sarcomas of the oral and maxillofacial region: a review of 32 cases in 25 years. *Clin Oral Investig* 2004;8:52–55.
44. Ogunlewe MO, Ajayi OF, Adeyemo WL, Ladeinde AL, James O. Osteogenic sarcoma of the jaw bones: a single institution experience over a 21-year period. *Oral Surg Oral Med Oral Pathol Oral Radiol Endod* 2006;101:76–81.
45. Kahn MF, Hayem F, Hayem G, Grossin M. Is diffuse sclerosing osteomyelitis of the mandible part of the synovitis, acne, pustulosis, hyperostosis, osteitis (SAPHO) syndrome? Analysis of seven cases. *Oral Surg Oral Med Oral Pathol* 1994;78:594–598.
46. Petrikowski CG, Pharoah MJ, Lee L, Grace MG. Radiographic differentiation of osteogenic sarcoma, osteomyelitis, and fibrous dysplasia of the jaws. *Oral Surg Oral Med Oral Pathol Oral Radiol Endod* 1995;80: 744–750.
47. Suei Y, Tanimoto K, Taguchi A, Yamada T, Yoshiga K, Ishikawa T, Wada T. Possible identity of diffuse sclerosing osteomyelitis and chronic recurrent multifocal osteomyelitis. One entity or two. *Oral Surg Oral Med Oral Pathol Oral Radiol Endod* 1995;80:401–408.
48. Lam DK, Sándor GK, Holmes HI, Evans AW, Clokie CM. A review of bisphosphonate-associated osteonecrosis of the jaws and its management. *J Can Dent Assoc* 2007;73:417–422.
49. Dore F, Filippi L, Biasotto M, Chiandussi S, Cavalli F, Di Lenarda R. Bone scintigraphy and SPECT/CT of bisphosphonate-induced osteonecrosis of the jaw. *J Nucl Med* 2009;50:30–35.
50. Bianchi SD, Scoletta M, Cassione FB, Migliaretti G, Mozzati M. Computerized tomographic findings in

51. bisphosphonate-associated osteonecrosis of the jaw in patients with cancer. *Oral Surg Oral Med Oral Pathol Oral Radiol Endod* 2007;104:249–258.
51. Bedogni A, Blandamura S, Lokmic Z, Palumbo C, Ragazzo M, Ferrari F, Tregnaghi A, Pietrogrande F, Procopio O, Saia G, Ferretti M, Bedogni G, Chiarini L, Ferronato G, Ninfo V, Lo Russo L, Lo Muzio L, Nocini PF. Bisphosphonate-associated jawbone osteonecrosis: a correlation between imaging techniques and histopathology. *Oral Surg Oral Med Oral Pathol Oral Radiol Endod* 2008;105:358–364.
52. Chiandussi S, Biasotto M, Dore F, Cavalli F, Cova MA, Di Lenarda R. Clinical and diagnostic imaging of bisphosphonate-associated osteonecrosis of the jaws. *Dentomaxillofac Radiol* 2006;35:236–243.
53. Waldron CA. Fibro-osseous lesions of the jaws. *J Oral Maxillofac Surg* 1993;51:828–835.
54. Eisenberg E, Eisenbud L. Benign fibro-osseous diseases: current concepts in historical perspective. *Clin Nor Am* 1997;9:551–562.
55. Jundt G. Fibrous dysplasia. Barnes L, Eveson J, Reichert P, Sidransky D, eds. *WHO Classification of Tumours, Oathology and Genetics of Tumours of the Head and Neck*. International Agency for Research on Cancer (IARC), Lyon 2005: pp 321–322.
56. MacDonald-Jankowski D. Fibrous dysplasia: a systematic review. *Dentomaxillofac Radiol* 2009;38:196–215.
57. Chapurlat RD, Orcel P. Fibrous dysplasia of bone and McCune-Albright syndrome. *Best Pract Res Clin Rheumatol* 2008;22:55–69.
58. Cohen MM Jr. The new bone biology: pathologic, molecular, and clinical correlates. *Am J Med Genet A* 2006;140:2646–2706.
59. Fahmy JL, Kaminsky CK, Kaufman F, Nelson MD Jr, Parisi MT. The radiological approach to precocious puberty. *Br J Radiol* 2000;73:560–567.
60. MacDonald-Jankowski DS, Li TK. Fibrous dysplasia in a Hong Kong community: the clinical and radiological features and outcomes of treatment. *Dentomaxillofac Radiol* 2009;38:63–72.
61. Slootweg PJ, Müller H. Differential diagnosis of fibro-osseous jaw lesions. A histological investigation on 30 cases. *J Craniomaxillofac Surg* 1990;18:210–214.
62. Schajowicz FM. Histological typing of bone tumours. 2nd ed. *WHO International Histological Classification of Tumours*. Springer-Verlag, London 1993: pp 39–40.
63. Eversole LR. Craniofacial fibrous dysplasia and ossifying fibroma. *Oral Maxillofacial Clin Nor Am* 1997;9:625–642.
64. MacDonald-Jankowski DS, Yeung R, Li TK, Lee KM. Computed tomography of fibrous dysplasia. *Dentomaxillofac Radiol* 2004;33:114–118.
65. Ricalde P, Horswell BB. Craniofacial fibrous dysplasia of the fronto-orbital region: a case series and literature review. *J Oral Maxillofac Surg* 2001;59:157–67; discussion 167–168.
66. Daly BD, Chow CC, Cockram CS. Unusual manifestations of craniofacial fibrous dysplasia: clinical, endocrinological and computed tomographic features. *Postgrad Med J* 1994;70:10–16.
67. Jacobsson S, Hallén O, Hollender L, Hansson CG, Lindströom J. Fibro-osseous lesion of the mandible mimicking chronic osteomyelitis. *Oral Surg Oral Med Oral Pathol* 1975;40:433–444.
68. Voytek TM, Ro JY, Edeiken J, Ayala AG. Fibrous dysplasia and cemento-ossifying fibroma. A histologic spectrum. *Am J Surg Pathol* 1995;19:775–781.
69. Yeow VK, Chen YR. Orthognathic surgery in craniomaxillofacial fibrous dysplasia. *J Craniofac Surg* 1999;10:155–159.
70. Sissons HA, Malcolm AJ. Fibrous dysplasia of bone: case report with autopsy study 80 years after the original clinical recognition of the bone lesions. *Skeletal Radiol* 1997;26:177–183.
71. Posnick JC. Fibrous dysplasia of the craniomaxillofacial region: current clinical perspectives. *Br J Oral Maxillofac Surg* 1998;36:264–273.
72. Posnick JC, Costello BJ. Discussion of Ricalde P, Horswell BB. Craniofacial fibrous dysplasia of the fronto-orbital region: a case series and literature review. *J Oral Maxillofac Surg* 2001;59:157–67. *J Oral Maxillofac Surg* 2001;59:167–168.
73. Cohen MM Jr. Merging the old skeletal biology with the new. I. Intramembranous ossification, endochondral ossification, ectopic bone, secondary cartilage, and pathologic considerations. *J Craniofac Genet Dev Biol* 2000;20:84–93.
74. Ruggieri P, Sim FH, Bond JR, Unni KK. Malignancies in fibrous dysplasia. *Cancer* 1994;73:1411–1424.
75. Daramola JO, Ajagbe HA, Obisesan AA, Lagundoye SB, Oluwasanmi JO. Fibrous dysplasia of the jaws in Nigerians. *Oral Surg Oral Med Oral Pathol* 1976;42:290–300.
76. Dorfman HD, Czerniak B. *Fibroosseous lesions. Bone Tumours*. Mosby, St. Louis 1998: pp 441–469.
77. Lustig LR, Holliday MJ, McCarthy EF, Nager GT. Fibrous dysplasia involving the skull base and temporal bone. *Arch Otolaryngol Head Neck Surg* 2001;127:1239–1247.
78. Slootweg PJ, Mofty SK. Ossifying fibroma. Barnes L, Eveson J, Reichert P, Sidransky D, eds. *WHO Classification of Tumours, Pathology and Genetics of Tumours of the Head and Neck*. International Agency for Research on Cancer (IARC), Lyon 2005: pp 319–320.
79. Kramer IRM, Pindborg JJ, Shear M. *Histological typing of odontogenic tumours*, 2nd ed. Springer-Verlag, London 1992: pp27–28.
80. Schajowicz FM. Histological typing of bone tumours. *WHO International Histological Classification of Tumours*, 2nd ed. Springer-Verlag, London 1998: pp40.
81. Slootweg PJ, Panders AK, Koopmans R, Nikkels PG. Juvenile ossifying fibroma. An analysis of 33 cases

with emphasis on histopathological aspects. *J Oral Pathol Med* 1994;23:385–388.
82. Brannon RB, Fowler CB. Benign fibro-osseous lesions: a review of current concepts. *Adv Anat Pathol* 2001;8:126–143.
83. Meister HP, Lufft W, Schlegel D. Differential diagnosis of fibro-osseous jaw lesions (fibrous dysplasia vs. ossifying fibroma). *Beitr Pathol* 1973;148:221–229.
84. Chen JD, Morrison C, Zhang C, Kahnoski K, Carpten JD, Teh BT. Hyperparathyroidism–jaw tumour syndrome. *J Intern Med* 2003;253:634–642.
85. MacDonald-Jankowski D. Ossifying fibroma: a systematic review. *Dentomaxillofac Radiol* 2009;38:495–513.
86. Sherman RS, Sternbergh WC. The roentgen appearance of ossifying fibroma of bone. *Radiology* 1948;50:595–609.
87. MacDonald-Jankowski DS, Li TK. Ossifying fibroma in a Hong Kong community: the clinical and radiological features and outcomes of treatment. *Dentomaxillofac Radiol* 2009;38:514–523.
88. Jee WH, Choi KH, Choe BY, Park JM, Shinn KS. Fibrous dysplasia: MR imaging characteristics with radiopathologic conelation. *AJRAm J Roentgenol* 1996;167:1523–1527.
89. Khanna JN, Andrade NN. Giant ossifying fibroma. Case report on a bimaxillary presentation. *Int J Oral Maxillofac Surg.* 1992;21:233–235.
90. Pindborg JJ, Kramer IRH, Torloni H. Histological typing of odontogenic tumours, jaw cysts and allied lesions. *WHO International Histological Classifications of Tumours*, No. 5. Geneva 1971: pp 31–34.
91. Slootweg PJ. Osseous dysplasia. Barnes L, Eveson J, Reichert P, Sidransky D, eds. *WHO Classification of Tumours, Pathology and Genetics of Tumours of the Head and Neck*. International Agency for Research on Cancer (IARC), Lyon 2005: pp 323.
92. Kramer IRM, Pindborg JJ, Shear M. *Histological typing of odontogenic tumours*, 2nd ed. Springer-Verlag, London 1992: pp 29–31.
93. Macdonald-Jankowski DS. Florid cemento-osseous dysplasia: a systematic review. *Dentomaxillofac Radiol* 2003;32:141–149.
94. MacDonald-Jankowski DS. Focal cemento-osseous dysplasia: a systematic review. *Dentomaxillofac Radiol* 2008;37:350–360.
95. Kawai T, Hiranuma H, Kishino M, Jikko A, Sakuda M. Cemento-osseous dysplasia of the jaws in 54 Japanese patients: a radiographic study. *Oral Surg Oral Med Oral Pathol Oral Radiol Endod* 1999;87:107–114.
96. Ariji Y, Ariji E, Higuchi Y, Kubo S, Nakayama E, Kanda S. Florid cemento-osseous dysplasia. Radiographic study with special emphasis on computed tomography. *Oral Surg Oral Med Oral Pathol* 1994;78:391–396.
97. Knutsen BM, Larheim TA, Johannessen S, Hillestad J, Solheim T, Koppang HS. Recurrent conventional cemento-ossifying fibroma of the mandible. *Dentomaxillofac Radiol* 2002;31:65–68.
98. Beylouni I, Farge P, Mazoyer JF, Coudert JL. Florid cemento-osseous dysplasia: Report of a case documented with computed tomography and 3D imaging. *Oral Surg Oral Med Oral Pathol Oral Radiol Endod* 1998;85:707–711.
99. Summerlin DJ, Tomich CE. Focal cemento-osseous dysplasia: a clinicopathologic study of 221 cases. *Oral Surg Oral Med Oral Pathol* 1994;78:611–620.
100. Su L, Weathers DR, Waldron CA. Distinguishing features of focal cemento-osseous dysplasias and cemento-ossifying fibromas: I. A pathologic spectrum of 316 cases. *Oral Surg Oral Med Oral Pathol Oral Radiol Endod* 1997;84:301–309.
101. Su L, Weathers DR, Waldron CA. Distinguishing features of focal cemento-osseous dysplasia and cemento-ossifying fibromas. II. A clinical and radiologic spectrum of 316 cases. *Oral Surg Oral Med Oral Pathol Oral Radiol Endod* 1997;84:540–549.
102. Waldron CA, Giansanti JS, Browand BC. Sclerotic cemental masses of the jaws (so-called chronic sclerosing osteomyelitis, sclerosing osteitis, multiple enostosis, and gigantiform cementoma. *Oral Surg Oral Med Oral Pathol* 1975;39:590–604.
103. Melrose RJ. The clinic-pathological spectrum of cemento-osseous dysplasia. *Oral Maxillofac Clin Nor Am* 1997;9:643–653.
104. Young SK, Markowitz NR, Sullivan S, Seale TW, Hirschi R. Familial gigantiform cementoma: classification and presentation of a large pedigree. *Oral Surg Oral Med Oral Pathol* 1989;68:740–747.
105. Waldron CA. Fibro-osseous lesions of the jaws. *J Oral Maxillofac Surg* 1993;51:828–835.
106. Toffanin A, Benetti R, Manconi R. Familial florid cemento-osseous dysplasia: a case report. *J Oral Maxillofac Surg* 2000;58:1440–1446.
107. Miyake M, Nagahata S. Florid cemento-osseous dysplasia. Report of a case. *Int J Oral Maxillofac Surg* 1999;28:56–57.
108. Ong ST, Siar CH. Florid cemento-osseous dysplasia in a young Chinese man. Case report. *Aust Dent J* 1997;42:404–408.
109. Abdelsayed RA, Eversole LR, Singh BS, Scarbrough FE. Gigantiform cementoma: clinicopathologic presentation of 3 cases. *Oral Surg Oral Med Oral Pathol Oral Radiol Endod* 2001;91:438–444.
110. Visnapuu V, Peltonen S, Ellilá T, Kerosuo E, Väánánen K, Happonen RP, Peltonen J. Periapical cemental dysplasia is common in women with NF1. *Eur J Med Genet* 2007;50:274–280.
111. Moshref M, Khojasteh A, Kazemi B, Roudsari MV, Varshowsaz M, Eslami B. Autosomal dominant gigantiform cementoma associated with bone fractures. *Am J Med Genet A* 2008;146A:644–648.
112. Kaplan I, Nicolaou Z, Hatuel D, Calderon S. Solitary central osteoma of the jaws: a diagnostic dilemma.

113. Larrea-Oyarbide N, Valmaseda-Castellón E, Berini-Aytés L, Gay-Escoda C. Osteomas of the craniofacial region. Review of 106 cases. *J Oral Pathol Med* 2008; 37:38–42.
114. Woldenberg Y, Nash M, Bodner L. Peripheral osteoma of the maxillofacial region. Diagnosis and management: a study of 14 cases. *Med Oral Patol Oral Cir Bucal* 2005;10:E139–142.
115. Jones AC, Prihoda TJ, Kacher JE, Odingo NA, Freedman PD. Osteoblastoma of the maxilla and mandible: a report of 24 cases, review of the literature, and discussion of its relationship to osteoid osteoma of the jaws. *Oral Surg Oral Med Oral Pathol Oral Radiol Endod* 2006;102:639–650.
116. van der Waal, I. Cementoblastoma. Barnes L, Eveson J, Reichert P, Sidransky D, eds. *WHO Classification of Tumours, Pathology and Genetics of Tumours of the Head and Neck*. International Agency for Research on Cancer (IARC), Lyon 2005: p 318.
117. MacDonald-Jankowski DS, Wu PC. Cementoblastoma in Hong Kong Chinese. A report of four cases. *Oral Surg Oral Med Oral Pathol* 1992;73:760–764.
118. Brannon RB, Fowler CB, Carpenter WM, Corio RL. Cementoblastoma: an innocuous neoplasm? A clinicopathologic study of 44 cases and review of the literature with special emphasis on recurrence. *Oral Surg Oral Med Oral Pathol Oral Radiol Endod* 2002;93: 311–320.
119. Dominguez FV, Frenandez LR, Luberti RF. Benign cementoblastoma: analysis of 25 cases. *Rev Asoc Odontol Argent* 2000;88:237–244. (in Spanish)
120. Praetorius P, Piatelli A. Odontoma, complex type. Barnes L, Eveson J, Reichert P, Sidransky D, eds. *WHO Classification of Tumours, Pathology and Genetics of Tumours of the Head and Neck*. International Agency for Research on Cancer (IARC), Lyon 2005: p310.
121. Praetorius P, Piatelli A. Odontoma, compound type. Barnes L, Eveson J, Reichert P, Sidransky D, eds. *WHO Classification of Tumours, Pathology and Genetics of Tumours of the Head and Neck*. International Agency for Research on Cancer (IARC), Lyon 2005: p311.
122. Chen Y, Li TJ, Gao Y, Yu SF. Ameloblastic fibroma and related lesions: a clinicopathologic study with reference to their nature and interrelationship. *J Oral Pathol Med* 2005;34:588–595.
123. MacDonald-Jankowski DS. Odontomas in a Chinese population. *Dentomaxillofac Radiol* 1996;25:186–192.
124. Takata T, Slootweg PJ. Calcifying epithelial odontogenic tumour. Barnes L, Eveson J, Reichert P, Sidransky D, eds. *WHO Classification of Tumours, Pathology and Genetics of Tumours of the Head and Neck*. International Agency for Research on Cancer (IARC), Lyon 2005: p302.
125. Philipsen HP, Reichart PA. Calcifying epithelial odontogenic tumour: biological profile based on 181 cases from the literature. *Oral Oncol* 2000(Jan);36(1): 17–26.
126. Kaplan I, Buchner A, Calderon S, Kaffe I. Radiological and clinical features of calcifying epithelial odontogenic tumour. *Dentomaxillofac Radiol* 2001;30: 22–28.
127. Struthers P, Shear M. Root resorption by ameloblastomas and cysts of the jaws. *Int J Oral Surg* 1976;5: 128–132.
128. Anavi Y, Kaplan I, Citir M, Calderon S. Clear-cell variant of calcifying epithelial odontogenic tumor: clinical and radiographic characteristics. *Oral Surg Oral Med Oral Pathol Oral Radiol Endod* 2003;95:332–329.
129. Philipsen HP, Nikal H. Adenomatoid odontogenic tumour. Barnes L, Eveson J, Reichert P, Sidransky D, eds. *WHO Classification of Tumours, Pathology and Genetics of Tumours of the Head and Neck*. International Agency for Research on Cancer (IARC), Lyon 2005: p304.
130. Philipsen HP, Reichart PA, Siar CH, Ng KH, Lau SH, Zhang X, Dhanuthai K, Swasdison S, Jainkittivong A, Meer S, Jivan V, Altini M, Hazarey V, Ogawa I, Takata T, Taylor AA, Godoy H, Delgado WA, Carlos-Bregni R, Macias JF, Matsuzaka K, Sato D, Vargas PA, Adebayo ET. An updated clinical and epidemiological profile of the adenomatoid odontogenic tumour: a collaborative retrospective study. *J Oral Pathol Med* 2007;36: 383–393.
131. Swasdison S, Dhanuthai K, Jainkittivong A, Philipsen HP. Adenomatoid odontogenic tumors: an analysis of 67 cases in a Thai population. *Oral Surg Oral Med Oral Pathol Oral Radiol Endod* 2008;105:210–215.
132. Leon JE, Mata GM, Fregnani ER, Carlos-Bregni R, de Almeida OP, Mosqueda-Taylor A, Vargas PA. Clinicopathological and immunohistochemical study of 39 cases of adenomatoid odontogenic tumour: a multicentric study. *Oral Oncol* 2005;41:835–842.
133. Dare A, Yamaguchi A, Yoshiki S, Okano T. Limitation of panoramic radiography in diagnosing adenomatoid odontogenic tumors. *Oral Surg Oral Med Oral Pathol* 1994;77:662–668.
134. Praetorius P, Ledesma-Montes C. Calcifying cystic odontogenic tumour. Barnes L, Eveson J, Reichert P, Sidransky D, eds. *WHO Classification of Tumours, Pathology and Genetics of Tumours of the Head and Neck*. International Agency for Research on Cancer (IARC), Lyon 2005: 313.
135. Ledesma-Montes C, Gorlin RJ, Shear M, Praetorius F, Mosqueda-Taylor A, Altini M, Unni K, Paes de Almeida O, Carlos-Bregni R, Romero de León E, Phillips V, Delgado-Azañero W, Meneses-García A. International collaborative study on ghost cell odontogenic tumours: calcifying cystic odontogenic tumour, dentinogenic ghost cell tumour and ghost cell odontogenic carcinoma. *J Oral Pathol Med* 2008;37:302–308.
136. Iida S, Fukuda Y, Ueda T, Aikawa T, Arizpe JE, Okura M. Calcifying odontogenic cyst: radiologic findings in

11 cases. *Oral Surg Oral Med Oral Pathol Oral Radiol Endod* 2006;101:356–362.
137. Li TJ, Yu SF. Clinicopathologic spectrum of the so-called calcifying odontogenic cysts: a study of 21 intraosseous cases with reconsideration of the terminology and classification. *Am J Surg Pathol* 2003;27:372–384.
138. Takeda Y, Tomich CE. Ameloblastic fibro-odontoma. Barnes L, Eveson J, Reichert P, Sidransky D, eds. *WHO Classification of Tumours, Pathology and Genetics of Tumours of the Head and Neck*. International Agency for Research on Cancer (IARC), Lyon 2005: p309.
139. McDonnell D. Dense bone island. A review of 107 patients. *Oral Surg Oral Med Oral Pathol* 1993;76:124–128.
140. MacDonald-Jankowski DS. Idiopathic osteosclerosis in the jaws of Britons and of the Hong Kong Chinese: radiology and systematic review. *Dentomaxillofac Radiol* 1999;28:357–363.
141. Petrikowski CG, Peters E. Longitudinal radiographic assessment of dense bone islands of the jaws. *Oral Surg Oral Med Oral Pathol Oral Radiol Endod* 1997;83:627–634.

Chapter 11
Maxillary antrum

Introduction

The maxillary antrum (also known as the "maxillary sinus") occupies a considerable part of the midface and is surrounded by important structures and organs. These are the orbits, the nasal and oral cavities, the ethmoid sinuses, the pterygopalatine and infratemporal fossae. Therefore, disease arising within it can involve these structures in addition to disease arising within them involving it in turn. The shape of the maxillary antrum forms an inverted pyramid with its apex set laterally at the root of the temporal process of the zygoma (Figure 11.1). Its central position within hemi-midface means that it is involved in most maxillary fractures as the three struts joining the maxilla with the skull base. These struts arising from two of the four angles of the antral cavity are the frontozygomatic the zygomaticotemporal and the frontomaxillary. They transmit the occlusal forces from the dentoalveolar process to the skull base.

The maxillary antrum has a communication with the nasal cavity via an ostium over half way up the medial wall above the attachment of the inferior turbinate (concha) (Figure 11.2). This is the sole or at least the main point of egress for the antral fluids. An accessory osteum may be present. Normal evacuation is dependent upon the integrity of the pseudostriated squamous epithelium that lines the lumen of the antrum.

The hard palate established the junction between the alveolar and basal process of the upper jaw, as did the mandibular canal for the mandible.[1] The profile of the hard palate is readily observed on lateral projections of the jaws, including panoramic radiographs (see Figure 1.24). The maxillary antrum frequently pneumatizes this process, particularly in the premolar region. Therefore diseases arising within this process or treatment for these diseases may involve the maxillary antrum. Extensive pneumatization of the alveolar process, as seen in Figure 11.3a), may indicate the presence of a lesion.

The maxillary antrum reaches adult size about 12 years of age. Chronic sinusitis during childhood has been suggested to be the cause of its failure to develop (aplasia) or its small size (hypoplasia) (Figures 11.4 and 11.5).[2,3] Pneumatization is reduced by red-marrow conversion during anemia.[2] Figure 11.6 exhibits a case of thalassemia, introduced in Chapter 9, affecting the maxillary antrum.

Although mucosal thickening of the maxillary antrum is common in symptom-free patients, it is considered normal if it is less than 4 mm (Figure 11.7).[2]

Sinusitis can be acute or chronic. A *de novo* acute sinusitis or an acute exacerbation of existing chronic sinusitis is generally painful. There may be a history of a recent upper respiratory tract infection. The intensity of the pain may vary with changes in position of the patient's head. If the maxillary antrum is infected, there may be tenderness of the anterior maxilla, and the vital premolar and molar teeth may be tender to percussion or biting. The initial diagnosis can be concluded on clinical evidence alone. Although conventional radiography is often unhelpful, radiographs should be taken in cases of a long-standing history of sinusitis. Unlike the *de novo* acute sinusitis, long-standing chronic sinusitis may present with thickening of the sinus's bony walls.[2] If sinusitis remains unresponsive to antibiotics and decongestants, it is necessary to exclude the possibility of other underlying pathology, the most important being a *squamous cell carcinoma* (SCC) (see Figure 18.20). This can achieve considerable dimensions prior to manifestation of symptoms, of which chronic sinusitis is one. The earlier the diagnosis, the better the prognosis.

Osteosarcoma affecting the maxillary antrum displays the similar features reported earlier (Chapter 10). Figure 11.8 displays substantial

---

Oral and Maxillofacial Radiology: A Diagnostic Approach, David MacDonald. © 2011 David MacDonald

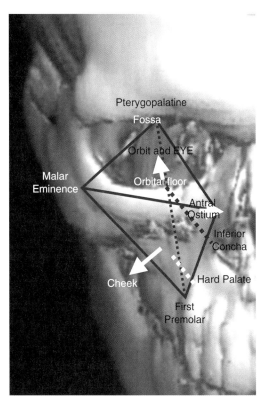

Figure 11.1. The anatomical parameters of the maxillary antrum.

soft-tissue involvement and swelling, which is not reflected by the extent of the radiologically apparent osseous (radiopaque) element of the disease.

The classical presentation of sinusitis in conventional radiography is thickening of the antral mucosa and the presence of fluid levels. The thickened mucosa in infective sinusitis presents as a radiopaque (of soft-tissue density) band parallel to the bony walls (see Figure 11.7). This may be accompanied with polyps (Figures 11.9, 11.10), particularly in allergy sinusitis. Fluid levels are best seen on a occipitomental projection with the patient sitting upright and with a horizontal beam. It is important to ensure that the petrous temporal bone does not overlie the inferior antrum (Figure 11.11a). The fluid nature of the fluid line can be confirmed by reradiographing following a 30° tilt of the head to one side. Occasionally, if the antral ostium is completely occluded a mucocele can occur resulting in the expansion and erosion of the antral walls. Mucoceles occur more frequently in the frontal and ethmoid sinuses where their close association with the cranium prompts a neurosurgical emergency.

Bacterial sinusitis can arise from a dental origin, for example, from a periapical lesion (see Figure 11.11) or from an oroantral fistula following extraction of a maxillary premolar or molar (see Figure 11.9) Pinhole fistulae are more likely to result in sinusitis rather than a wide fistula, which allows free drainage. Figure 11.12 displays new bone at the apex of a root-filled tooth.

Antroliths, calcifications within the maxillary antrum, are occasionally observed (Figure 11.13). Such radiopacities may also represent exostoses arising from the antral wall. The latter may be considered when one of the exostosis' margins appears diffuse rather than sharp (Figure 11.14) or if there has been no change in position over time (Figure 11.15).

Other lesions such as neoplasms and cysts affecting the maxillary antrum are either intrinsic to the antrum (arise within it) or are extrinsic to it (arise outside it) and invade it secondarily. The most frequent intrinsic lesions are *mucosal antral cysts* (MACs) (Figures 11.16–18). As the MAC represents an accumulation of fluid within the antral mucosa, but without an epithelial lining, it may be referred to as a "pseudocyst." They vary in frequency in different communities depending upon local climate and culture. They are more frequent in Hong Kong[4] than in inner-city London,[5] at least during late summer to early autumn when the radiographs were taken. They may also vary within the community according to the seasons. They are dome-shaped and on the panoramic radiography appear to rise most frequently from the antral floor. On *helical computed tomography* (HCT) they are observed arising from other walls, particularly the lateral wall, which is outside the panoramic radiograph's focal trough. Although features of periapical pathosis are classically absent in the alveolar bone subjacent to the MACs (Figures 11.16–18), in such situations where a periapical pathosis is present, the antral phenomenon is more likely to represent a mucositis induced by the underlying periapical inflammation rather than to be a MAC.

The MAC, although a soft-tissue structure, is apparent on conventional radiography by virtue of the silhouette effect (see Figure 11.17). This allows the surface contour of soft-tissue structures to

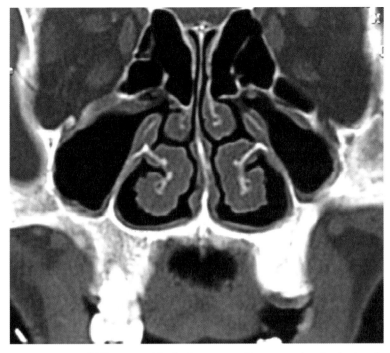

Figure 11.2. Normal maxillary antrum (MA) coronal helical computed tomography displaying the midface at the level of the osteum of the MA. The ostium's position, above the inferior concha is toward the roof of the MA. The inferior orbital canal is laterally positioned anteriorly to the lumen of the hypoplastic MA. **Note:** The white rim represents the mucosal surface enhancing postcontrast. The darker part is the soft-tissue of the turbinates.

become visible by contrasting them against air-filled spaces. For this reason the outlines of the heart, the aorta, and the pulmonary veins are made visible because they are silhouetted against the translucent air-filled lung fields on the chest radiograph.

Other lesions, such as cysts and neoplasms that arise within the alveolus, can expand upward into the maxillary antrum to create a domelike structure. The difference is that they have a cortex at their periphery. This cortex is the now upwardly displaced floor of the antrum.

The MAC or any other soft-tissue structure silhouetted against an air-filled space may also appear to have a radiopaque periphery. This is a result of the Mach band effect and is accompanied by a black line immediately surrounding the entire MAC. This is more fully discussed in Chapter 3.

Another aspect of lesions arising from the maxillary alveolus and expanding into the maxillary antrum is that the portion of the lesion appearing inferior to the image of the hard plate on panoramic radiographs appears radiolucent, whereas that portion appearing above it and thus completely enveloped by the air-filled space of the maxillary antrum is radiopaque (see Figure 11.17).

Furthermore, if the lesion arising from the alveolus becomes infected, the cortex representing the upwardly expanded floor of the maxillary antrum can be lost. The lesion may become indistinguishable from the MAC or other pathology intrinsic to the maxillary antrum except for the radiolucency within the alveolus (Figure 11.19) or "hyperpneumatization" (see Figure 11.3a). This loss of the upwardly displaced floor of the maxillary antrum can be readily appreciated in infected *keratocystic odontogenic tumors* (KCOT) (Figure 11.20).[6] In addition to the loss of the antral floor much of the maxillary tuberosity and posterior antral wall was absent strongly suggesting presence of an antral malignancy, more specifically a carcinoma.

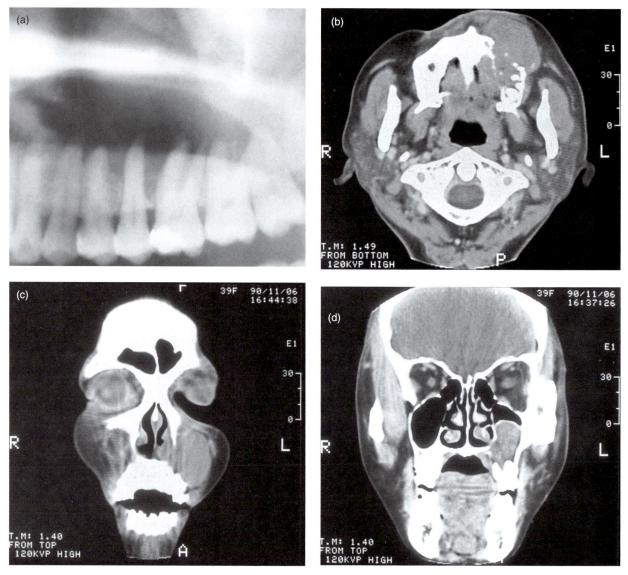

Figure 11.3. The panoramic radiograph and three soft-tissue window computed tomographs (CT) of a case of non-Hodgkin's lymphoma (NHL) that arose from lymphoid tissue within the alveolus. Although the majority of NHLs arise within Waldeyer's ring, a number can arise outside it. They are generally bone sparing, unless they arise in bone itself. The breaching of the buccal and palatal cortices and substantial soft-tissue mass are reminiscent of a squamous cell carcinoma, but both the substantial palatal expansion and concave upward expansion of the floor of the maxillary antrum are more typical of a benign neoplasm or cyst. (a) The panoramic radiograph displays extensive loss of bone around the roots of maxillary molars in addition to premolars. This degree of bone loss far exceeds normal pneumatization of the alveolus. (b) Axial CT, at the level of the first cervical vertebra. The alveolar bone around the roots on the left side appear to have been resorbed. The buccal cortex is absent, whereas the palatal cortex is expanded, but it eroded with some perforations. The buccal soft-tissue mass has extended medially between the surface of the anterior alveolus and the skin of the upper lip. (c) Coronal CT, at the level of the first premolars, almost occludes the left antral and nasal cavities, leaving a residual air space just below the medial portion of the orbital floor. The medial and lateral bony walls, in addition to a considerable portion of the alveolar process are absent. The soft-tissue mass of the lesions has expanded medially into the nasal cavity. (d) Coronal CT, at the level of the first molars, has perforated the palatal cortex and expanded into the palatal submucosa, whereas the lateral wall appears intact. The lesion is separated from the floor of the orbit by a residual air space. The dome-shaped surface of the lesion is delineated by a well-defined uniformly thick cortex. Reprinted with permission from Li TK, MacDonald-Jankowski DS. An unusual presentation of a high-grade non Hodgkin's lymphoma in the maxilla. *Dentomaxillofacial Radiology* 1991;20:224–226.

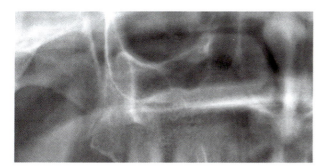

Figure 11.4. Panoramic radiograph displaying a very hypoplastic maxillary antrum.

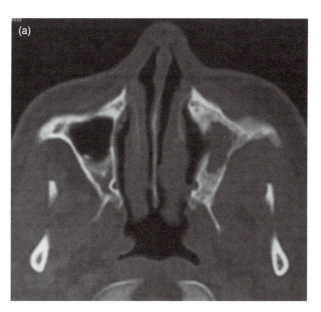

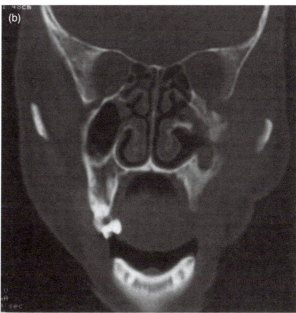

Figure 11.5. Computed tomography (CT; bone windows) displaying a hemimaxilla which has been affected by osteomyelitis. This had resulted in loss of teeth and marked hypoplasia of the maxilla and also of the maxillary antrum. Some of the features of this lesion were secondary to the surgery required to remove sequestra and teeth. (a) Axial CT displays the hypoplasia of the residual maxilla and of the maxillary antrum. The latter cavity has been obturated with soft tissue. (b) Coronal CT displays the above. The floor of the orbit on the affected side is thicker. The inferior concha is not attached to the affected maxilla in this section. **Note:** Osteomyelitis affecting the maxilla usually affects children and is spread to the maxilla by the blood (hematogenous).

The carcinoma, in particular the SCC, affecting the maxillary antrum extensively destroys all bony structures and invades directly adjacent soft-tissue structures (see Figure 18.20). These require further evaluation by computed tomography with contrast and *magnetic resonance imaging* (MRI) to fully appreciate the extent of their invasion. Successful treatment required complete surgical ablation of the neoplasm. Not all malignant neoplasms behave in this manner nor require such radical treatment. There is a small synthesis of cases of *non-Hodgkin's lymphoma* (NHL) that arise within the alveolar process. One such case displayed an upward displacement both of the floor and of the posterior lateral wall of the maxillary antrum very much in the manner of a benign neoplasm or cyst (see Figure 11.3). NHLs are generally treated by radiotherapy rather than by surgery.[7]

Odontogenic neoplasms arising in the subjacent alveolus can involve the maxillary antrum secondarily. The most frequent of these are three of the most important—the ameloblastoma, KCOT, and odontogenic myxoma—because of their propensity to recur.

Although a systemic review revealed that only 9% of ameloblastomas affect the maxilla and 82% of that affected the posterior sextant, in the detailed Hong Kong series within the same report the majority were solid (or multilocular) ameloblastomas. These maxillary cases first presented later in life than the average mandibular ameloblastoma within the same report and are more likely to affect the anterior sextant wholly or in part.[8,9] This pattern has now been substantially confirmed in another Chinese report.[10] If these ameloblastomas become sufficiently large they can affect the maxillary antrum (Figure 11.21). Solid ameloblastomas

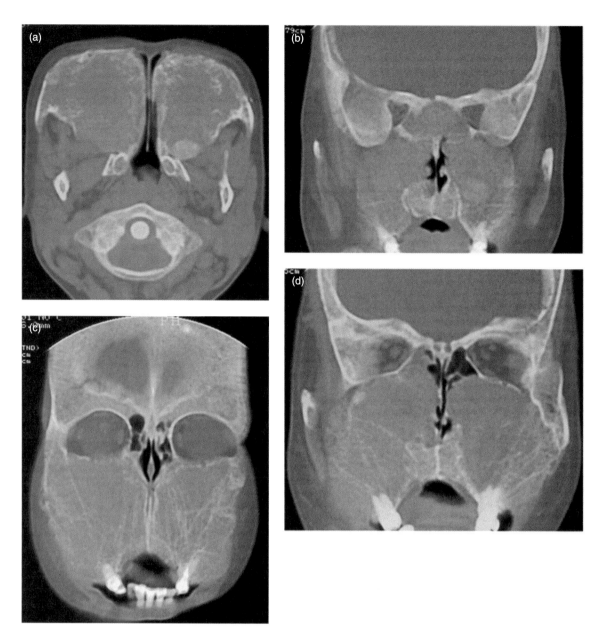

Figure 11.6. Computed tomography (bone windows) of thalassemia. See also Figures 9.5 and 17.21 for other images from the same patient. (a) Axial section, through the atlantoaxial articulation, reveals the complete absence of a maxillary antrum. The entire normal bone has been replaced by thickened and coarse trabeculae. There is a well-defined ovoid sclerosis on the posterior wall of the right maxilla. The cortex is either diffusely thickened or absent. Both maxillae are expanded. (b) Coronal section confirms the complete absence of the maxillary antrum observed in (b). The pattern of the background radiodensity is peau d'orange. (c) Coronal section through the middle of the globe (eyeball) displays the above. There are fewer trabeculae save for a few vertical trabeculae. The floor of the orbit is diffusely thickened or absent, whereas the roof is distinct and intact. (d) Coronal section, just behind the globe (note the rectus muscles and optic nerve). It displays the same pattern as (c). There is a small well-defined ovoid sclerosis just below the orbit on the right maxilla.

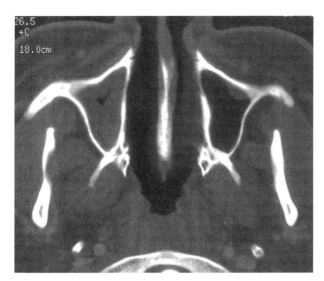

Figure 11.7. This axial computed tomograph (bone window) displays the normal triangular shape of the maxillary antrum (MA). Its degree of pneumatization, size, and shape are generally symmetrical. Note that the posteriolateral wall is sigmoid shaped. Neoplastic lesions generally expand the posterior half of this wall. Lumen of the MA on one side is nearly obturated by substantial expansion of the antral mucosa, leaving only a residual air space in the center. The other side reveals the presence of a shallow dome-shaped opacity of soft-tissue radiodensity arising from the anterior wall of the MA.

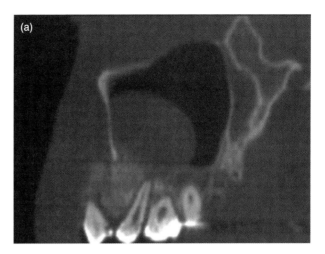

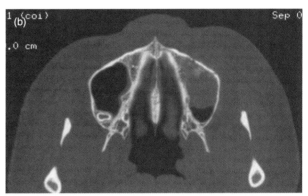

Figure 11.8. Computed tomography (CT) of osteosarcoma affecting the anterior sextant of the maxilla and extending into the maxillary antrum (MA). (a) Sagital CT (bone window) exhibiting a poorly defined radiopacity displacing the roots of the canine and lateral incisor. Striae are present at the periphery of the lesion. The intraluminal mass is largely soft tissue with deeper sunburst striae. (b) Axial CT (bone window) displays a loss of the trabeculae and cortex at the anteriomedial angle of the MA and an invasion by a soft-tissue mass and striae deeper within it. The lateral surface of the medial wall of the MA presents with striae arising from the anterior two-thirds of its length. Just anterior to the anterior (facial) wall of the MA is a small area of periosteal reaction with faint striae. The affected side exhibits substantial swelling of the face.

require resection with a margin to minimize recurrence. If untreated or inadequately treated they can spread to the orbit or even the skull base (Figure 11.22) and may cause death.[11]

The KCOT affects the maxilla as frequently as the mandible. Although most mandibular KCOTs are multilocular, most of those that affect the posterior maxilla are not only unilocular (Figure 11.23), but they present earlier in life.[12] Although this earlier age of presentation is similar to that observed for syndromic KCOTs (*nevoid basal cell carcinoma syndrome*; NBCCS) the latter can be distinguished by its other stigmata, which may include multiple KCOTs (Figure 11.24), already discussed in Chapter 9. A reason for the unilocular presentation and buccolingual expansion of the KCOT arising subjacent to the maxillary antrum could be that the buccal and lingual cortices of the maxillary alveolus and the cortex of the floor of the maxillary antrum are considerably thinner than those of the body of the mandible, where expansion, if it occurs, is minimal.[13]

KCOTs affecting the maxillary antrum generally present with substantial expansion during adolescence or early adulthood. If presenting with unerupted teeth, particularly third molars, they should be distinguished from dentigerous cysts.

Dentigerous cysts affecting the maxillary antrum can also arise from canines (Figure 11.25)

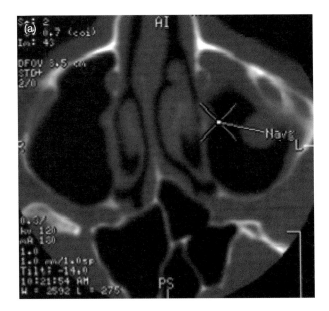

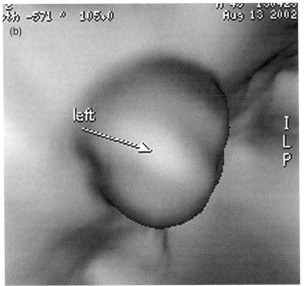

Figure 11.9. Mucositis and antral polyps can occur secondary to an oroantral fistula. Computed tomography of the maxillary antrum (a). This virtual antroscopy (b) displays the MAC suspended from the roof of the antral cavity. Figure (b) reprinted with permission from MacDonald-Jankowski DS, Li TK. Computed tomography for oral and maxillofacial Surgeons. Part 1: Spiral computed tomography. *Asian Journal of Oral Maxillofacial Surgery* 2006;18: 68–77.

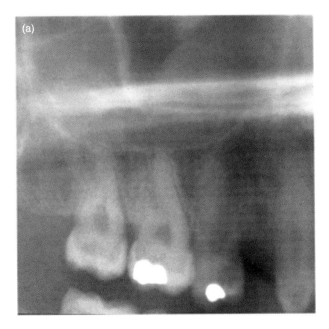

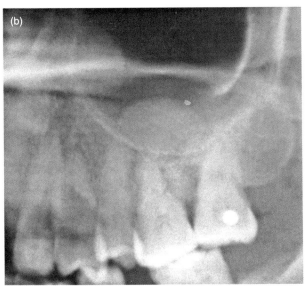

Figure 11.10. Panoramic radiographs from cases displaying polyps arising from the floor of the maxillary antrum.

and supernumeraries (Figure 11.26). Another cyst that should be considered in the differential diagnosis is the *orthokeratinized odontogenic cyst* (OOC) (Figure 11.27), which frequently is associated with an unerupted tooth.[14] A recent systematic review revealed the paucity of detailed reports of the clinical and radiological features of this lesion.[14] The OOC potentially accounts for 10% of all formerly designated odontogenic keratocysts.[14] It is important to distinguish it from the odontogenic

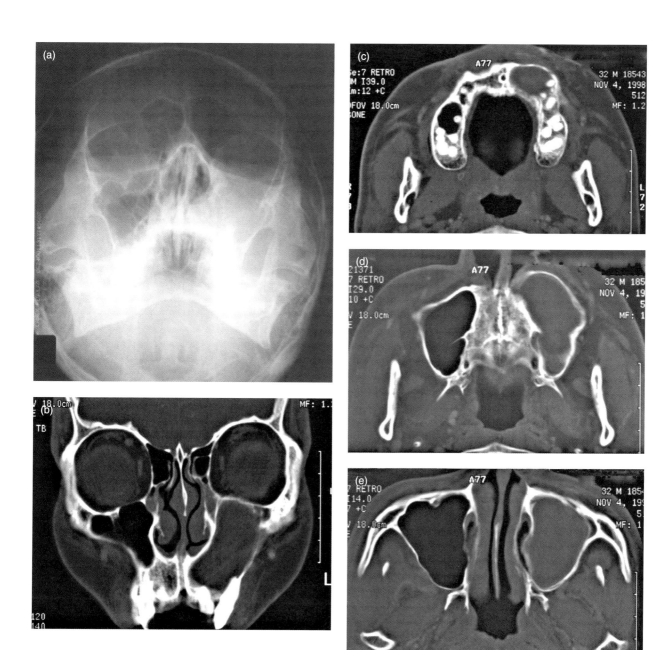

Figure 11.11. Conventional radiography and computed tomography (CT) of a radicular cyst arising in the maxilla's anterior sextant affecting the maxillary antrum (MA). (a) 30° occipitomental projection displaying complete radiopacity of the left MA. Note that the petrous temporal bone is correctly below the floor of the antrum. There is no obvious buccal expansion, but the medial wall appears to have been displaced into the left nasal cavity. (b) Coronal CT (bone window), at the level of the maxillary canines, exhibits displacement of the lateral and medial walls and the roof of the MA. (c) Axial CT (bone window), through the alveolus, displaying a radiolucency occupying the left anterior region. It is associated with buccal expansion. (d) Axial CT (bone window), through the hard palate, showing an opacification of the entire MA. The anterior wall has been substantially expanded. (e) Axial CT (bone window), at the level of the neck of the condyle, displaying substantial obturation of the antrum by a corticated lesion containing a homogeneous light radiopaque content. The medial wall has been expanded. There is a residual air space at the anteriolateral angle of the MA.

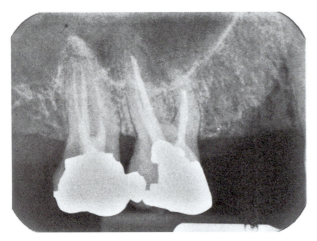

Figure 11.12. Periapical radiograph displaying a root-filled tooth, which exhibits new bone formation at the apex, analogous to the involucrum observed in Figure 10.14.

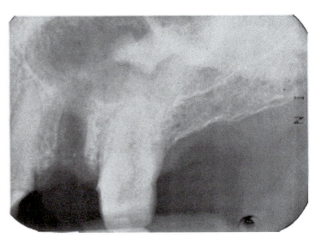

Figure 11.13. A periapical radiograph displaying as a nonroot–appearing antrolith within the maxillary antrum. Adjacent to it is an extraction socket.

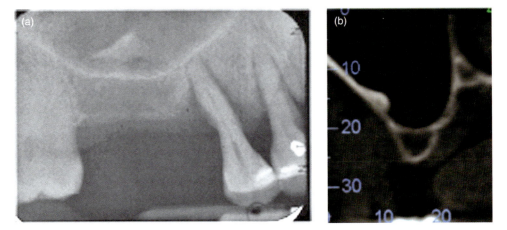

Figure 11.14. Periapical radiography (a) and cone-beam computed tomograph (CBCT; b) of an endostosis within the lateral wall of the maxillary antrum (MA). (a) The superior surfaces are well defined, but the inferior margin is not. (b) The exostosis is on the lateral wall of the MA. The lumen of the MA is otherwise completely normal; that is, the mucosa is not thick enough to be visible and there is no fluid.

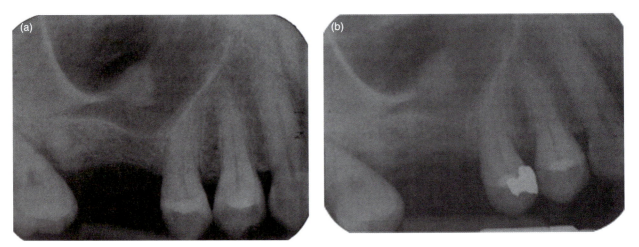

Figure 11.15. Periapical radiographs (a and b) taken 3 years apart. They display an antrolith, which has not changed in site and shape during that period. (a) A well-defined radiopacity is sited just distal and superior to the apex of the second premolar. (b) Three years later, the second premolar has acquired an amalgam restoration, but the antrolith has remained unchanged in site and in shape.

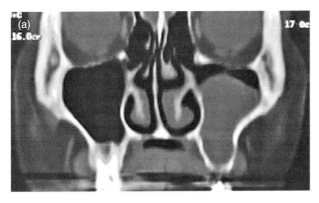

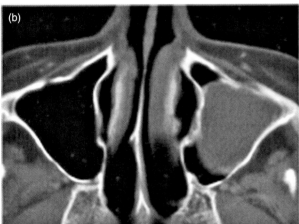

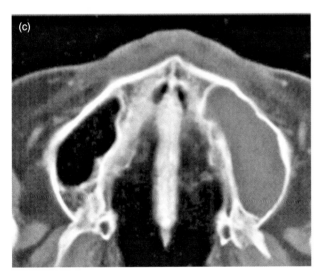

Figure 11.16. Postcontrast computed tomography (bone window) of a mucosal antral cyst (or pseudocyst). The cyst presents as a soft-tissue density structure partially filling the maxillary antrum. The white "rim" denotes contrast "enhancement" of the surface mucosal rather than a bony cortex as observed in Figure 11.3(d). The surface mucosa of the turbinates are also enhanced (compare with Figure 11.2). Therefore, this is a large mucosal antral cyst rather than a lesion arising within the alveolus (compare with Figure 11.11).

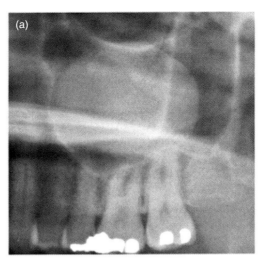

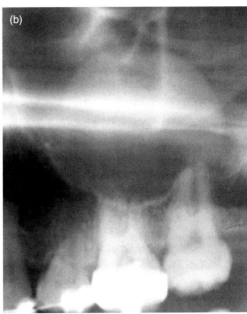

Figure 11.17. The panoramic radiographs of two different mucosal antral cysts (MAC; also called *pseudocysts*). (a) This MAC is centered upon the maxillary first molar tooth. Although this tooth has a moderate restoration, the likelihood of pulpal necrosis is low. (b) This MAC's subjacent teeth are undergoing orthodontic treatment and therefore presumably noncarious. There is agenesis of the second premolar. Reprinted with permission from MacDonald-Jankowski DS. Fibro-osseous lesions of the face and jaws. *Clinical Radiology* 2004;59:11–25.

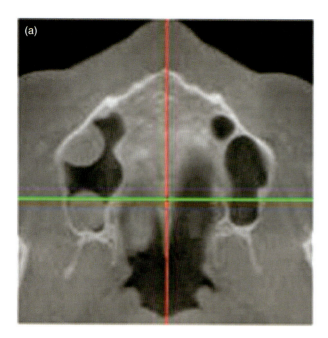

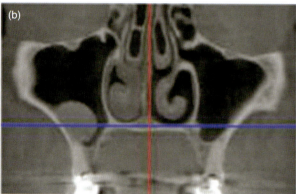

Figure 11.18. Cone-beam computed tomography (CBCT) of the maxillary antrum (MA). (a) Axial CBCT, at the level of the hard palate, displaying the nasopharynx; the mucosal surface silhouetted against the air-filled space. One antrum exhibits two dome-shaped opacities of a similar radiodensity as that of the nasopharyngeal tissues. These dome-shaped radiopacities are likely to be mucosal antralcysts or polyps. (b) Coronal CBCT section through the more anterior radiopacity. The mucosa of the ipsilateral inferior concha is swollen in comparison to the contralateral. The middle conchae and ethmoid air cells appear normal. **Note 1:** Inferior orbital canal in the floor of the orbit. **Note 2:** The osteum of the left MA. Figures courtesy of Dr. Babak Chehroudi, Faculty of Dentistry, University of British Columbia.

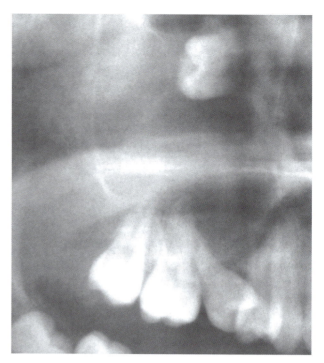

Figure 11.19. Panoramic radiograph displaying a keratocystic odontogenic tumor affecting the maxillary antrum (MA). The MA displays a soft-tissue radiopacity. A tooth, which appears like a premolar, has been displaced upwardly into the MA almost to the floor of the orbit. The root of a premolar has been displaced distally. The alveolus is translucent.

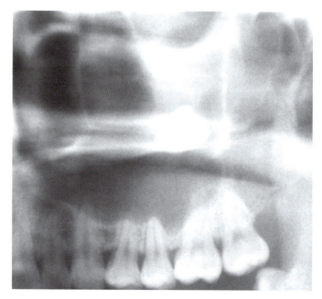

Figure 11.20. Panoramic radiograph displaying a keratocystic odontogenic tumor affecting the maxillary antrum (MA). The MA displays a soft-tissue radiopacity separated from a residual air-filled space anterior by a cortex. A tooth, which is likely a third molar, has been displaced upwardly into the MA to the level of the hard palate. The alveolus appears to be hyperpneumatized. There is no cortex at the maxillary tuberosity and adjacent posterior antral wall.

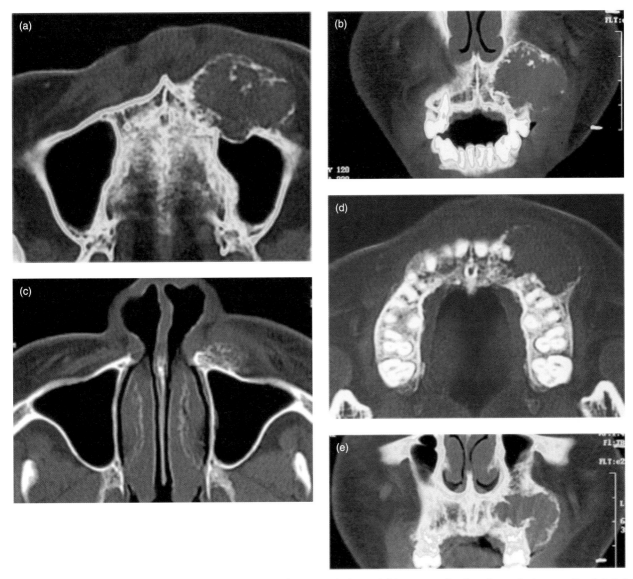

Figure 11.21. Computed tomography (CT) of a solid (multilocular) ameloblastoma affecting the anterior maxilla. (a) Axial CT (bone window), at the level of the hard palate, revealing an ovoid lesion largely expanding out into the soft tissue of the face. The cortex on this surface has been substantially eroded and perforated in places. There are some radiopaque features extending partly into the lesion. The lesion has eroded and displaced the medial aspect of the anterior wall of the maxillary antrum backward into the lumen. Multiple radiopacities are observed in the anterior aspect of the lesion. (b) Coronal CT (bone window) reveals most of the features observed in (a). The multiple radiopacities are localized to the superior aspect of the lesion. (c) Axial CT (bone window), through the inferior concha, displaying the most superior extent of this lesion. It lies upon the surface of the anterior wall of the maxillary antrum. (d) Axial CT (bone window), through the level of the hamulus, displaying the deepest reach of the lesion into the alveolus. The cortex at its anterior periphery appears substantially absent. (e) Coronal CT (bone window) displays both the marked buccal expansion and its extent palatally into the alveolus. It also displays the septa within the lesion and the marked erosion and absence of its buccal cortex toward the gingival margin.

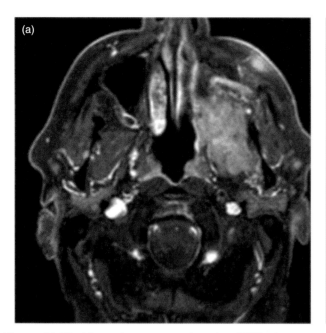 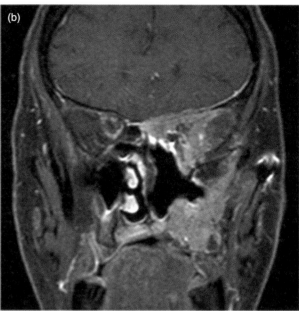

Figure 11.22. Magnetic resonance imaging (MRI) of a recurrent maxillary ameloblastoma invading the brain. This patient has had numerous surgeries including a left maxillectomy; the surgical defect was filled with a radial forearm reconstruction flap. (a) Axial postgadolinium T1-weighted MRI, at the level of the base of the sigmoid notch, displaying an enhanced lesion occupying the site of the maxillary antrum and the infratemporal space. (b) Coronal postgadolinium T1-weighted MRI, at the level of the condyle, displays enhancement of a lesion invading a reconstruction flap used to rebuild the inferior and lateral walls of the maxillary antrum. Above the maxillary antrum, the ameloblastoma has invaded the ethmoid air-cells, the orbital apex, the dura mater of the anterior cranial fossa and the anterior part of the temporal lobe. Figure courtesy of Dr. Montgomery Martin, British Columbia Cancer Agency.

neoplasms because it displays little tendency to recur and has little if any association with NBCCS.

Seventy-five percent of the cases of odontogenic myxoma in a systematic review affect the posterior sextant.[15] The odontogenic myxoma can achieve substantial dimensions obturating the antrum and expanding and eroded its walls (Figure 11.28).[16] The computed tomograph in Figure 11.28 displays striae that resemble a sunray pattern. Although the odontogenic myxoma arising primarily within the anterior sextant is not yet of substantial dimensions, it can still involve the maxillary antrum without the expansion observed in Figure 11.29. The odontogenic myxoma requires resection to minimize recurrence.

Fibrous dysplasia (FD) affects the maxillary antrum either from its origin within the alveolus or within the base of the skull. The latter is obviously more likely to affect vision. A recent systemic review has revealed that the literature based on cases of FD arising within the alveolus affecting a specific community very unlikely to cause ocular problems.[17] Therefore, with the exception of the occasionally sporadic case the patient's vision is unlikely to be threatened by FD arising within the alveolus. Fifty-eight percent of FDs included by the systematic review affect the maxilla.[17] FD affecting the maxilla enlarges it, sometimes grotesquely, although retaining a broad resemblance to its original (normal) shape (Figure 11.30a). The predominant pattern of radiodensity on conventional radiography is generally ground glass.[18] Computed tomography may display a wider range of radiodensities (Figures 4.5, 11.30–11.36). All maxillary cases in a recent case series affected the maxillary antrum.[18] In a series of cases investigated by computed tomography,[18] in one case the lateral wall displayed the fusiform expansion classically observed in the mandible (Figure 11.32). As the lesion grows, the lumen of the maxillary antrum is gradually obliterated.[19] The vertical dimension of the maxilla is increased both vertically and horizontally, which gives rise to the clinically apparent grossly expanded maxilla already observed in Figure 11.30. In most cases of monostotic fibrous dysplasia, affecting the jaws, arising within the

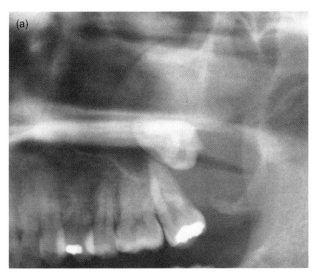
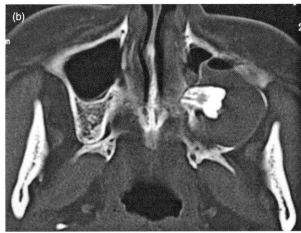
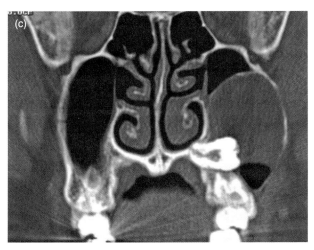

Figure 11.23. A panoramic radiograph (a), coronal (b), and axial (c) computed tomography (CT) (bone windows) exhibit a keratocystic odontogenic tumor (KCOT) within the lumen of the maxillary antrum. The KCOT presents on the panoramic radiograph as a radiopacity above the hard palate and as a radiolucency below it, lying within the alveolus. The radiopacity arises because its soft tissue attenuated the X ray beam more than the air-filled antrum. The KCOT has obturated most of the antral lumen. The radiolucency in the third permanent molar site of the alveolar bone arises from the KCOT, removing the normal bone for this site as it grows. The most obvious feature displayed is that of the unerupted third molar. This is transversely impacted with its roots embedded into the attachment of the hard palate with the alveolus. Root formation is still incomplete. The overall presentation is that of a dentigerous cyst. The only clue that this may not be so is the attachment of the lesion to the root surface. It is apical to the cementoenamel junction.

alveolus or at least close to it, the pterygoid body and plates are not generally affected. The anteriorly apparent expansion of the maxillary FD appears to use the pterygoid bone as a base to push off against (Figures 11.32–34). Although occasionally the pterygoid bone may be displaced backward, in other cases it remains undisplaced but with some posterior-laterally expanded dysplastic maxilla lying against the lateral surface of the lateral pterygoid plate (Figures 11.33, 11.34). The vertical expansion raises the floor of the orbit, but infrequently to such an extent to occasion ocular signs, such as proptosis as observed in Figure 11.30. The more usual upward expansion of FD arising within the jaws is exhibited in Figures 11.31 and 11.35a.

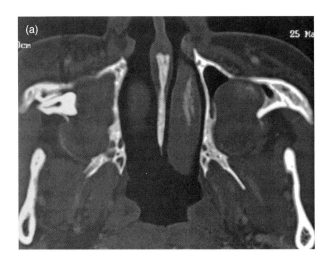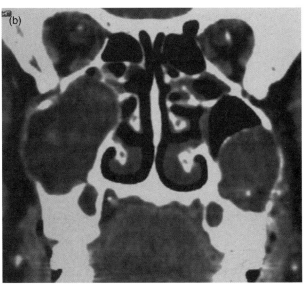

Figure 11.24. Computed tomography (CT) of keratocystic odontogenic tumor (KCOT) affecting both maxillary antra (MA) in a case of nevoid basal cell carcinoma syndrome (NBCCS). (a) Axial CT (bone window) displays KCOTs obturating both MAs. The right MA has been completely obturated, whereas the left has been only partially obturated. The presence of the residual air space (anteriorly) suggests that the MA is draining normally. Each MA displays an expansion of the posterior aspect of the posteriolateral wall. The right KCOT contains an unerupted tooth the apex of which has been displaced through the posteriolateral wall close to the MA's zygomatic angle. (b) Coronal CT (soft-tissue window) with intravenous contrast (note enhancement of the ophthalmic, facial, and right greater palatine arteries) displays complete obturation and buccolingual expansion of the left MA (the medial wall has also been displaced medially). The right MA exhibits a residual air space.

Polyostotic fibrous dysplasia can, in addition to the jaws, affect other facial and skull bones, including the base of the skull (see Figures 11.35, 11.36). Involvement of the last can threaten vision by compressing the optic nerve. Other cases of polyostotic fibrous dysplasia affecting the base of the skull are displayed in Figures 17.23–17.25.

Radiopaque lesions such as ossifying fibroma[20,21] and the odontoma[22] can grow upward into the maxillary antrum. Seventeen percent of ossifying fibromas affect the posterior sextant of the maxilla.[22] Three of the four maxillary ossifying fibromas affected the antrum, of which two obturated it.[20]

The ossifying fibroma[20,21] and complex odontomas[22] are generally indistinguishable on conventional radiographs. According to a synthesis of case reports on large odontomas,[22] most are likely to grow and occupy the entire vertical dimension of the maxillary antrum (Figure 11.37). The tooth it prevented from erupting is frequently reported as being displaced upward to the floor of the orbit (Figure 11.38).

Dense bone islands (idiopathic osteosclerosis) are infrequently found in the maxilla.[23] Osseous dysplasia (OD) can arise in the posterior alveolus.[24]

The posterior sextant was affected in 73% of all cases with florid osseous dysplasia (FOD) (Figure 11.39a–d),[24] but in only 9% of cases of focal osseous dysplasia (FocOD).[25] Although the OD lesions may achieve substantial dimensions and protrude into the maxillary antrum as in Figure 11.39e, they are infrequently associated with symptoms.

Occasionally other rarer cysts and neoplasms may affect the maxillary antrum; Figure 11.40 exhibits one such lesion, the ameloblastic fibroodontoma. Figure 11.40 displays the bone windows; the corresponding soft-tissue windows were reported by Piette et al.[26]

Radiopacities, such as Figure 11.41, associated with the floor of the antrum that may convinc-

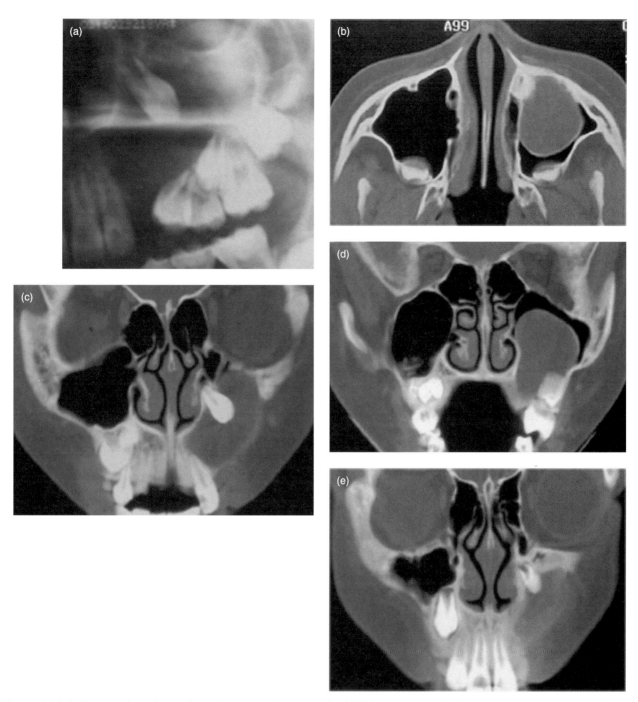

Figure 11.25. Panoramic radiograph and computed tomography (CT; bone windows) of a dentigerous cyst arising from a maxillary canine obturating most of the maxillary antrum (MA). (a) The panoramic radiograph displays the unerupted canine with incompletely formed roots sited high within the MA. The second premolar, second molar, and third molar are unerupted. (b) Axial CT, above the hard palate, displays the canine lodged at the MA's medioanterior angle. It has expanded into the anterior nasal cavity. The cystic cavity has obturated the anterior three-fourths of the MA's lumen and has a well-defined cortex. The posterior fourth is air-filled. The most posterior aspect of each MA contains the developing third molar (this is obvious in (a). (c) Coronal CT, at the level of the nasopalatine duct, exhibits the unerupted canine impacted into the medial wall of the MA apex first. The cyst appears to be attached to the canine at the cementoenamel junction. The cyst has expanded the buccal wall of the maxillary antrum and adjacent alveolus. (d) Coronal CT, at the level of the unerupted second premolar, displays the cyst filling the inferior four-fifths of the MA. The upwardly displaced former floor of the antrum is clearly present as a well-defined cortex. The residual air-filled space occupies the superior fifth of the MA's lumen. The medial wall of the MA has been displaced medially. (e) Coronal CT, at the level of the lacrymal ducts. The buccal cortex, though largely eroded and indistinct, is clearly expanded.

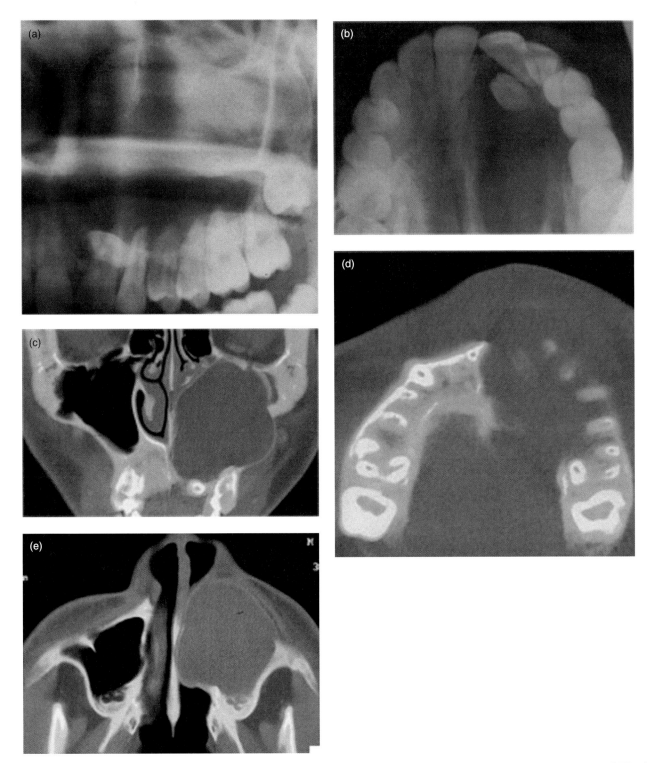

Figure 11.26. Conventional radiographs (panoramic radiograph and anterior occlusal and computed tomographs [CT]) of a dentigerous cyst (DC) arising from a supernumerary (mesiodens) affecting the maxillary antrum (MA). (a) Panoramic radiograph displaying the DC on the mesiodens obturating the MA. There is marked hyperpneumatization of the alveolus. The DC has also caused substantial root resorption. (b) Anterior occlusal showing the mesiodens between the roots of the central incisors. (c) Coronal CT taken at the level the inferior orbital nerve begins its course from the inferior orbital canal/groove to the infraorbital foramen. It shows the DC obturating the entire MA and expanding its buccal and lingual cortices widely. (d) Axial CT (bone window), through the alveolus, exhibiting substantial root resorption. (e) Axial CT, at the level of the infraorbital foramen, exhibiting substantial expansion anteriorly and medially.

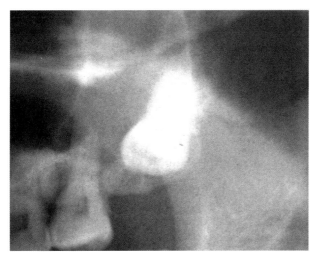

Figure 11.27. Panoramic radiograph displaying an orthokeratinized odontogenic cyst (OOC) affecting the maxillary antrum (MA). The posterior MA is radiopaque, whereas the anterior MA is still translucent. The radiopacity is bounded by a cortex anteriorly. The root of associated unerupted third molar appears to be in intimate contact with the OOC. The tooth's follicular space is discernible over the entire crown and is unlikely to be continuous with the OOC.

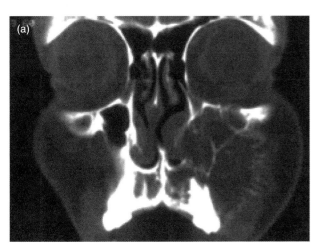
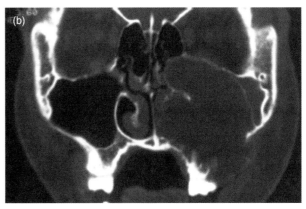

Figure 11.28. Computed tomography (CT: bone windows) of odontogenic myxoma affecting the maxilla. It has extensively involved the maxillary antrum (MA). (a) Coronal CT displays the substantial expansion, medially and laterally, at the level of the maxillary premolars and infraorbital foramen/canal. The medial wall of the MA has been displaced and eroded, whereas the lateral wall has been replaced by striae lined up in the direction of the expansion, presenting as a sunray spicular pattern. A number of large "soap bubbles" are visible medially. The pattern between the anterior teeth appears like the honeycomb pattern. (b) Coronal CT at the level of the first maxillary molar displaying almost complete obliteration of the MA with the exception of the superiolaterally remnant of the lumen, which is no longer air-filled. The inferior aspect of the MA has been expanded and eroded. The floor of the orbit has been displaced upward and the medial wall medially. Part of the bone of the inferior concha has been preserved within the lesion. The bone around the roots of the first molar has been resorbed. (c) Axial CT, at the level of the molar apices, displaying complete obliteration of the MA. The OM has expanded anteriorly medially and laterally. The periphery of the lesion has parallel striae and loculi, whereas the center of the OM is devoid of structure. (d) Axial CT, at the level of the infraorbital foramen, showing almost complete obliteration of the left MA with the exception of a space (no longer air-filled) on the anteriolateral aspect of the lesion, extending from the widened infraorbital foramen to the anteriolateral angle. Much of the posteriolateral wall is absent. There is a soft-tissue expansion of the lesion through this perforation. The posterior aspect of this penetration is covered by displaced cortex, which is juxtapositioned to the lateral aspect of the lateral pterygoid plate. It has a smooth contour and its edges overlap the lateral aspect of the perforated bone. The medial wall has been displaced toward the nasal septum. The medial bony wall is largely intact although substantially eroded. There are two bone density opacities within the lesion at the anteriomedial angle of the MA. Figures (a), (b), and (d) reprinted with permission from MacDonald-Jankowski DS. Fibro-osseous lesions of the face and jaws. *Clinical Radiology* 2004;59:11–25.

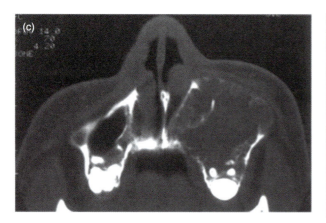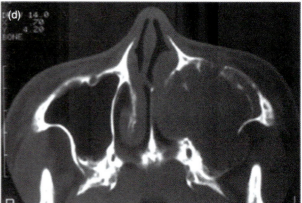

Figure 11.28. (*Continued*).

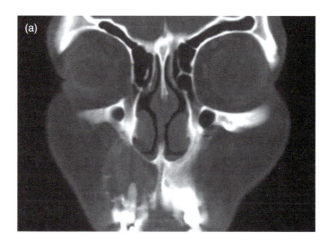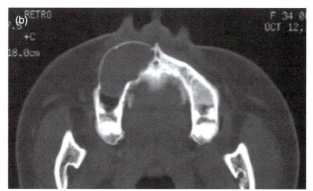

Figure 11.29. Computed tomography (bone windows) of an odontogenic myxoma arising from the anterior sextant and involving the maxillary antrum (MA). (a) Coronal CT, at the level of the maxillary canines, displays a well-defined multilocular radiolucency expanding the buccal cortex. (b) Axial CT, at the level of the anterior nasal spine, shows a radiolucency, which has a largely smooth expanded buccal cortex, but a multiloculated medial border. The buccal surface has a small septum on its internal aspect. The odontogenic myxoma's posterior corticated margin is extending back into the lumen of the MA.

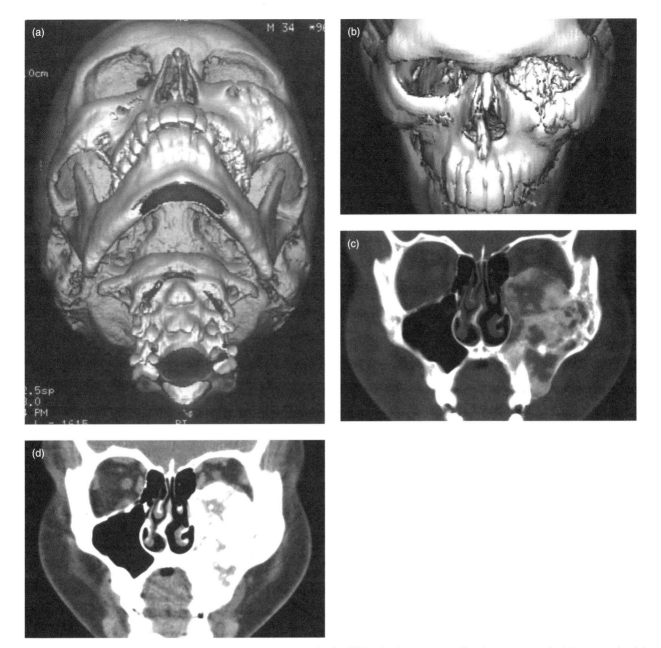

Figure 11.30. Computed tomography (CT) of fibrous dysplasia (FD) affecting the maxilla that presented with proptosis. (a) A three-dimensional reconstruction of a case of fibrous dysplasia affecting the maxilla displays the affected side as a coarse overgrowth of the normal bony profile. (b) A three-dimensional reconstruction of a case of fibrous dysplasia affecting the maxilla displays an upward displacement of the orbital floor. (c) A coronal CT section behind the globe (eyeball). This bone window exhibits a substantial soft-tissue element. The upward expansion is substantial and appreciably constricts the orbital cavity in comparison to the contralateral orbit. (d) A coronal CT section behind the globe (eyeball). This soft-tissue window more clearly displays the optic nerve and four rectus muscles of the eye.

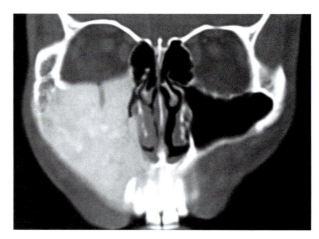

Figure 11.31. Computed tomograph (bone window) of fibrous dysplasia (FD) affecting the maxilla. It has obliterated the maxillary antrum (MA). The FD has expanded the buccal aspect of the maxilla and slightly upwardly displaced the floor of the orbit. The vertically—directed canal carries the inferior orbital nerve toward the infraorbital foramen.

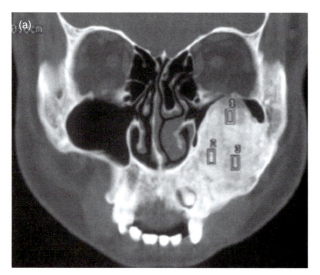

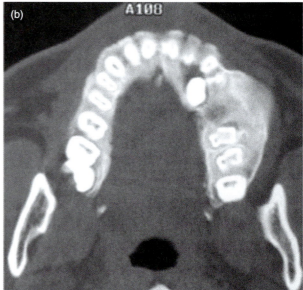

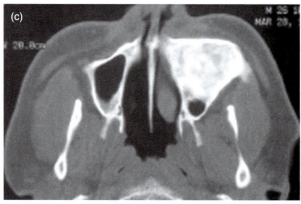

Figure 11.32. Computed tomography (CT; bone windows) of fibrous dysplasia (FD) affecting the left maxilla and the maxillary antrum (MA). (a) Coronal CT, at the level of the maxillary canines. The FD extends from the left malar to the midline. Although the FD has also almost completely obliterated the MA in the coronal plane there are two air-filled spaces superiolaterally (zygomatic recess) and medially. The rounded outline of the antral part of the lesion is unusual for FD and is more indicative of a benign neoplasm, whereas the poorly defined medial margin is indicative of FD. The floor of the orbit is slightly displaced. The general radiopacity is a ground-glass pattern. The cortex of the inferior aspect of the lateral wall is not apparent. There is an unerupted tooth within the FD and the follicular space is clear visible. The "boxes" represent the density measurement sites (in Hounsfield units). (b) Axial CT, of the maxillary alveolus, displays the follicle of the above-mentioned unerupted tooth. The follicle has no lamina dura. The buccal aspect of the alveolus has been substantially expanded. (c) Axial CT, at the level of the base of the sigmoid notch. It exhibits the substantial expansion observed above. The bone of the posterior MA appears normal, and the associated antral lumen is still patent and air-filled.

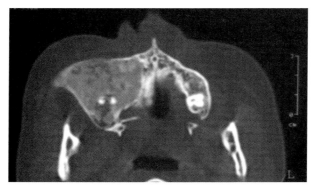

Figure 11.33. Axial computed tomograph of fibrous dysplasia completely obturating the maxillary antrum and in apposition with the pterygoid body, which has not been displaced.

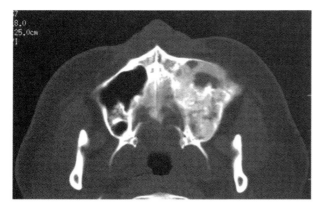

Figure 11.34. Computed tomography of FD in Figure 11.30 obturating the maxillary antrum (MA). There is a translucent zone anteriorly. The FD is firmly juxtaposed with regard to the pterygoid plates and is also partly apposed to the lateral aspect of the lateral plate. Reprinted with permission from MacDonald-Jankowski DS, Yeung R, Li TK, Lee KM. Computed tomography of fibrous dysplasia. *Dentomaxillofacial Radiology* 2004;33:114–118.

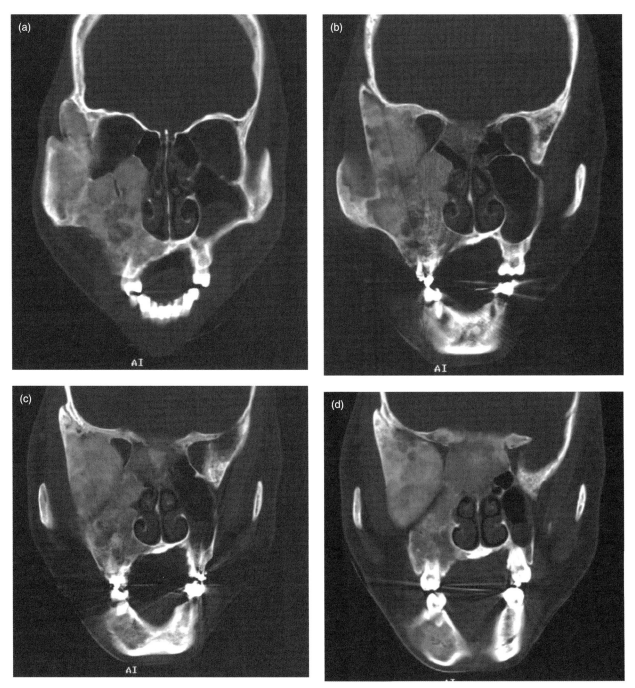

Figure 11.35. These coronal computed tomographs (CT; bone window) display fibrous dysplasia (FD) affecting the mandible, the maxilla, the zygoma, and the sphenoid. (a) FD affecting the hemimaxilla and contiguous zygoma. The maxillary antrum has been completely obturated. Although the predominant pattern of radiodensity is the ground glass pattern, areas of sclerosis and rarefaction are also apparent. The infraorbital canal running toward the infraorbital foramen is still patent as a vertical slit. Although the affected hemimaxilla (and also the zygoma) displays substantial buccal expansion and appreciable medially directed expansion of the lateral wall of the nasal cavity, upward expansion of the floor of the orbit is minimal. The affected alveolus displays buccolingual expansion. The lack of a cortex on the buccal aspect of the alveolus should provoke consideration that this may have been biopsied. (b) Medially directed expansion of the lateral wall of the orbital cavity. The ethmoid air cells are being obturated by the FD affecting the sphenoid. The buccal cortex is intact. (c) FD affecting the ipsilateral hemimandible and the greater wing and body of the sphenoid. The ethmoid air cells have been completely obturated. (d) The ethmoid air cells have been completely obturated. The superior and inferior orbital fissures, although narrower than the contralateral unaffected side are nevertheless still patent. The affected hemimandible exhibits substantial buccolingual expansion.

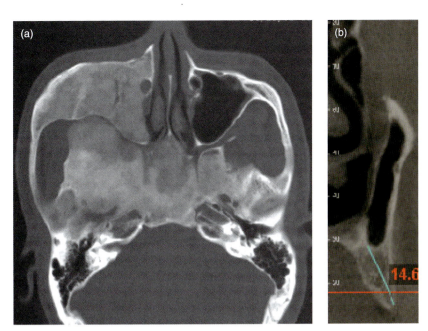

Figure 11.36. Computed tomography (CT) of polyostotic fibrous dysplasia affecting a maxillary antrum (MA) on one side and the adjacent base of the skull. (a) Axial of another case of polyostotic fibrous dysplasia. The MA of the affected side has been completely obliterated by the FD expanding outward both the anterior wall of the maxilla and the malar bone. Nevertheless, the inferior orbital canal has been preserved as has the diameter of the nasolacrimal duct. The latter has lost most of its cortex. The anterior portion of the lateral wall of the nasal cavity, although retaining its cortex has been displaced medially. Although the base of the skull has also been extensively affected, the ipsilateral zygomatic arch has not. Unlike the above canal and duct the FD has completely obliterated the foramen ovale and spinosum. Also it has crossed the midline. The downwardly expanded base of the skull and the maxilla are still separated by the pterygomaxillary fossa now both narrowed and lengthened. (b) Coronal cone-beam computed tomography (CBCT) displays the nasolacrimal canal in a normal patient.

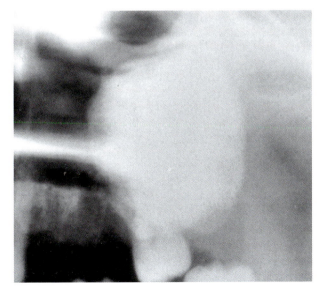

Figure 11.37. This panoramic radiograph displays a complex odontoma presenting as a well-defined radiopacity completely obturating the posterior part of the lumen of the maxillary antrum. A capsule is apparent as radiolucent spaces on the superior and anterioinferior aspects of the lesion. A cortex is clearly observed on the lesions anteriosuperior aspect. The radiodensity of a substantial portion of the lesion is similar to that of teeth. It has displaced the root of the erupted maxillary third molar mesially. Reprinted with permission from MacDonald-Jankowski DS. Fibro-osseous lesions of the face and jaws. *Clinical Radiology* 2004;59:11–25.

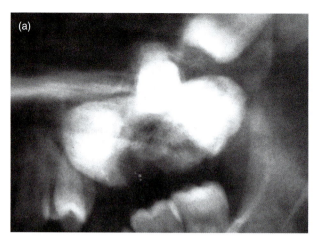
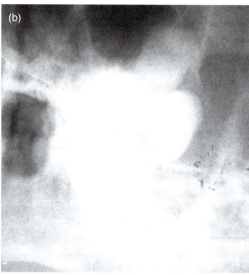

Figure 11.38. A panoramic radiograph (a) and a posterioanterior projection (b) display a well-defined radiopacity within the maxillary antrum. Upon removal it was found histopathologically to be a complex odontoma. It has prevented the eruption of a molar tooth.

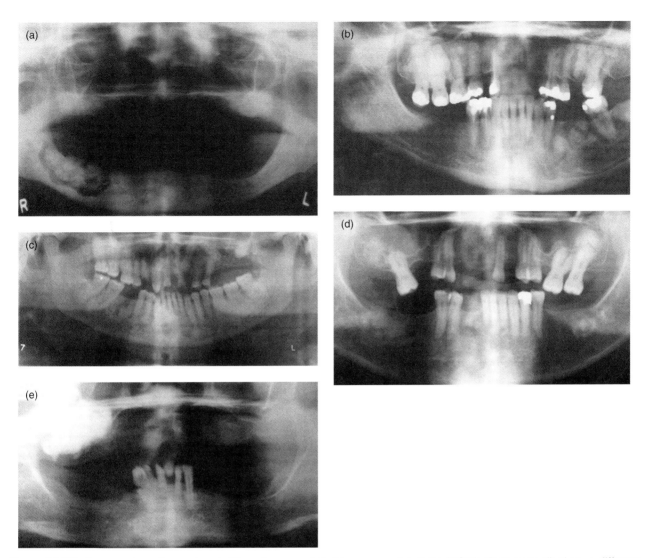

Figure 11.39. Panoramic radiographs of different cases of florid osseous dysplasia (FOD). Each case displays a different presentation of the FODs sited in the posterior maxillary sextants and the maxillary antrum (MA). A radiolucent line separating the FOD from the floor of the MA can be discerned wholly in (b), (c), and (d) and partially in (a) and (e). Figure (a) reprinted with permission from MacDonald-Jankowski DS. Florid osseous dysplasia in the Hong Kong Chinese. *Dentomaxillofacial Radiology* 1996;25:39–41. Figure (c) reprinted with permission from MacDonald-Jankowski DS. Fibro-osseous lesions of the face and jaws. *Clinical Radiology* 2004;59:11–25.

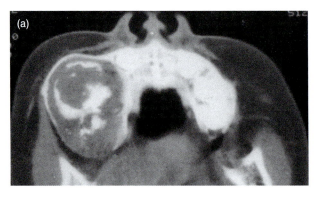
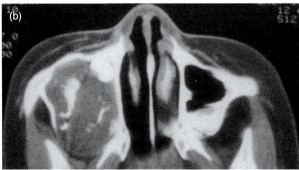
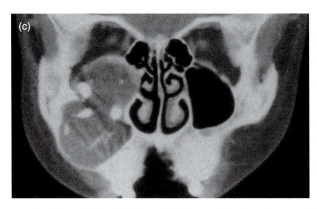

Figure 11.40. Computed tomography of ameloblastic fibro-odontoma affecting the maxillary antrum (MA). (a) Axial CT (bone window) through the maxillary alveolus, displaying a well-defined egg-shaped corticated radiolucent area containing central opacities. There is substantial buccal expansion. (b) Axial CT (bone window) through the middle of the nasal cavity, displaying a well-defined corticated radiolucent area containing central opacities. The posterior aspect of the posteriolateral wall of the MA has been outwardly displaced. There is a tooth crown impacted into the anteriomedial angle of the MA. (c) Coronal CT (bone window) through the maxillary canine, displaying a well-defined corticated radiolucent area containing central opacities. The floor of the orbit has been upwardly displaced.

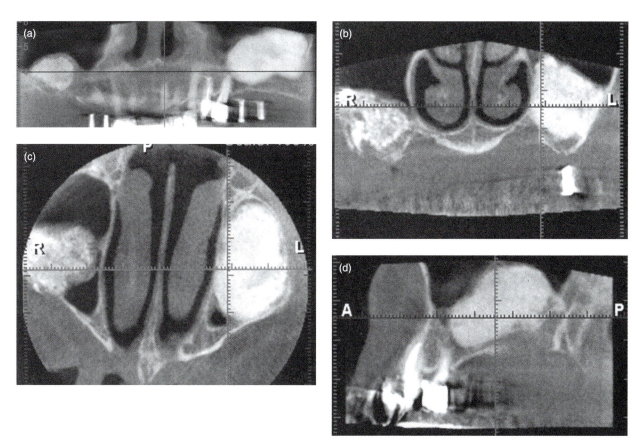

Figure 11.41. Cone-beam computed tomography (CBCT) of bilateral sinus-lift grafts. (a) A panoramic reconstruction displaying the sinus-lift grafts are a bilateral pair of well-defined radiopacities within the posterior maxillary sextants. (b) A CBCT coronal reconstruction displaying both radiopacities. The left sinus-lift graft is larger and more dense. (c) A CBCT axial reconstruction of both sinus-lift grafts. (d) The sagittal reconstruction of the left sinus-lift graft showing the formation of a new bony antral floor over it.

ingly look like pathology may be iatrogenic; they are sinus-lift grafts.

References

1. MacDonald-Jankowski DS. Fibro-osseous lesions of the face and jaws. *Clin Radiol* 2004;59:11–25.
2. Eggesbø HB. Radiological imaging of inflammatory lesions in the nasal cavity and paranasal sinuses. *Eur Radiol* 2006;16:872–888.
3. Kim HY, Kim MB, Dhong HJ, Jung YG, Min JY, Chung SK, Lee HJ, Chung SC, Ryu NG. Changes of maxillary sinus volume and bony thickness of the paranasal sinuses in longstanding pediatric chronic rhinosinusitis. *Int J Pediatr Otorhinolaryngol* 2008;72:103–108.
4. MacDonald-Jankowski DS. Mucosal antral cysts in a Chinese population. *Dentomaxillofac Radiol* 1993;22: 208–210.
5. MacDonald-Jankowski DS. Mucosal antral cysts observed within a London inner-city population. *Clin Radiol* 1994;49:195–198.
6. MacDonald-Jankowski DS. The involvement of the maxillary antrum by odontogenic keratocysts. *Clin Radiol* 1992;45:31–33.
7. Li TK, MacDonald-Jankowski DS. An unusual presentation of a high-grade, non-Hodgkin's lymphoma in the maxilla. *Dentomaxillofac Radiol* 1991;20:224–226.

8. MacDonald-Jankowski DS, Yeung R, Lee KM, Li TK. Ameloblastoma in the Hong Kong Chinese. Part 1: systematic review and clinical presentation. *Dentomaxillofac Radiol* 2004;33:71–82.
9. MacDonald-Jankowski DS, Yeung R, Lee KM, Li TK. Ameloblastoma in the Hong Kong Chinese. Part 2: systematic review and radiological presentation. *Dentomaxillofac Radiol.* 2004;33:141–151.
10. Luo HY, Li TJ. Odontogenic tumors: A study of 1309 cases in a Chinese population. *Oral Oncol* 2009 (Jan 13) [Epub ahead of print].
11. Nastri AL, Wiesenfeld D, Radden BG, Eveson J, Scully C. Maxillary ameloblastoma: a retrospective study of 13 cases. *Br J Oral Maxillofac Surg* 1995;33:28–32.
12. MacDonald-Jankowski DS, Li TKL. Keratocystic odontogenic tumour in a Hong Kong community; the clinical and radiological presentations and the outcomes of treatment and follow-up. *Dentomaxillofac Radiol* 2010;39:167–175.
13. MacDonald-Jankowski DS. Keratocystic odontogenic tumour; a systematic review. *Dentomaxillofac Radiol* 2011;40:1–23.
14. MacDonald-Jankowski DS. Orthokeratinized odontogenic cyst; a systematic review. *Dentomaxillofac Radiol* 2010;39:455–467.
15. MacDonald-Jankowski DS, Yeung R, Lee KM, Li TK. Odontogenic myxomas in the Hong Kong Chinese: clinico-radiological presentation and systematic review. *Dentomaxillofac Radiol* 2002;31:71–83.
16. MacDonald-Jankowski DS, Yeung R, Lee KM, Li TK. Computed tomography of odontogenic myxoma. *Clin Radiol* 2004;59:281–287.
17. MacDonald-Jankowski DS. Glandular odontogenic cyst: a systematic review. *Dentomaxillofac Radiol* 2010;39:127–139
18. Macdonald-Jankowski DS, Li TK. Fibrous dysplasia in a Hong Kong community: the clinical and radiological features and outcomes of treatment. *Dentomaxillofac Radiol* 2009;38:63–72.
19. MacDonald-Jankowski DS, Yeung R, Lee KM, Li TK. Computed tomography of fibrous dysplasia. *Dentomaxillofac Radiol* 2004;33:114–118.
20. Macdonald-Jankowski DS. Ossifying fibroma: systematic review. *Dentomaxillofac Radiol* 2009;38:495–513.
21. Macdonald-Jankowski DS, Li TK. Ossifying fibroma in a Hong Kong community: the clinical and radiological features and outcomes of treatment. *Dentomaxillofac Radiol* 2009;38:514–523.
22. MacDonald-Jankowski DS. Odontomas in a Chinese population. *Dentomaxillofac Radiol* 1996;25:186–192.
23. MacDonald-Jankowski DS. Idiopathic osteosclerosis in the jaws of Britons and of the Hong Kong Chinese: radiology and systematic review. *Dentomaxillofac Radiol* 1999;28:357–363.
24. MacDonald-Jankowski DS. Florid cemento-osseous dysplasia: a systematic review. *Dentomaxillofac Radiol* 2003;32:141–149.
25. MacDonald-Jankowski DS. Focal cemento–osseous dysplasia: a systematic review. *Dentomaxillofac Radiol* 2008;37:350–360.
26. Piette EM, Tideman H, Wu PC. Massive maxillary ameloblastic fibro-odontoma; case report with surgical management. *J Oral Maxillofac Surg* 1990;48:526–530.

Chapter 12
Temporomandibular joint

The *temporomandibular joint* (TMJ) has three basic components, the condylar head (Figure 12.1), joint space, (Figure 12.2), and glenoid fossa and articular eminence of the temporal bone (Figure 12.1). On *conventional radiography*, including panoramic radiography, conventional tomography, bone window *helical computed tomography* (HCT) (Figure 12.3a), and *cone-beam computed tomography* (CBCT) the joint space is visualized as a radiolucent structure by virtue of its wholly soft-tissue content. The only radiopacities that may be observed occasionally within this space by the aforementioned modalities are *articular loose bodies* (Figure 12.2). These radiopacities range from innocent joint mice (isolated bone fragments from the condyle or temporal bone), *synovial chondromatosis*,[1] and *pseudogout (chondrocalcinosis)*.[2] The last two diseases can erode through the skull base. Yokota et al. had included *chondrosarcoma* and *osteosarcoma* in the differential diagnosis of a case of synovial chondromatosis.[3]

The anatomical components and disease characteristics of the soft tissue of the joint space are displayed by soft-tissue window HCT and especially by *magnetic resonance imaging* (MRI) (see Figure 12.3b, c). Pereira et al. pioneered an inconclusive study using *ultrasound*.[4]

The first image usually taken of a patient presenting with symptoms or signs indicating a TMJ problem is the panoramic radiograph. This provides the clinician with a lateral view of the condylar head and neck. Although the width of the focal plane of the panoramic radiograph is likely to include the whole width of the condylar head, the shape of the head can vary between patients. Nevertheless, if the condyles are symmetrical in shape and size, it is reasonable to assume they are normal, particularly in the absence of symptoms. The significance of flattened condyles, erosions, and osteophytes (Figure 12.4) are considered later.

The size of the condylar heads can be assessed first by ensuring that the patient has been properly positioned within the panoramic radiographic unit prior to exposure. This may be readily assessed by comparing the width of the vertical rami and the molar teeth of both sides. Any difference should then be compared to the patient. If the patient displays no difference, the image has been distorted due to incorrect positioning.

After it has been determined that one side is indeed larger, the clinician needs to determine which side is abnormal, because the other, smaller side could be hypoplastic. This can be appreciated by an abnormally shaped vertical ramus and an obtuse gonial angle (the angle formed by the lower border of the mandible and the posterior margin of the vertical ramus) (see Figure 10.30). Hypoplasia of one side can result from a developmental accident such as hemorrhage of the stapedial artery *in utero* with disruption of adjacent tissues including the condylar growth center. It can also occur in infancy due to radiotherapy. The midline of the mandible is skewed toward the hypoplastic side (Figure 12.5b). An increase in size could be due to hyperplasia, neoplasia, or dysplasia.

Lesions affecting the condyle may arise either primarily within it, such as the osteoma (Figure 12.5) and chondrosarcoma, or arise elsewhere in the mandible and subsequently involve it, such as osteogenic sarcoma (Figure 12.6) and *fibrous dysplasia* (FD) (Figure 12.7).

One report on cases of FD affecting an East Asian community included 13 mandibular cases, of which 12 affected the posterior sextant including the ramus.[5] Of these 12, 6 cases affected the condyle. This high proportion of condylar involvement was not observed for most other lesions with a predilection for the posterior mandible affecting the same community. Of the 31 cases out of 36 ameloblastomas affecting the posterior sextant of the mandible, 2 involved the condyle.[6] Both were of the unicystic variant. One of these cases is featured in Figure 1.34. It has not replaced entirely the normal trabeculae of the condyle. The

---

Oral and Maxillofacial Radiology: A Diagnostic Approach, David MacDonald. © 2011 David MacDonald

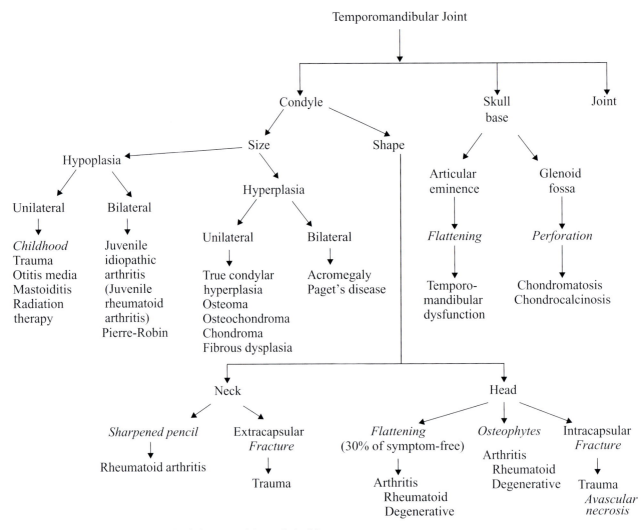

Figure 12.1. Temporomandibular joint—condyle and skull base.

condyle was involved by only 1 of the 13 mandibular KCOTs, which affected the posterior sextant.[7] Only 1 of 6 mandibular cases of odontogenic myxoma approached the condyle.[8] These results would suggest that odontogenic neoplasms are unlikely to reach the condyle even though the majority arise within the posterior sextant.

Although the TMJ may be spared in most odontogenic neoplasms, certainly in the Southern Chinese community from which the above reports[6-8] were derived, nevertheless it is likely to be challenged by the change in jaw dynamics occasioned by these large lesions and their subsequent management.

In the population at large it is not uncommon for individuals to experience pain and stiffness within one or both temporomandibular joints at least once in their life. These are the clinical hallmarks of the most common TMJ disease, *temporomandibular dysfunction* (TMD).

Temporomandibular Joint Disorder

TMD has been subject throughout the decades to almost the whole gamut of imaging and surgical modalities. MRI was proposed a decade ago as the "gold standard" for TMJ imaging but nevertheless remains controversial.[9] Despite the existence of a substantial body of literature in its support, some have urged caution. Among them are Larheim[10] and Limchaichana et al.[11] The latter's systematic review revealed that the published work was not

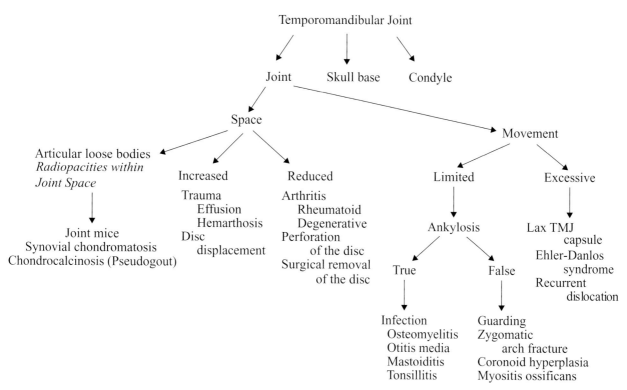

Figure 12.2. Temporomandibular joint—joint.

sufficiently rigorous and that not one report had a high level of evidence.

Limchaichana et al. stated that a *verification bias* was created because the reports were composed of patients in need of surgery and thereby sicker than the representative population that would be investigated upon the report's findings.[11] Another reason is that they lack a "gold standard." This means that if the test and reference methods co-vary, "the sensitivity and specificity will often be too high." Finally, no report considered "diagnostic thinking efficacy or therapeutic efficacy."[11]

According to Larheim's consideration of the literature on TMJ, abnormalities have so far not been adequately assessed by clinical examination.[10] Larheim makes it very clear that a thorough physical examination should be first performed in order to identify the data that only imaging can provide, before any treatment can be planned and provided.[10]

Another of Limchaichana et al.'s major criticisms was that the reports varied too much in their methodology, rendering any meta-analysis not possible.[11] Such variation speaks to the still unsettled nature of many of the fundamental principals upon which management of TMD is based. A major feature of Klasser and Greene's address "The changing field of temporomandibular disorders: What dentists need to know" was the continued uncertainty about the theoretical basis of TMD.[12] Nevertheless, they maintained that patients can be effectively diagnosed and treated by dentists taking a "low-tech and high-prudence" therapeutic approach. Invasive, irreversible and aggressive treatments should, therefore, be avoided.[12]

Although a panoramic radiograph gives an overview of the condyle and adjacent structures, markedly more joint components were accessible with sagittal cross-sectional tomography.[13] Unfortunately, tomography becomes time-consuming, particularly if a bilateral investigation is required. CBCT has been shown to produce similar accuracy but in a much shorter time while imparting perhaps lower radiation dose to the patient.[14,15] The other advantage is that CBCT images "may be very 'reader-friendly' and easy to become familiar with." Both modalities were ineffective when required to identify flattening of the condylar head, defects, and osteophytes.[14]

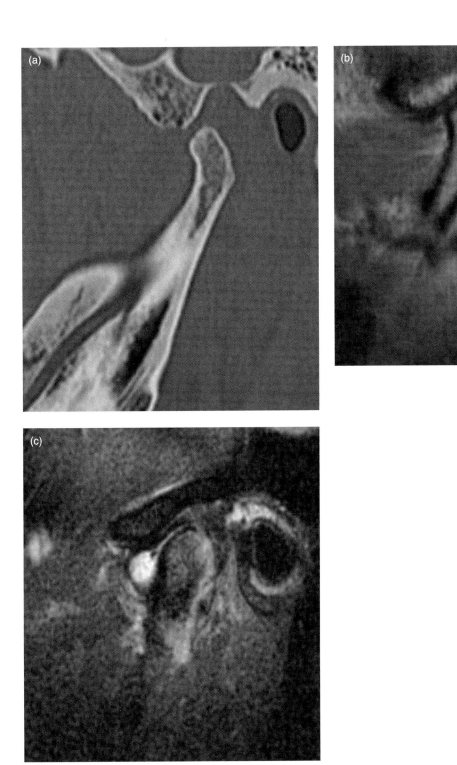

Figure 12.3. The bone window computed tomograph (a) and magnetic resonance images (T1-weighted (b), and fat-saturated T2-weighted (c) are of the same patient investigated for a painful joint after trauma). (a) This displays exquisite bone detail. The soft tissue in this bone window is a uniform gray, punctuated only by the air-filled external acoustic meatus. (b) The hypointensity of the cortices of the condyle and glenoid fossa is similar to that of the air-filled external acoustic meatus. The T1-weighted MR image displays the soft-tissue anatomy in great detail. The isointense (grey) muscle fasiculae of the lateral pterygoid muscle are clearly obvious inserting into the pterygoid fossa. The interarticular disc appears as a hypointense (dark) structure in a normal relationship to the condylar head. (c) The hyperintense (bright white) area represents an effusion into the joint space and is indicative of ongoing inflammation. The hyperintense condylar marrow, which appeared relatively hyperintense in (b) is now hypointense in the fat-saturated (fat-suppression) T2-weighted image. Reprinted with permission from MacDonald-Jankowski DS, Li TK, Matthew I. Magnetic resonance imaging for oral and maxillofacial Surgeons. Part 2: Clinical applications. *Asian Journal of Oral Maxillofacial Surgery* 2006;18:236–247.

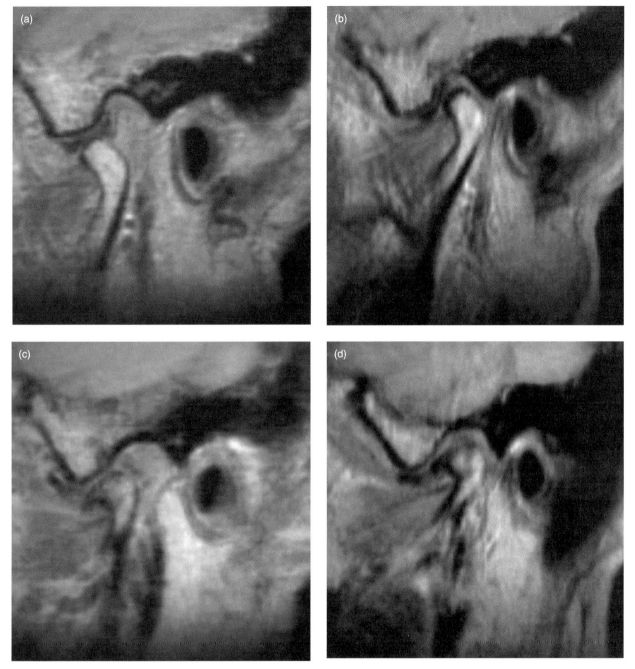

Figure 12.4. These T1-weighted magnetic resonance images (MRI) display the side displaying partial disc displacement displayed normal disc position in the open mouth position (a) and anterior disc displacement in the closed mouth position (b), whereas the side displaying complete disc displacement displayed anterior disc displacement both in the open (c) and closed(d) mouth positions. The marrow of the condyle is normally fat-filled and is hyperintense on T1-weighted MRI. The abnormal side exhibits an osteophyte, which gives the abnormal condyle the appearance of a bird's head. The normal side displays the meniscus (hypointense disc) properly interposed between the condyle and articular surface in both open and closed positions. The abnormal side displays anterior disc displacement. Reprinted with permission from MacDonald-Jankowski DS, Li TK, Matthew I. Magnetic resonance imaging for oral and maxillofacial Surgeons. Part 2: Clinical applications. *Asian Journal of Oral Maxillofacial Surgery* 2006;18:236–247.

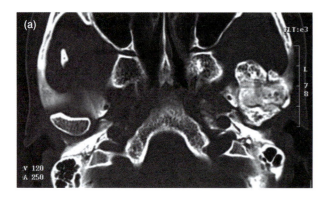

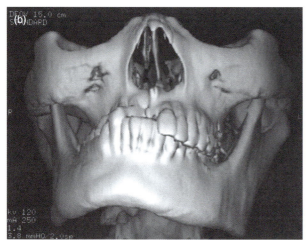

Figure 12.5. The computed tomography of an osteoma affecting the condylar head. (a) An axial computed tomograph (bone window) displaying the osteoma. In comparison with the normal ovoid-shaped contralateral condylar head, the affected condyle has substantially expanded in all directions. It has substantially reduced the joint space. (b) A three-dimensional reconstruction of the osteoma displaying it as an expansion mainly of the medial pole of the condylar head. There is a midline shift to the unaffected side and marked facial asymmetry.

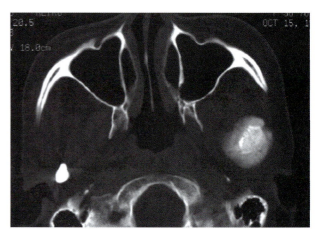

Figure 12.6. Axial computed tomograph (bone window) of osteosarcoma affecting the condylar head. The primary site in the body of the mandible recurred twice after wide resections (See Figure 10.12). It has now spread to the condyle. The axial section displays an outline of a normal-sized and -shaped condylar head surrounded by a sunburst halo.

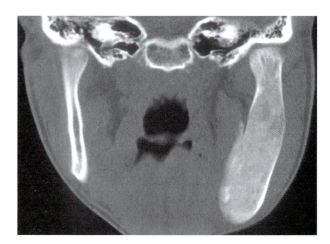

Figure 12.7. This coronal computed tomograph (bone window) displays fibrous dysplasia affecting the mandible involving the condyle. The superior or articular portion of the condyle appears sclerotic and the neck is wider. Note the lingula.

The appearance of the above osseous changes of flattening of the condylar head and small osteophytes, but associated with a smooth cortical outline, should be interpreted not as osteoarthritis but as remodeling.[10] Some flattening of the condyle has been reported in 35% of symptom-free patients.[16] No erosions or osteophytes were observed in such patients.[16]

For the assessment of internal derangement of the TMJ, MRI is the modality of choice.[17] These will now be briefly overviewed. MRI investigations largely used T1-weighted image (see Figures 12.3b and 12.4) to determine the position of the articular disc.[18] An imaging protocol, categories, and distributions of TMJ disc position in the mouth-closed position are set out by Larheim and Westesson.[19]

Although disc displacement is significantly more frequently prevalent in TMD patients than in symptom-free volunteers, it can also appear in the latter group. Because this does not occur in preschool children, it can be surmised that disc displacement is an acquired phenomenon that may develop early in life. Disc displacement on its own is only part of the TMD story. The presence of joint effusions and marrow abnormalities optimally displayed by MRI, suggest a subset of cases, which have more severe joint pathology. Fifteen percent of TMD patients display joint effusion and 30% display bone marrow abnormalities.[10]

Joint effusion, which is an excess of fluid in the joint space (see Figure 12.3c) may be associated with joint pain and inflammatory changes[17] and is not displayed by T1-weighted images; indeed, it is not even readily visualized on T2-weighted images. To display a joint effusion, fat suppression is necessary, which in turn reduces the scanning time and, therefore, reduces the risk of movement artifacts.

Fluid in the synovial spaces is displayed as a hyperintense signal on T2-weighted MRI (see Figure 12.3c). It has been observed in half of symptom-free volunteers ranging from just dots or as a line to a "moderate" amount. The amount of fluid within the joint space must be a "marked" accumulation, such as evident in Figure 12.3c, to be described as an effusion. Larheim reported that this effusion in two-thirds of cases was found exclusively or predominantly in the upper compartment of the anteriolateral recess. Although the discs of both joints of most patients were displaced, effusion was predominantly unilateral. The amount of effusion varied over time and correlated well with the severity of pain.[10]

The bone marrow appears normally as a homogeneous bright signal on proton density images and a homogeneous intermediate signal on T2-weighted images. Edema of the marrow presents as a reduced signal on proton density and an increased signal on T2-weighted images.[10]

The most reliable presentation of osteonecrosis on MRI is the appearance of edema and sclerosis of the marrow. Although the designation of osteonecrosis has been awarded to this presentation as it is observed with regard to osteonecrosis of the hip, this condition in the TMJ appears to be less aggressive. Currently there is no published evidence to suggest that patients presenting with this phenomenon should be treated differently.[10]

Orhan et al. suggested that MRIs of TMJs should also be examined for evidence of inflammatory disease, such as *otomastoiditis*, in adjacent structures.[20]

Single photon emission computed tomography fused with computed tomography (SPECT/CT) images, permit attenuation correction of the SPECT data for TMD by means of anatomical mapping. However, the diagnostic efficiency is limited by the 10 mm wide CT slices, which reduce the spatial resolution.[21]

Petersson overviewed TMJ imaging and related it to the 1992 edition of the *Research Diagnostic Criteria for Temporomandibular Disorders* (RDC/TMD).[22] This overview extended from panoramic and conventional radiography through MRI. Although he cautiously identified MRI as the better but expensive modality, he concluded that currently there is "no clear evidence for when (his italics)" it should be used. He suggested that "there is a need for high-quality studies on the diagnostic efficacy of MRI that incorporates accepted methodological criteria."[22]

Juvenile Idiopathic Arthritis

Juvenile idiopathic arthritis (JIA, also called "juvenile rheumatoid arthritis") is an autoimmune oligo- or polyarticular disease affecting 1:500 children.[23] Fifty-five percent of cases in a Greek region suffered uveitis, which can result in blindness.[24] Arvidsson et al. reported the craniofacial growth disturbances related to TMJ abnormality in Norwegian JIA patients that were followed up for 27 years.[25] TMJ involvement was 40% of JIA cases, similar to that in other Scandinavian communities. "Microstomia occurred with bilateral TMJ involvement only and in 27% in the entire series of patients." Nevertheless, "growth disturbances did not always follow TMJ involvement, not even when affected early."[25]

Simard et al. have just reported in their nationwide study based on the five registries, which included Swedish National Patient Registry, Swedish Cancer Registry, and The Causes of Death Registry.[26] They found a significantly higher risk of malignancy, particularly lymphoproliferative, among those JIA patients diagnosed in the last 2 decades in comparison to the general population. They suggested that this could be due to the newer treatments, which may affect immune function.[26]

References

1. Yu Q, Yang J, Wang P, Shi H, Luo J. CT features of synovial chondromatosis in the temporomandibular joint. *Oral Surg Oral Med Oral Pathol Oral Radiol Endod* 2004;97:524–528.
2. Barthelemy I, Karanas Y, Sannajust JP, Emering C, Mondie JM. Gout of the temporomandibular joint: pitfalls in diagnosis. *J Craniomaxillofac Surg* 2001;29: 307–310.
3. Yokota N, Inenaga C, Tokuyama T, Nishizawa S, Miura K, Namba H. Synovial chondromatosis of the temporomandibular joint with intracranial extension. *Neurol Med Chir (Tokyo)* 2008;48:266–270.
4. Pereira LJ, Gaviao MB, Bonjardim LR, Castelo PM. Ultrasound and tomographic evaluation of temporomandibular joints in adolescents with and without signs and symptoms of temporomandibular disorders: a pilot study. *Dentomaxillofac Radiol* 2007;36:402–408.
5. MacDonald-Jankowski DS, Li TK. Fibrous dysplasia in a Hong Kong community: the clinical and radiological features and outcomes of treatment. *Dentomaxillofac Radiol* 2009;38:63–72.
6. MacDonald-Jankowski DS, Yeung R, Lee KM, Li TK. Ameloblastoma in the Hong Kong Chinese. Part 2: systematic review and radiological presentation. *Dentomaxillofac Radiol* 2004;33:141–151.
7. MacDonald-Jankowski DS, Li TK. Keratocystic odontogenic tumor in a Hong Kong community: the clinical and radiological presentations and the outcomes of treatment and follow-up. *Dentomaxillofacial Radiol.* 39:167–175.
8. MacDonald-Jankowski DS, Yeung R, Lee KM, Li TK. Odontogenic myxomas in the Hong Kong Chinese: clinico-radiological presentation and systematic review. *Dentomaxillofac Radiol* 2002;31:71–83.
9. Schmitter M, Gabbert O, Ohlmann B, Hassel A, Wolff D, Rammelsberg P, Kress B. Assessment of the reliability and validity of panoramic imaging for assessment of mandibular condyle morphology using both MRI and clinical examination as the gold standard. *Oral Surg Oral Med Oral Pathol Oral Radiol Endod* 2006;102: 220–224.
10. Larheim TA. Role of magnetic resonance imaging in the clinical diagnosis of the temporomandibular joint. *Cells Tissues Organs* 2005;180:6–21.
11. Limchaichana N, Petersson A, Rohlin M. The efficacy of magnetic resonance imaging in the diagnosis of degenerative and inflammatory temporomandibular joint disorders: a systematic literature review. *Oral Surg Oral Med Oral Pathol Oral Radiol Endod* 2006;102: 521–536.
12. Klasser GD, Greene CS. The changing field of temporomandibular disorders: what dentists need to know. *J Can Dent Assoc* 2009;75:49–53.
13. Hintze H, Wiese M, Wenzel A. Comparison of three radiographic methods for detection of morphological and tomographic examination. *Dentomaxillofac Radiol* 2009;38:134–140.
14. Hintze H, Wiese M, Wenzel A. Cone beam CT and conventional tomography for the detection of morphological temporomandibular joint changes. *Dentomaxillofac Radiol* 2007;36:192–197.
15. Hussain AM, Packota G, Major PW, Flores-Mir C. Role of different imaging modalities in assessment of temporomandibular joint erosions and osteophytes: a systematic review. *Dentomaxillofac Radiol* 2008;37:63–71.
16. Brooks SL, Westesson PL, Eriksson L, Hansson LG, Barsotti JB. Prevalence of osseous changes in the temporomandibular joint of asymptomatic persons without internal derangement. *Oral Surg Oral Med Oral Pathol* 1992;73:118–122.
17. Mori S, Kaneda T, Lee K, Kato M, Motohashi J, Ogura I. T2-weighted MRI for the assessment of joint effusion: comparative study of conventional spin-echo and fast spin-echo sequences. *Oral Surg Oral Med Oral Pathol Oral Radiol Endod* 2004;97:768–774.
18. Gossi DB, Gallo LM, Bahr E, Pallo S. Dynamic intraarticular space variation in clicking TMJs. *J Dent Res* 2004;83:480–484.
19. Larheim TA, Westesson P-L. TMJ imaging. In Laskin DM, Greene CS, Hylander WL, eds. TMDs. *An Evidence-based Approach to Diagnosis and Treatment*. Chicago Quintessence Publishing Co, Inc., Chicago 2006: pp 149–179.
20. Orhan K, Nishiyama H, Tadashi S, Shumei M, Furukawa S. MR of 2270 TMJs: prevalence of radiographic presence of otomastoiditis in temporomandibular joint disorders. *Eur J Radiol* 2005;55:102–107.
21. Coutinho A, Fenyo-Pereira M, Dib LL, Lima EN. The role of SPECT/CT with 99mTc-MDP image fusion to diagnose temporomandibular dysfunction. *Oral Surg Oral Med Oral Pathol Oral Radiol Endod* 2006;101:224–230.
22. Petersson A. What you can and cannot see in TMJ imaging—an overview related to the RDC/TMD diagnostic system. *J Oral Rehabil* 2010;37:771–778.
23. Boros C, Whitehead B. Juvenile idiopathic arthritis. *Aust Fam Physician* 2010;39:630–636.
24. Asproudis I, Felekis T, Tsanou E, Gorezis S, Karali E, Alfantaki S, Siamopoulou-Mauridou A, Aspiotis M. Juvenile idiopathic arthritis-associated uveitis: Data from a region in western Greece. *Clin Ophthalmol* 2010; 4:343–347.
25. Arvidsson LZ, Flatø B, Larheim TA. Radiographic TMJ abnormalities in patients with juvenile idiopathic arthritis followed for 27 years. *Oral Surg Oral Med Oral Pathol Oral Radiol Endod* 2009;108:114–123.
26. Simard J, Neovius M, Hagelberg S, Askling J. Juvenile idiopathic arthritis and risk of cancer: A nationwide cohort study. *Arthritis Rheum* 2010 (Sep 8) [Epub ahead of print].

Chapter 13
Imaging of the salivary glands

Introduction

Imaging of the salivary glands includes almost the entire range of imaging modalities, from conventional imaging through *computed tomography* (CT), *magnetic resonance imaging* (MRI), *ultrasound* (US), and *positron emission tomography* (PET). Figure 13.1 overviews the main glandular lesions and their imaging strategies. The classical imaging technique for salivary gland disease is *sialography* (Figure 13.2). Although it temporarily slipped into abeyance, it has experienced a renaissance due to its central role in interventional sialography, a therapeutic modality for the conservative (or minimal surgical) treatment of obstructive glandular disease. This conservative treatment can also be achieved under US guidance and may include lithotripsy. The basic principles of US were introduced in introduced in Chapter 8.

Diseases affecting salivary glands can affect one or more glands. Affected glands generally appear swollen or enlarged. This may be accompanied by pain, particularly if acute onset, and perhaps an alteration in saliva flow rate. The last is usually reduced. Those arising from obstructive disease or neoplasia most frequently affect one gland, whereas those caused by systemic disease affect more than one and frequently present bilaterally. Although the smaller major salivary glands (submandibular and sublingual) can be affected by the latter it is usually swelling of the larger parotid glands that is most clinically obvious.

Bilateral Swelling

Bilateral swelling of the salivary gland can be of rapid onset or chronic. The most frequent cause for rapid onset of swelling and pain is mumps, whereas the chronic swellings are features of Sjogren's syndrome and *human immunodeficency virus* (HIV) infection.

MUMPS

Acute sialadenitis, such as in mumps, are generally readily diagnosed by their classical presentation and so quickly self-resolving that there is little role for radiology, particularly when mumps presents within an epidemic and affects young patients. Nevertheless, presentation in the older patient may require appropriate imaging, such as US. Radiology itself may provoke "iodide mumps," a rare response to intravenous contrast.

Hitherto, mumps was a significant public health issue prior to the *measles, mumps and rubella* (MMR) vaccine. As a result incidence of mumps plummeted from the 1960s, only to reemerge recently among adolescents and young adults[1] and in many schools and universities in North America and Europe.[1-3] The morbidity is very high: cerebrospinal fluid pleocytosis (occurs in 50% of all cases), orchitis (up to 30%), and spontaneous abortion (27%).[2] There may not be a single cause; a lack of vaccination, an incomplete vaccination, or a decline efficiency of vaccination have been variously proposed.

US is generally not indicated, unless the mumps manifests unilaterally. On US, mumps-affected parotid has a rounded shape "with a convex lateral surface and a hypoechogenic structure."[4] Such a case is displayed in Figure 13.3. This case was a middle-aged male, who presented with a rapid onset of bilateral swelling of the submandibular and parotid glands. Early diagnosis and intravenous immunotherapy may minimize complications of mumps.[2]

HIV-ASSOCIATED SALIVARY GLAND DISEASE

Although all salivary glands are affected by HIV, the parotid glands are most affected due to their

Oral and Maxillofacial Radiology: A Diagnostic Approach, David MacDonald. © 2011 David MacDonald

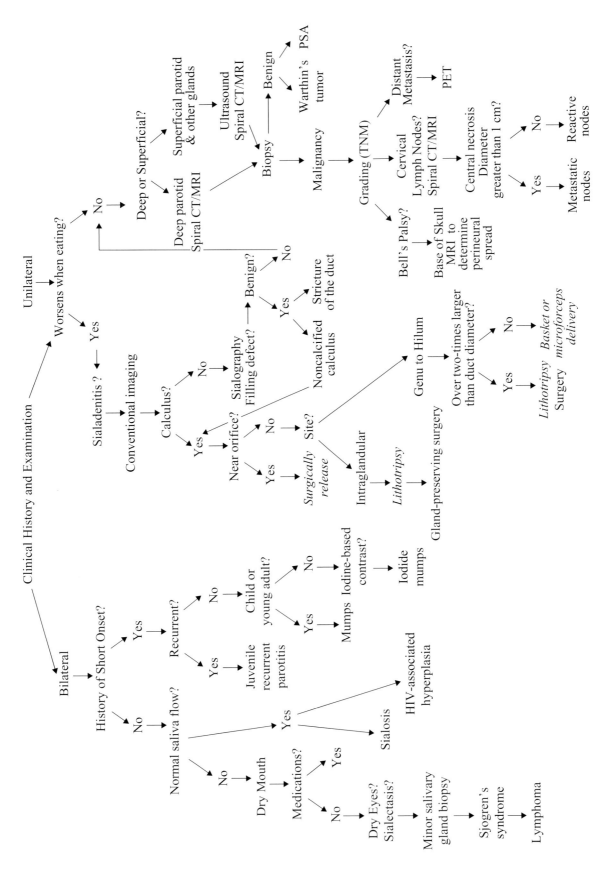

Figure 13.1. Overview of the principal glandular lesions and their imaging strategies. PET, positron emission tomography; PSA, pleomorphic salivary adenoma.

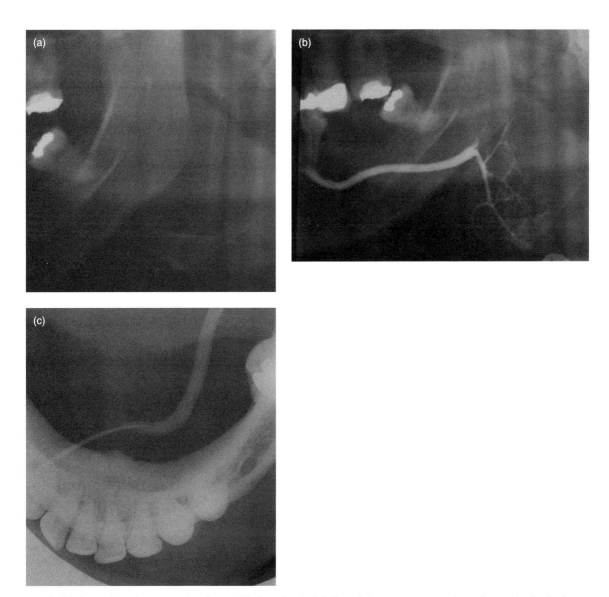

Figure 13.2. Sialography of a normal submandibular gland. (a) Pre-sialogram panoramic radiograph displaying no radiopacities. (b) Panoramic radiograph showing a normal submandibular gland. The "blush" around the terminal ducts represents some overfilling of the acini. The submandibular gland is of uniform width and displays no filling defects. It follows a horizontal route until it reaches the posterior margin of the myohyoid muscle and turns sharply downward to the hilum of the submandibular gland. (c) A true occlusal image displaying the cannula inserted into the submandibular duct through the submandibular papilla.

inclusion of lymphoid tissue. Three types of lesions may be encountered: inflammatory reaction to the virus, lymphomas, and infection.[5]

The inflammatory reaction is expressed by benign lymphoepithelial lesions and AIDS-related cysts. The benign lymphoepithelial lesions appear on CT, MRI, and US as dilated cystic ducts within hyperplastic lymphoid tissue. They frequently cause the painless swellings of the parotid. This may be bilateral in up to a fifth of cases. On the other hand AIDS-related cysts appear on the above modalities as multiple lesions, which are bilateral in 80% of cases. These benign lesions are generally not treated and may regress spontaneously.[5]

Primary salivary gland lymphomas can occur in any gland and present as a painless mass. Severe infections of the salivary gland can occur. Lymph node hyperplasia affects half of the patients. The

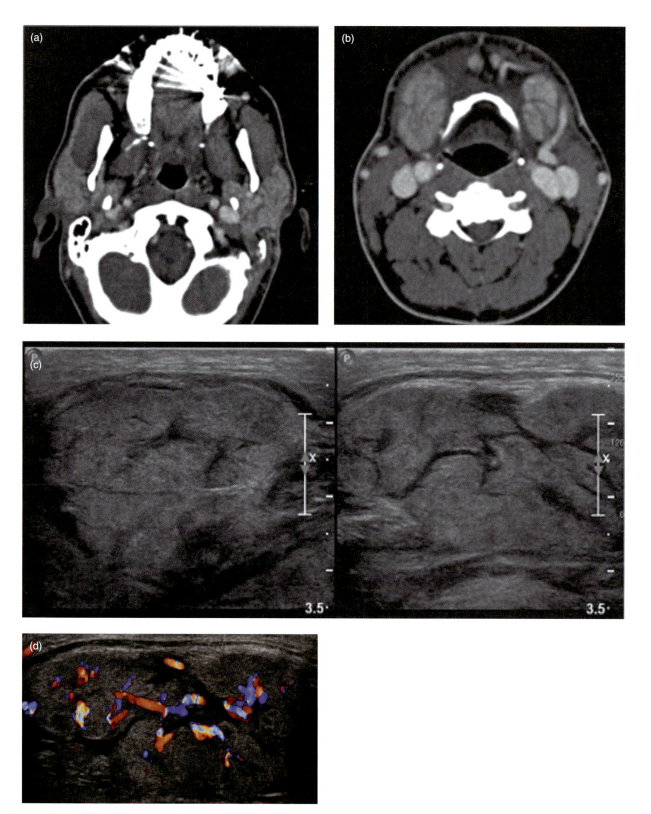

Figure 13.3. Adult mumps. (a) Contrast enhanced soft-tissue window axial computed tomograph at the level of the first cervical vertebra. Bilateral swollen parotid glands are displayed. The accessory lobe and the duct of the right parotid gland are displayed. (b) Contrast enhanced soft-tissue window axial computed tomograph at the level of the third cervical vertebra. Bilateral swollen submandibular glands are displayed. The facial vein is observed running on the lateral surface of the left submandibular gland. (c) The grayscale ultrasound displays diffuse almost symmetrical swelling of both submandibular glands. The parenchyma has a diffuse heterogeneous hypoechoic presentation, which is consistent with an acute sialadenitis. There are no hyperechoic foci suggestive of sepsis. (d) Color Doppler ultrasound reveals hyperemia, which is consistent with acute sialadenitis. Figures courtesy Dr. Eli Whitney; Oral Medicine, Faculty of Dentistry, UBC.

posterior cervical nodes are most frequently affected. Needle biopsy in addition to imaging is often required to determine whether the node is reactive, infected or neoplastic. The last may represent Karposi's sarcoma, Hodgkin's lymphoma, and non–Hodgkin's lymphoma.[5]

SIALOSIS

Sialosis is a noninflammatory nonneoplastic phenomenon, which manifests as recurrent painless, usually bilateral, swelling, principally of the parotid glands. US displays hyperechoic parotid glands. Sialosis is associated with endocrine and deficiency diseases. It also is associated with alcoholism, malnutrition, and cirrhosis.

SJOGREN'S SYNDROME

The 5-step American-European classification criteria for Sjogren's syndrome is set out and discussed by Ellis.[6] Three-quarters of cases are in females in the fourth to seventh decades. The parotid is affected in 90%. Although bilateral disease is typical, one side may be more severely affected. The recurring progressive swelling can be accompanied occasionally by discomfort or even pain. Although the lobar architecture is preserved, the progressive parenchymal damage eventually results in the typical sialectasis obvious on sialography (Figure 13.4). it should be noted that sialectasis is not pathognomonic for Sjogren's syndrome because it can be secondary to infection of the salivary glands.

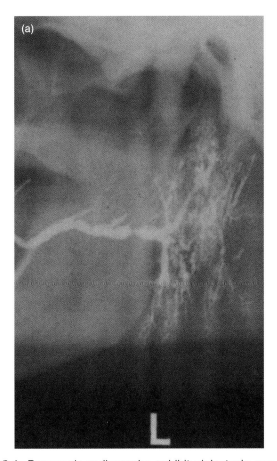
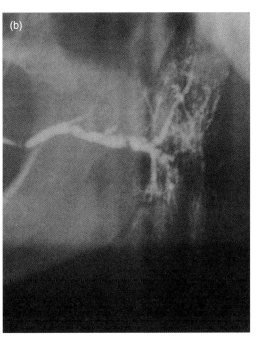

Figure 13.4. Panoramic radiographs exhibit sialectasis secondary to Sjogren's syndrome. (a) The sialographic image displaying irregular ducts and sialectasis or dilations within the substance of the parotid gland. There is a constriction of the main duct distal to the ducts for the accessory parotid lobe. (b) The emptying film (the cannula is removed) demonstrates no emptying, meaning that the constrictions of the main are preventing drainage. The emptying film demonstrates that there is also a constriction close to the parotid papilla.

Poul et al. reported that high spatial resolution US is more sensitive than sialography for the investigation of Sjogren's syndrome. Furthermore, they reported that accuracy is increased when US is carried out in conjunction with sialography.[7]

Although neoplasms of the salivary glands have long been displayed by MRI; (see also Kinoshita et al.'s pictorial essay[8]), preliminary work by a Japanese team presents MRI sialography as an alternative to conventional sialography. They compared normal and Sjögren's syndrome patients[9] with stimulated[10] and unstimulated salivary flow.[11]

Bacterial Sialadenitis

Acute sialadenitis, caused by bacteria rather than a virus as in mumps are generally unilateral. Its presentation on US may be similar to that observed in Figure 13.3c and d, but may display hyperechoic foci suggestive of air within the gland and are indicative of sepsis. An example of this is Howlett's Figure 15.[12] nevertheless, such foci are only occasionally observed in such an infection. Gritzmann et al. advise that in such situations a meticulous grayscale US is required in order to detect moving debris within an abscess. This abscess can then be aspirated under US guidance.[13]

Chronic sialadenitis affecting the parotid is common and manifests as recurrent often painful swellings with purulent discharge.[14] Chronic sialadenitis manifests on sialography as dilated irregular ducts and loss of normal saliva production (manifest by reduced or no emptying following removal of the cannula (Figure 13.5). Choi et al.[15] reported a correlation between the grades of inflammation of the sialographic images and the amount of retention of the contrast medium, and those with the degree of salivation.

A cause of chronic sialadentitis is obstructive glandular disease caused by calculi (Figures 13.6 and 13.7) and stenoses (see Figure 13.4). Marchal et al.[16] published under the auspices of the European Salivary Gland Society the following sialographic classification of salivary duct pathologies, endoscopic classification of salivary lithiasis, endoscopic classification of salivary stenosis, and endoscoptic classifications of dilatations.

According to Ngu et al.,[17] many of the 36% of patients reporting the symptoms of obstructive salivary disease suffered only a single episode. As a result they do not display any abnormality on sialography, Therefore, Ngu et al. recommend that such cases should be initially investigated by US in order to avoid unnecessary irradiation.

Sialolithiasis accounts for half of major salivary gland disease. According to Iro et al.[18] it accounts for 60 cases per million per year. Its peak occurs early in the fourth decade.

Although calculi are most frequently calcified and visible on conventional radiology and on HCT, some calculi, particularly of the parotid gland, are noncalcified and therefore translucent. They may be visualized only by US or by sialography as a filling defect (see Figure 13.7).

According to McGurk et al.[19] the submandibular gland is most vulnerable to sialolithiasis. It accounts for 63–94% of calculi in contrast to the parotid's 6–21%. The sublingual gland is infrequently affected. The distribution of calculi is 48.5% at the hilum of the submandibular gland, 29% in the submandibular duct, 9.6% in the parotid duct, 7.6% in the submandibular gland, 3.6% in the parotid gland and 1.8% at its hilum. Simultaneous bilateral calculi occur in 1%. McGurk et al. also report that the size of submandibular calculi is about 8.5 mm in diameter, in contrast to 6.6 mm for the parotid.[19] They affirm that size impacts management. Smaller calculi are more easily eradicated by both lithotripsy and, if mobile, basket delivery.

Traditionally, calculi lying anteriorly in the duct within the floor of the mouth have been released by a simple incision, whereas those more posteriorly placed have frequently necessitated adenectomy (removal of the gland) with its attendant morbidity. Current developments in technology and techniques in minimal invasive management of calculi and stenoses have afforded oral and maxillofacial radiologists a unique opportunity to move from their traditional diagnostic role into direct patient treatment. This was made possible by interventional sialography using wire baskets. This has since been joined by microendoscopy (and microforceps) and lithotripsy. Retrieval of calculi by basket or microforceps is the treatment of those calculi with a diameter less than 7 mm.[18]

Recently an European multicenter report on minimally invasive management of salivary of 4,691 patients over 14 years has confirmed its reliability in calculi elimination; 80% of the calculi had been successfully removed leaving a functionally normal salivary gland.[18] It should be noted that

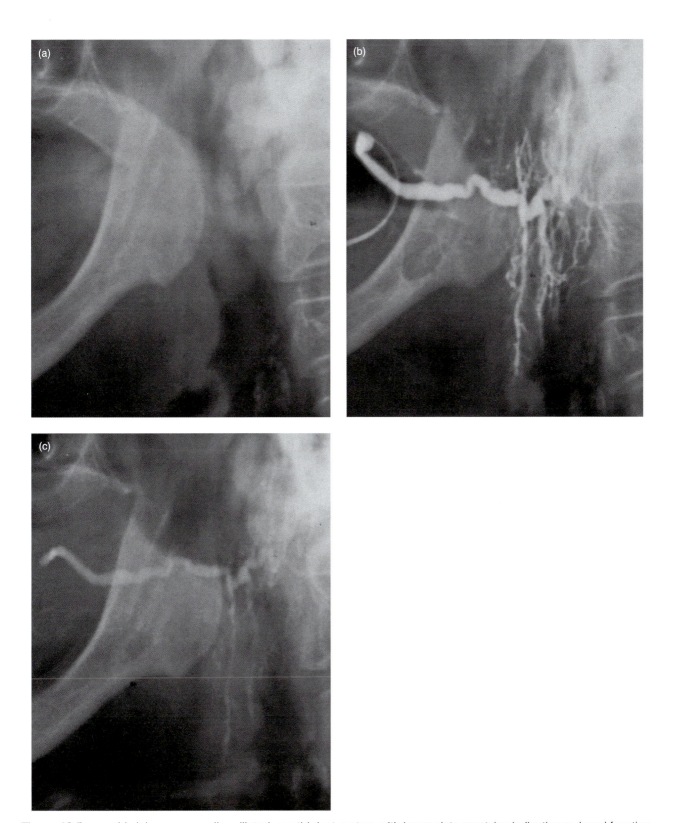

Figure 13.5. parotid sialogram revealing dilated parotid duct system with incomplete emptying indicating reduced function. (a) Pre-sialogram indicating no calculus or radiolucency. (b) Sialogram displaying extensive dilation of the parotid duct system. It also reveals a well-defined radiolucency within the vertical ramus not apparent in (a). This artifact arose from the delineation of escaped contrast medium of an air bubble within the posterior lingual sulcus. (c) Post-sialogram or emptying image exhibits incomplete emptying. The artifact in (b) has now disappeared.

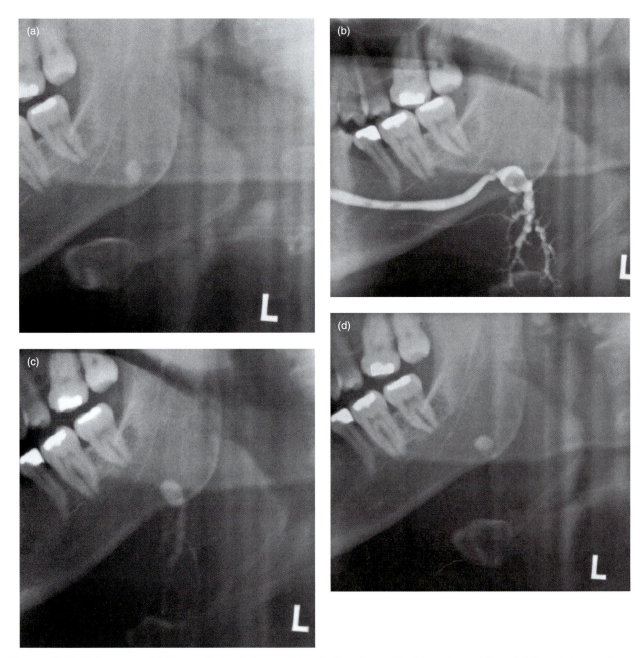

Figure 13.6. Sialogram exhibiting a dilated intraglandular duct system with delayed emptying. (a) Pre-sialogram image displaying a calculus. (b) Sialogram displaying a dilated intraglandular duct system. There is also a dilation within the extraglandular duct just at the genu at the posterior margin of the myelohyoid muscle. Although the position of this dilation is consistent with the position of the calculus, the central lucency within this dilation suggests that only part of the calculus is calcified. (c) Immediate emptying shows incomplete emptying. (d) Emptying 5 minutes later shows complete emptying, returning the image to its precontrast state observed in (a). **Note:** This case demonstrates the importance of taking a precontrast image,

this recovery of function was not to its original level. Nevertheless, only 3% of the gland ultimately had to be removed. Although the remaining 17% of patients retained fragments of calculi they are free of symptoms. Lithotripsy of the parotid render 60% calculus-free and relieved a further 30% of their symptoms even although they still retained fragments.

Strictures of the salivary gland, according to Ngu et al.[17] are perhaps more common, and more

Chapter 13: Imaging of the salivary glands

important than previously perceived. This has also been observed by those using microendoscopy.[20] The reason could be because it is not possibly to distinguish between strictures and calculi on the basis of clinical presentation. This phenomenon is best appreciated by sialography because it will "simulate the meal-time scenario where copious saliva is produced rapidly." This is not possible with other imaging modalities such as ultrasound, CT, and MRI. Once the stricture has been identified, it can then be dilated by balloons.[21]

Warthin's Tumor

Warthin's tumor (papillary cystadenoma lymphomatosum), the second most common tumor of the parotid gland, is a benign epithelial tumor composed of glandular and cystic elements. Its stroma has lymphoid tissue.[22] The ultrasound, MRI, and HCT images of a case are compared in Figure 17.18. Warthin's tumor's ability to concentrate pertechnetate (99mTc) results in positive scintigraphy.[6]

Conservative surgery may have a 5–12% recurrence, whereas total or subtotal parotectomy has no recurrence.[23] Warthin's tumor is discussed further in Chapter 17.

Neoplasms

Salivary gland neoplasms account for less than 3% of all tumors.[24] Although the central role of needle biopsy in the differentiation between malignant and benign neoplasms is almost unanimously proclaimed by the literature,[24–26] there are clear supporting roles for advanced imaging modalities; these are discussed in relation to pleomorphic salivary adenoma (MRI in Chapter 6 and HCT in Chapter 17) and malignant neoplasms of the salivary glands in Chapter 18.

Needle biopsies are fine needle (aspiration) and core needle; the former provides a specimen for cytology, whereas the latter delivers a core of tissue for histopathological evaluation. Recent reports came down strongly in favor of the core needle method.[27,28] Pratap et al.[27] preferred US-guided core needle biopsy because it is very safe (only one case with complications, a subclinical hematoma). They found that only 4% of the core biopsies were not diagnostic in contrast to 26% for fine needle aspiration cytology.

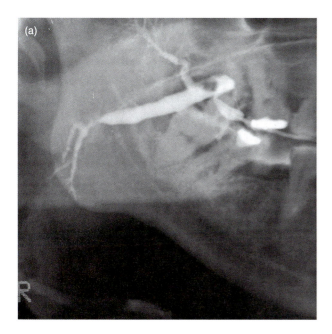
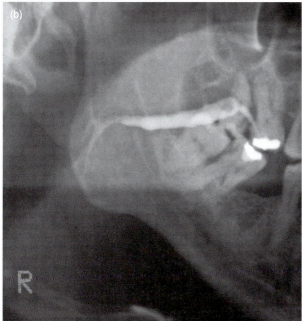

Figure 13.7. Sialogram showing a filling defect and substantial duct dilation, which remain unchanged after emptying. (a) Sialogram displays a filling defect within the main parotid duct. Both the main and accessory ducts of the parotid display marked dilation. Little of the normal arboreal pattern of the parotid gland is obvious. (b) Only the duct to the accessory lobe of the parotid has cleared, indicating that it has normally functioning acini.

Breeze et al. using core biopsies acquired under US guidance,[29] cautioned that because the samples are usually taken from the center of the lesion, the capsule is usually absent. This may lead to a wrong diagnosis, because infiltration of the capsule in an otherwise histopathological benign lesion would have been missed.

Palpation and MRI were superior to US for prediction of a tumor's location according to de Ru et al.[26] It is important to add at this point that the quality of US and its interpretation is very dependent upon the excellence of the equipment and the technical and interpretative abilities of the operator. Furthermore, the use of US guided needle biopsies by surgeons with or without input by the radiologist is likely to continue to grow because it is more likely to be more readily available than an MRI, which at best gives the predictions as to whether a lesion is benign or malignant.[30]

The literature is mixed in its reporting of the relative effectiveness of fine needle biopsies and MRI. Generally both are similarly predictive of benign and malignant lesions. Paris et al. reported that both together reduced false negatives.[31] US-guided needle biopsy was highly specific for malignancy,[32] enabling more reliable preoperative patient counseling and reduced pathological surprises at operation and after the surgical specimen had been acquired. Inohara et al. added that MRI should be reserved until the needle biopsy "shows the indication for surgical intervention."[30]

PLEOMORPHIC SALIVARY ADENOMA

Pleomorphic salivary adenoma (PSA), also more simply called *pleomorphic adenoma*, is the most common salivary gland neoplasm. Although it is benign, its capsule is not always intact and it has satellite micronodules. Early diagnosis and treatment is important because it may undergo malignant transformation, particularly if it has persisted for over 10 years.[33]

Although formerly large PSAs were detected as filling defects in sialograms in modern practice, it may be best to first evaluate it by US, if available, and more definitively by MRI (see Figures 6.2–4, 11 and 17.17).

Although the superficial parotid and other salivary glands are readily accessible to investigation by US and needle biopsy, this is less true for the deep extension of the parotid. Brunese et al. observed that multiphasic CT with an 8-minute acquisition permitted differentiation of all PSA from Warthin's tumors and malignant salivary gland neoplasms.[33] The PSA is further discussed in Chapter 17.

References

1. Shanley JD. The resurgence of mumps in young adults and adolescents. *Cleve Clin J Med* 2007;74:42–44,47, 48.
2. Senanayake SN. Mumps: a resurgent disease with protean manifestations. *Med J Aust* 2008;189:456–459.
3. Watson-Creed G, Saunders A, Scott J, Lowe L, Pettipas J, Hatchette TF. Two successive outbreaks of mumps in Nova Scotia among vaccinated adolescents and young adults. *CMAJ* 2006;175:483–488.
4. Gritzmann N, Rettenbacher T, Hollerweger A, Macheiner P, Hübner E. Sonography of the salivary glands. *Eur Radiol* 2003;13:964–975.
5. Marsot-Dupuch K, Quillard J, Meyohas MC. Head and neck lesions in the immunocompromised host. *Eur Radiol* 2004;3:E155–167.
6. Ellis GL. Lymphoid lesions of salivary glands: malignant and benign. *Med Oral Patol Oral Cir Buccal* 2007; 12:E479–485.
7. Poul JH, Brown JE, Davies J. Retrospective study of the effectiveness of high resolution ultrasound compared with sialography in the diagnosis of Sjogren's syndrome. *Dentomaxillofac Radiol* 2008;37:392–397.
8. Kinoshita T, Ishii K, Naganuma H, Okitsu T. MR imaging findings of parotid tumors with pathologic diagnostic clues: a pictorial essay. *Clin Imaging* 2004;28: 93–101.
9. Morimoto Y, Habu M, Tomoyose T, Ono K, Tanaka T, Yoshioka I, Tominaga K, Yamashita Y, Ansai T, Kito S, Okabe S, Takahashi T, Takehara T, Fukuda J, Inenaga K, Ohba T. Dynamic magnetic resonance sialography as a new diagnostic technique for patients with Sjogren's syndrome. *Oral Dis* 2006;12:408–414.
10. Morimoto Y, Ono K, Tanaka T, Kito S, Inoue H, Shinohara Y, Yokota M, Inenaga K, Ohba T. The functional evaluation of salivary glands using dynamic MR sialography following citric acid stimulation: a preliminary study. *Oral Surg Oral Med Oral Pathol Oral Radiol Endod* 2005;100:357–364.
11. Ono K, Morimoto Y, Inoue H, Masuda W, Tanaka T, Inenaga K. Relationship of the unstimulated whole saliva flow rate and salivary gland size estimated by magnetic resonance image in healthy young humans. *Arch Oral Biol* 2006;51:345–349.
12. Howlett DC. High resolution ultrasound assessment of the parotid gland. *Brit J Radiol* 2003;76:271–277.
13. Gritzmann N, Rettenbacher T, Hollerweger A, Macheiner P, Hübner E. Sonography of the salivary glands. *Eur Radiol* 2003;13:964–975.

14. Wang S, Marchal F, Zou Z, Zhou J, Qi S. Classification and management of chronic sialadenitis of the parotid gland. *J Oral Rehabil* 2009;36:2–8.
15. Choi JW, Lee SS, Huh KH, Yi WJ, Heo MS, Choi SC. The relationship between sialographic images and clinical symptoms of inflammatory parotid gland diseases. *Oral Surg Oral Med Oral Pathol Oral Radiol Endod* 2009;107:e49–56.
16. Marchal F, Chossegros C, Faure F, Delas B, Bizeau A, Mortensen B, Schaitkin B, Buchwald C, Cenjor C, Yu C, Campisi D, Eisele D, Greger D, Trikeriotis D, Pabst G, Kolenda J, Hagemann M, Tarabichi M, Guntinas-Lichius O, Homoe P, Carrau R, Irvine R, Studer R, Wang S, Fischer U, Van der Poorten V, Saban Y, Barki G. Salivary stones and stenosis. A comprehensive classification. *Rev Stomatol Chir Maxillofac* 2008;109:233–236.
17. Ngu RK, Brown JE, Whaites EJ, Drage NA, Ng SY, Makdissi J. Salivary duct strictures: nature and incidence in benign salivary obstruction. *Dentomaxillofac Radiol* 2007;36:63–67.
18. Iro H, Zenk J, Escudier MP, Nahlieli O, Capaccio P, Katz P, Brown J, McGurk M. Outcome of minimally invasive management of salivary calculi in 4,691 patients. *Laryngoscope* 2009;119:263–268.
19. McGurk M, Escudier MP, Thomas BL, Brown JE. A revolution in the management of obstructive salivary gland disease. *Dent Update* 2006;33:28–30,33–36. Erratum in *Dent Update* 2006;33:83.
20. Nahlieli O, Shacham R, Yoffe B, Eliav E. Diagnosis and treatment of strictures and kinks in salivary gland ducts. *J Oral Maxillofac Surg* 2001;59:484–490; discussion, 490–492.
21. Drage NA, Brown JE, Escudier MP, Wilson RF, McGurk M. Balloon dilatation of salivary duct strictures: report on 36 treated glands. *Cardiovasc Intervent Radiol* 2002(Sep-Oct);25:356–359.
22. Jung SM, Hao SP. Warthin's tumor with multiple granulomas: a clinicopathologic study of six cases. *Diagn Cytopathol* 2006;34:564–567.
23. Klussmann JP, Wittekindt C, Florian Preuss S, Al Attab A, Schroeder U, Guntinas-Lichius O. High risk for bilateral Warthin tumor in heavy smokers—review of 185 cases. *Acta Otolaryngol* 2006;126:1213–1217.
24. Lee YY, Wong KT, King AD, Ahuja AT. Imaging of salivary gland tumours. *Eur J Radiol* 2008;66:419–436.
25. Alphs HH, Eisele DW, Westra WH. The role of fine needle aspiration in the evaluation of parotid masses. *Curr Opin Otolaryngol Head Neck Surg* 2006;14:62–66.
26. de Ru JA, van Leeuwen MS, van Benthem PP, Velthuis BK, Sie-Go DM, Hordijk GJ. Do magnetic resonance imaging and ultrasound add anything to the preoperative workup of parotid gland tumors? *J Oral Maxillofac Surg* 2007;65:945–952.
27. Pratap R, Qayyum A, Ahmed N, Jani P, Berman LH. Ultrasound-guided core needle biopsy of parotid gland swellings. *J Laryngol Otol* 2008;122:1–4.
28. Pfeiffer J, Kayser G, Ridder GJ. Diagnostic effectiveness of sonography-assisted cutting needle biopsy in uncommon cervicofacial lesions. *Oral Surg Oral Med Oral Pathol Oral Radiol Endod* 2009;107:173–179.
29. Breeze J, Andi A, Williams MD, Howlett DC. The use of fine needle core biopsy under ultrasound guidance in the diagnosis of a parotid mass. *Br J Oral Maxillofac Surg* 2009;47:78–79.
30. Inohara H, Akahani S, Yamamoto Y, Hattori K, Tomiyama Y, Tomita Y, Aozasa K, Kubo T. The role of fine-needle aspiration cytology and magnetic resonance imaging in the management of parotid mass lesions. *Acta Otolaryngol* 2008;128:1152–1158.
31. Paris J, Facon F, Pascal T, Chrestian MA, Moulin G, Zanaret M. Preoperative diagnostic values of fine-needle cytology and MRI in parotid gland tumors. *Eur Arch Otorhinolaryngol* 2005;262:27–31.
32. Bajaj Y, Singh S, Cozens N, Sharp J. Critical clinical appraisal of the role of ultrasound guided fine needle aspiration cytology in the management of parotid tumours. *J Laryngol Otol* 2005;119:289–292.
33. Brunese L, Ciccarelli R, Fucili S, Romeo A, Napolitano G, D'Auria V, Collina A, Califano L, Cappabianca S, Sodano A. Pleomorphic adenoma of parotid gland: delayed enhancement on computed tomography. *Dentomaxillofac Radiol* 2008;37:464–469.

Chapter 14
Fractures of the face and jaws

Fractures of the facial skeleton occur due to assault,[1] traffic accidents,[2] and sporting accidents.[3] The last has become increasingly important primarily due to the increased popularity of skiing and snowboarding.[3] Although snowboarding injury cases are twice as likely to produce facial fractures than those caused by skiing, skiing is more likely to result in more than one facial fracture.[3] The traffic accident cases, despite regulations with regard to seat belts, helmets, and child seats, account for 46% of all facial fractures in a recent Brazilian urban report, whereas assault and sports account for 26% and 6%, respectively.[2]

Cross-sectional imaging is frequently required for fractures of the skeleton of the midface. These fractures are generally complex (Figures 14.1–14.3), reflecting the midface's complex anatomy, which contains the eyes and nose. The classical fracture patterns of fractures of the middle third of the face are the LeFort I, II, and III, (Figure 14.1) and the zygomatic (also called *malar*) fractures (Figure 14.2).

Of the fractures caused by violence reported by Salonen et al., the most common was the fracture of the nasal bones (35%). LeFort and zygomatic fractures account for 8% and 18%, respectively.[1] Twenty-six of their 48 cases of LeFort fractures were asymmetrical. Six of the 22 symmetrical cases each displayed bilateral LeFort I, II, and III fractures. Figure 14.3 displays a 3-D reconstruction of a violent assault case displaying all three LeFort fractures, fractures of the nasoethmoid complex and the mandible.

Fractures of the mandible are usually simple (Figures 1.12, 14.3–14.5) because of the mandible's shape. The mandible may be described as a bent long bone with a synovial joint at each end. Nevertheless, complex fractures of the mandible can also occur (Figure 14.6.). Furthermore, substantial swelling, which makes the clinical evaluation of the patient difficult, can quickly follow severe facial trauma. Thai et al. reported that the clinical examination was accurate in only two-thirds of mandibular fractures,[4] thus emphasizing the importance of radiology to the assessment of the fractured mandible.

The panoramic radiograph had been held by some to be an "unofficial gold standard" for imaging fractures of the mandible.[5] Roth et al. reviewed both the panoramic radiographic and HCT images of 217 patients.[6] They found that HCT identified more fractures, particularly those of the angle, ramus, and condylar neck, than the panoramic images.[6] Preda et al. reported that many of their patients with complex maxillofacial fractures benefited from the HCT's short scan time because of their multiple trauma and possible damaged organs that were not yet fully stabilized.[7]

Dean et al. reported that although the incidence of life-threatening hemorrhage of facial fractures is low (0.33%), when it does occur the risk of death is high.[8] Nearly a quarter of their 19 cases died after receiving a mean of 9.5 units of red blood cells. Although other injuries contributed to the deaths in most cases, they reported no deaths in the cases treated by arterial ligation.[8] Nevertheless, due to ligation's higher morbidity, Dean et al. advise the use of radiological embolization (by angiography) as a better alternative. Their protocol (see their Figure 10) clearly indicates that radiologists should perform this.[8]

The traumatic force that caused the facial injury is more likely to cause serious injury to the eye, the cervical spine, and the brain than a severe hemorrhage. Salonen et al. reported that 44% of assault victims with facial fractures had not only sustained multiple noncontiguous facial fractures (likely to result from repeated blows), but 26% had orbital fractures and 6% had base-of-the-skull fractures.[1] Although a suggestion that HCT could be considered routinely for facial injuries would be in conflict with the need for clear clinical indications for radiography, the decision to prescribe CT may

---

Oral and Maxillofacial Radiology: A Diagnostic Approach,
David MacDonald. © 2011 David MacDonald

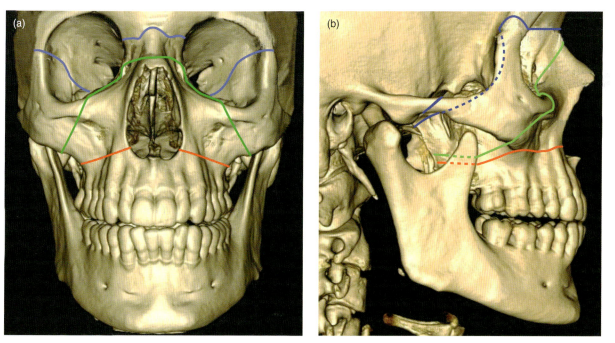

Figure 14.1. The LeFort fractures of the midface. Red—LeFort I. Blue—LeFort II. Green—LeFort III (fracture through the base of the skull and therefore a neurosurgical referral). Acknowledgment: Bruce McCaughey, Senior Photography/Audio-Visual technician; Faculty of Dentistry; University of British Columbia.

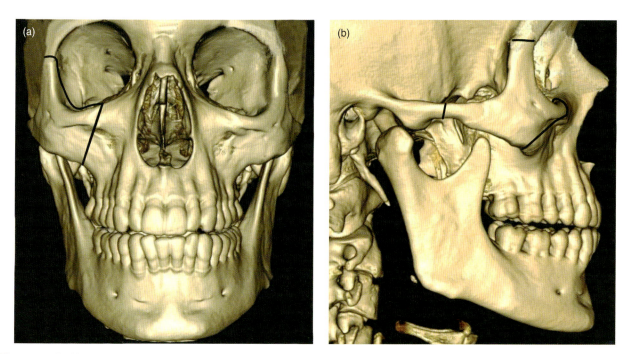

Figure 14.2. Zygomatic fracture, also called malar fracture. Acknowledgment: Bruce McCaughey, Senior Photography/Audio-Visual technician; Faculty of Dentistry; University of British Columbia.

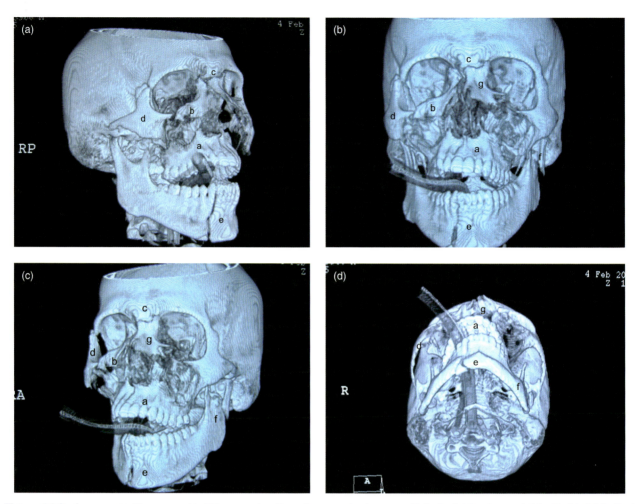

Figure 14.3. A computed tomographic 3D reconstruction of a case features multiple facial fractures. (a) LeFort I; (b) LeFort II; (c) LeFort III; (d) zygomatic fracture; (e) fractured mandible—midline; (f) fractured mandible—condyle; (g) Nasal complex fracture. **Note:** Left zygomatic bone articulations with adjacent frontal and temporal bones are still intact. Acknowledgment: Dr. Ian Matthew; Faculty of Dentistry; University of British Columbia. Figure 14.3c reprinted with permission from MacDonald-Jankowski DS, Li TK. Computed tomography for oral and maxillofacial Surgeons. Part 1: Spiral computed tomography. *Asian Journal of Oral Maxillofacial Surgery* 2006;18:68–77.

have been made to address the imaging needs of other head and neck specialists, who may have a more pressing interest in the same patient. Holmgren et al. found that 84% of those head trauma cases investigated by HCT did not require a further HCT to investigate concurrent facial trauma.[9]

Jamal et al. suggest that an ophthalmic examination should be undertaken preoperatively of all zygomaxillary fractures, because 10% of these fractures are associated with major or blinding injuries to the eyeball itself and a further 6% with traumatic optic neuropathy.[10]

Lee et al. reported that fractures of the orbital floor and medial wall each accounted for a third of all orbital fractures.[11] In addition to the blow-out fracture in which the orbital contents expand into adjacent cavities, there are also blow-in fractures.[11] Enophthalmos is indicative of a severe orbital blow-out.[12] HCT permits better appreciation of injury to the optic nerve at its passage through the optic canal at the time of first presentation.[11] Early use of CT in the diagnosis of nasoethmoid orbital fractures in conjunction with aggressive treatment will optimize the success of the outcome, minimizing later postoperative deformities.[13]

Cervical spine fractures accompany 6.7% of all maxillofacial fractures in the United States of America.[14] Elahi et al. reported that the 6 and 7 cervical vertebrae were affected in 41% of their

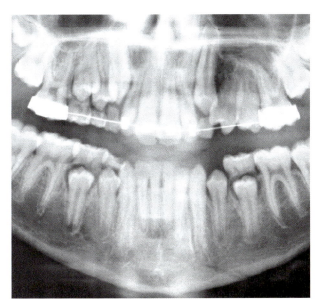

Figure 14.4. Panoramic radiograph displaying an undisplaced fracture through the paramedian mandible of this 12-year-old. The two fracture lines indicate fracture of the buccal and lingual cortex rather than comminution.

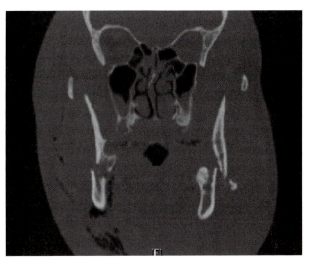

Figure 14.6. Computed tomography of comminuted fracture of the mandible. This fracture is accompanied by significant facial and neck swelling. Figure courtesy of Dr. Martin Aidelbaum, Faculty of Dentistry, University of British Columbia.

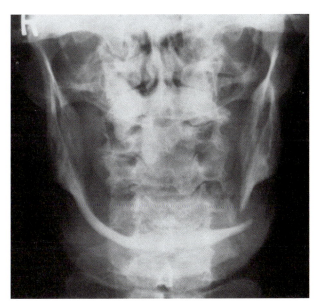

Figure 14.5. Posterioanterior projection of the mandible showing a fracture on the left. The mandible is pencil-thin.

cases of cervical spine injury.[15] Therefore all seven cervical vertebrae should be included in any imaging protocol of the cervical spine. Mithani et al. report that fractures of the upper face are associated with injury to the middle and lower cervical spine, and unilateral mandibular fractures are associated with injury to the upper cervical spine.[16]

Fractures of the middle face are associated with base-of-skull fractures and intracranial injury.[16] Bilateral middle facial fractures are more associated with death.[14]

The use of CBCT technology (see Chapter 5) has been successfully transferred to the operating room or theater by incorporation of a C-arm. Pohlenz et al. reported that two-thirds of their 177 patients imaged with such a unit were facial and mandibular fracture cases.[17] Postoperatively this technology allowed immediate revision of the surgery. Although at the time of writing (2010) there appears to be no other objective reporting of the application of CBCT to the diagnosis and management of facial fractures, Shintaku et al., using HCT literature on facial fractures, have endeavored to illustrate how CBCT may be applied to facial fractures.[18]

References

1. Salonen EM, Koivikko MP, Koskinen SK. Violence-related facial trauma: analysis of multidetector computed tomography findings of 727 patients. *Dentomaxillofac Radiol* 2010;39:107–113.
2. Leles JL, dos Santos EJ, Jorge FD, da Silva ET, Leles CR. Risk factors for maxillofacial injuries in a Brazilian emergency hospital sample. *J Appl Oral Sci* 2010;18: 23–29.

3. Tuli T, Haechl O, Berger N, Laimer K, Jank S, Kloss F, Brandstätter A, Gassner R. Facial trauma: how dangerous are skiing and snowboarding? *J Oral Maxillofac Surg* 2010;68:293–299.
4. Thai KN, Hummel RP III, Kitzmiller WJ, Luchette FA. The role of computed tomographic scanning in the management of facial trauma. *J Trauma* 1997;43:214–217; discussion 217–218.
5. Wilson IF, Lokeh A, Benjamin CI, Hilger PA, Hamlar DD, Ondrey FG, Tashjian JH, Thomas W, Schubert W. Prospective comparison of panoramic tomography (zonography) and helical computed tomography in the diagnosis and operative management of mandibular fractures. *Plast Reconstr Surg* 2001;107:1369–1375.
6. Roth FS, Kokoska MS, Awwad EE, Martin DS, Olson OT, Holler LH, Hollenbeak CS. The identification of mandible fractures by helical computed tomography and panorex tomography. *J Craniofac Surg* 2005;16:394–399.
7. Preda L, La Fianza A, Di Maggio EM, Dore K, Schifino MR, Mevio F, Campani R. Complex maxillofacial trauma: diagnostic contribution of multiplanar and tridimensional spiral CT imaging [In Italian]. *Radiol Med (Torino)* 1998;96:178–184.
8. Dean NR, Ledgard JP, Katsaros J. Massive hemorrhage in facial fracture patients: definition, incidence, and management. *Plast Reconstr Surg* 2009;123:680–690.
9. Holmgren EP, Dierks EJ, Homer LD, Potter BE. Facial computed tomography use in trauma patients who require a head computed tomogram. *J Oral Maxillofac Surg* 2004;62:913–918.
10. Jamal BT, Pfahler SM, Lane KA, Bilyk JR, Pribitkin EA, Diecidue RJ, Taub DI. Ophthalmic injuries in patients with zygomaticomaxillary complex fractures requiring surgical repair. *J Oral Maxillofac Surg* 2009;67:986–989.
11. Lee HJ, Jilani M, Frohman L, Baker S. CT of orbital trauma. *Emerg Radiol* 2004;10:168–172.
12. Kelley P, Hopper R, Gruss J. Evaluation and treatment of zygomatic fractures. *Plast Reconstr Surg* 2007;120:5S–15S.
13. Sargent LA. Nasoethmoid orbital fractures: diagnosis and treatment. *Plast Reconstr Surg* 2007;120:16S–31S.
14. Roccia F, Cassarino E, Boccaletti R, Stura G. Cervical spine fractures associated with maxillofacial trauma: an 11-year review. *J Craniofac Surg* 2007;18:1259–1263.
15. Elahi MM, Brar MS, Ahmed N, Howley DB, Nishtar S, Mahoney JL. Cervical spine injury in association with craniomaxillofacial fractures. *Plast Reconstr Surg* 2008;121:201–208.
16. Mithani SK, St-Hilaire H, Brooke BS, Smith IM, Bluebond-Langner R, Rodriguez ED. Predictable patterns of intracranial and cervical spine injury in craniomaxillofacial trauma: analysis of 4786 patients. *Plast Reconstr Surg* 2009;123:1293–1301.
17. Pohlenz P, Blessmann M, Blake F, Heinrich S, Schmelzle R, Heiland M. Clinical indications and perspectives for intraoperative cone-beam computed tomography in oral and maxillofacial surgery. *Oral Surg Oral Med Oral Pathol Oral Radiol Endod* 2007;103:412–417.
18. Shintaku WH, Venturin JS, Azevedo B, Noujeim M. Applications of cone-beam computed tomography in fractures of the maxillofacial complex. *Dent Traumatol* 2009;25:358–366.

Chapter 15
Osseointegrated implants

Introduction

Prosthodontics has become one of the more exciting areas in dentistry due to the development of two disruptive technologies: *osseointegrated implants* and *cone-beam computed tomography* (CBCT). Although the American Academy of Oral and Maxillofacial Radiology (AAOMR) recommends that some form of cross-sectional imaging be used for preimplant assessment[1] not all jurisdictions demand this; the European Association for Osseointegration (EAO) does not.[2] It declares that the "choice of technique is based on the lowest dose giving the required diagnostic information." Nevertheless, this report did not include CBCT compelling the *European Academy of Dental and Maxillofacial Radiology* (EADMFR) to issue its "Basic principles for the use of dental cone beam CT."[3]

In many communities, the osseointegrated implant, which will now be simply referred to as the "implant," has become the treatment of choice for replacement of one or more missing teeth. Current implant placement can be successful when the cases have been properly selected and the procedure properly executed. A recent systematic review reported survival of implants placed in the completely edentulous mandibular and maxillary arches.[4] The results are summarized in Table 15.1.

Although training for implants is clearly a postgraduate activity in most jurisdictions, this training includes an in-depth understanding in the field of oral and maxillofacial radiology, dentoalveolar surgery, and the principles of modern prosthodontics. Because different specialties are involved, a team approach is commonly required to optimize planning.

Success of the dental implant depends largely on the thorough preoperative assessment of the patient's medical and oral condition. With regard to the latter, the oral and maxillofacial radiologist's role is superlative. This requires assessment of the alveolar bed on and around the prospective implant. The anatomy, quantity, and quality of alveolar bone determines the suitability for implant placement and the appropriate size, length, and position of the implant (Figures 15.1, 15.2). All implant planning is made with some form of radiographic investigation. The range of imaging methods for implant assessment varies considerably. It can be anything from simple intraoral *periapical radiography* to *helical computed tomography* (HCT) and CBCT (Table 15.2). Other imaging methods such as *magnetic resonance imaging* (MRI) and *ultrasonography* (US) have been explored with limited success and will not be considered further.

Because conventional intraoral and panoramic radiographs can provide only anatomical information in the 2-dimensional (2-D) mesiodistal plane, some form of tomographic or 3-dimensional (3-D) imaging is needed to demonstrate anatomy in the buccolingual plane. This enables measurement of the width of the alveolar bed. These images can be obtained in an analog (film) or in a digital format. Each has its own measuring and calibration system to ensure relatively accurate measurement. Projection magnification, which varies widely, particularly for panoramic radiography, has to be accounted for.

To maximize the chance for successful osseointegration, implants have to be placed accurately in the alveolar bed in a 3-D space. The objective is to have an adequate area of contact between the implant and sound alveolar bone, while avoiding anatomical hazards such the as mandibular canal, incisive canal, maxillary antrum, and submandibular fossa. Although Vazquez et al. have shown that panoramic radiography can be considered a safe preimplant assessment modality for routine posterior mandibular implants "if a safety margin of at least 2 mm above the mandibular canal is

---

Oral and Maxillofacial Radiology: A Diagnostic Approach, David MacDonald. © 2011 David MacDonald

Table 15.1. Summary of the survival of implants from the systematic review of implants in completely edentulous arches

	Mandible		Maxilla	
	Fixed	Removable	Fixed	Removable
5-year survival	97%	96%	88%	
10-year survival	91%	95%	81%	
15-year survival	82%		70%	

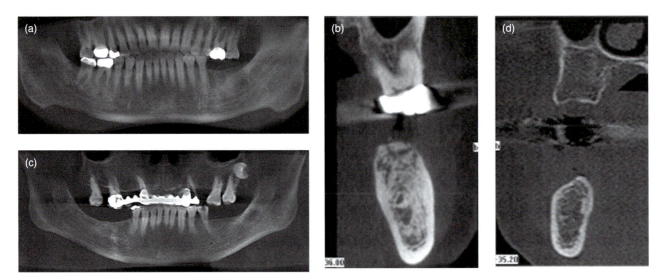

Figure 15.1. The panoramic and cross-sectional reconstructions of cone-beam computed tomography of two cases (a,b and c,d) display edentulous alveolar processes with good bone height and quality for both jaws.

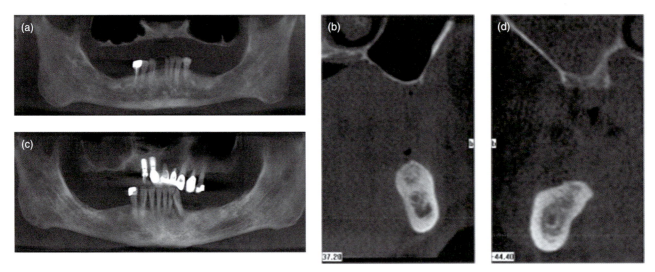

Figure 15.2. The panoramic and cross-sectional reconstructions of cone-beam computed tomography of two cases (a,b and c,d) display edentulous alveolar processes with reduced bone height and quality for both jaws. Note that for both cases the floor of the maxillary antrum is separated from the oral mucosa by only a cortex, This is particularly apparent in b.

Table 15.2. A comparison of the main features of 6 imaging modalities that can be employed in preimplant assessment

Imaging Modality	Convenience, Footprint, and Price	Cross-section (Y/N)	Spatial Resolution	Magnification and Distortion	Patient Positioning	Software (Y/N)	EPR (Y/N)
Intraoral radiography	+++++	N	++++	None if paralleling technique is used	Upright	N	Y
Panoramic radiography	+++++	N	+++	++++	Upright	N	Y
Linear tomography	+++	Y	++	++++	Upright	N	N
Complex motion tomography	+++	Y	+++	+++	Upright	N	N
Helical computed tomography	+	Y	+	+	Supine	Y	?
Cone-beam computed tomography	++	Y	+++	+	Upright	Y	Y

Key: EPR, electronic patient record; N, no; Y, yes.
The number of + (+ to +++++) indicates an increased facilitation of quality of a feature.

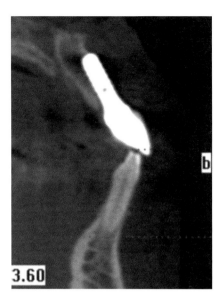

Figure 15.3. Cone-beam computed tomography (CBCT) displaying perforation (indeed, replacement) of the labial cortex by the implant. This implant has failed.

respected,"[5] buccal or lingual cortical plate perforation can occur easily without adequate appreciation of the buccolingual dimension (Figure 15.3). Therefore, this objective can be achieved only when the anatomy of the alveolar bed is clearly visualized in a multiplanar or a 3-D format. Many anatomical hazards, mandibular and incisive canals, foramina, concavity of cortical plate, and thin alveolar ridges cannot be demonstrated with *conventional radiography*; these will be *considered* at the end of this chapter. Therefore, cross-sectional imaging is required. The cross-sectional imaging modalities that have been applied to preimplant assessment are *linear tomography, complex-motion tomography*, HCT, and CBCT.

LINEAR TOMOGRAPHY

This type of tomography is not recommended by the AAOMR for implant imaging.[1] Linear tomography has been reported to be subject to distortion.[6] Some dental panoramic machines include linear tomography. Linear tomography demonstrates the anatomy of a section of the jaws in a flat buccolingual plane. Typical slice thickness is about 3–6 mm and each examination series would consist of three to four tomographic cuts covering a one- to two-teeth region. There is typically a magnification of 1.4. Special calibrated rulers or magnified overlay transparence are used to measure the exact dimension of the jawbone on the film.

The general spatial resolution of the images is poor (Figure 15.4), but it still provides an image of sufficient quality to show obvious anatomical features, such as the maxillary sinus and the submandibular fossa. When no other superior imaging method is available, linear tomography is superior to conventional 2-D imaging alone.

The panoramic radiographic machine has to be adjusted to perform linear tomography with a custom-made head positioning device such as a jaw registration block. The procedure is complicated and technique-sensitive because the imaging plane is determined by the exact head positioning. Unless very experienced, most clinicians find this examination troublesome. The inferior image quality and complicated procedure have limited its utility.

COMPLEX MOTION TOMOGRAPHY

Complex motion tomography includes spiral and hypocycloidal tomography. Bou Serhal et al. demonstrated that their spiral conventional tomographic unit displayed very good localization of the mandibular canal "if the clinician takes into consideration the maximal overestimation encountered with this technique and if precautions are taken during the radiographic procedure, e.g., positioning of the head and immobility during scanning."[7] Hypocycloidal conventional tomography is reported to create the best blurring of objects outside the focal trough. Spiral conventional tomography has been demonstrated to provide very good display of the mandibular canal. A more recent report indicated that spiral tomography significantly changed presurgical treatment plans.[8]

Complex motion tomography was most frequently used for preimplant assessment, prior to the advent of CBCT. The Scanora (Figure 15.5) and Cranex both use spiral tomography, whereas the CommCat uses hypocycloidal motion. Their complex motion produces very even blurring of objects outside the focal trough, thus enhancing the clarity of the structure of interest. The Scanora uses a rotating anode X-ray tube, which has a much higher loading factor and a smaller focal spot. The quality of these sectional images is significantly better than the linear ones. With a larger swing angle, the slice thickness can be reduced to 2 mm. A typical examination series would cover a

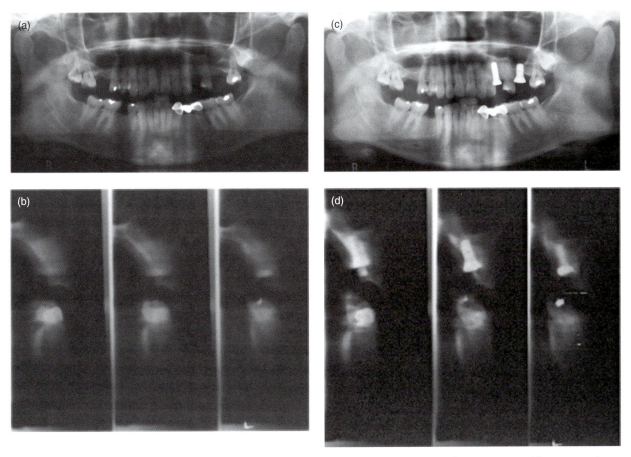

Figure 15.4. Linear tomography of implant site preimplant (a,b) and postimplant (c,d) assessment. Views a and c are panoramic radiographs. Views b and d are linear tomographs. The spatial resolution is poor.

one- to two-teeth region. The magnification factor for Scanora in buccolingual tomography is 1.7. A typical series of projections for a one- to two-teeth region takes around 2 minutes. The spiral movement of the film cassette holder in front of the patient's face can induce dizziness in sensitive patients. Patients are advised to close their eyes during exposure because any patient movement during exposure would cause unacceptable movement artifacts.

Both linear and complex-motion (especially spiral) tomography can be acquired using an analog (film) or a digital system. The radiation dose for one or two sections is similar to one panoramic radiograph. Multiple implant sites require multiple exposures. A full mouth assessment would take up to 15 projections and would be very time consuming. Therefore, complex motion tomography is best used for planning of relatively simple implant cases that need only one or two implants.

HELICAL COMPUTED TOMOGRAPHY

HCT requires a dedicated high amperage power line in order to sustain the high tube current and long exposure. Special cooling of the CT room is needed to remove the heat generated by the tube head. Therefore, it is best situated in a hospital. The basics of HCT are discussed in Chapter 4.

The major advancements in HCT are in large part due to improved computer power and upgraded image sensors. The scan time can be very rapid. The scan for the maxillofacial region for implant planning in a 64-slice machine is around 5–10 seconds. The reconstruction time is only a couple of seconds; by the time the patient has left the HCT

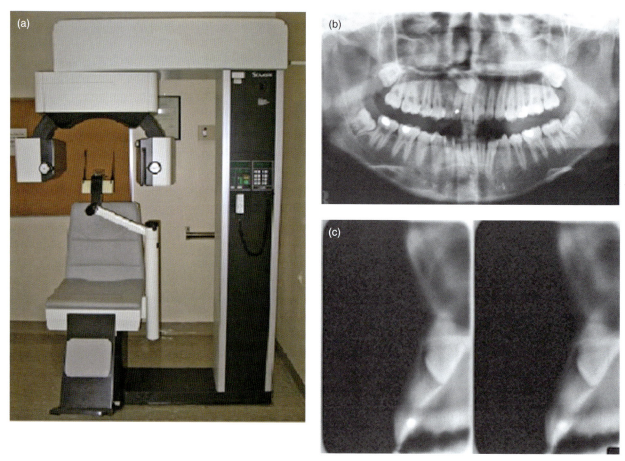

Figure 15.5. Scanora unit (a). The panoramic radiograph (b) and two transverse sections (c) of an unerupted maxillary tooth.

table, the images are ready for viewing. The workstation of the machine, besides storing patient files, is also loaded with a wide range of image viewing software, such as multiplanar reformatting, 3D reconstruction using surface shade display, or volume rendering methods. Most of these are introduced in Chapter 4.

The original viewing software for implant planning is called Dentascan (General Electric, Pittsfield, USA), which is still extensively used. This is a form of multiplanar reformatting, which reformats sectional images perpendicular to the dental arches. The results are high-quality diagnostic images that are calibrated to life size and printed on films. The clinician can make measurements directly on films using a standard metric ruler to obtain accurate dimensional measurement of the alveolar bed and jawbones. Because the image is set to be perpendicular to the occlusal plane, measurement in an oblique plane is not possible.

All image data from HCT is stored in *Digital Imaging and Communications in Medicine* (DICOM) format. It was created by the *National Electrical Manufacturers Association* (NEMA, USA) to aid the distribution and viewing of medical images, such as CT, MRI, and US. This image file format is not compatible with ordinary personal computers. Special DICOM viewer software has to be included to view the files.

Some clinicians still prefer to have the CT images printed on analog format (films) for easy viewing. However, images once printed cannot be adjusted for window width and window level (see chapter 4). There is a loss of information in the transfer of data from the scanning computer to the printer. Unless frequently calibrated, the radiographic printer can lose its dimensional accuracy after repeated use. In time, the volume of CT films can grow considerably, complicating storage and easy file retrieval. Digital diagnostic images are best viewed digitally, that is using a computer monitor. Clinic layout must be changed to facilitate easy assess to a large, high-quality computer monitor both for dentist and patient viewing (see Chapter 2.).

Modern HCT provides high-quality sectional images with good spatial resolution. Its benefit is somewhat offset by relatively high radiation dose and high equipment costs. The typical radiation dose for an HCT is at least 100 times that of similar examination by linear or spiral tomography.[2] Considering that implant treatment is for rehabilitation of dentition rather than for detecting life-threatening pathology, the high radiation dose imparted by HCT is rightly a concern for the clinicians and their patients.

Most HCT facilities are operated by medical radiologists and radiographers and give priority to medical patients. Some dental practitioners may find that the medical CT center has difficulty understanding the special needs of dental implant surgery. Unless there is good communication and understanding between the medical radiologist and referring dental practitioner, the full benefit of HCT may not be realized. For the best results, there should be zero gantry tilt and the patient's occlusal plane should be vertical. It is difficult to assess and align the occlusal plane in an HCT supine setting. The head holder is rigid and not adjustable.

CONE-BEAM COMPUTED TOMOGRAPHY

CBCT scanners have been available for craniofacial imaging since 1999 in Europe and 2001 in the United States. At present, all major manufacturers of dental X-ray equipment have CBCT. Some machines are cone-beam only equipment (see Figure 5.5a); others are add-on extras onto standard panoramic radiographic machines (see Figure 5.5b). The hybrid machine offers the choice of a standard panoramic radiographic and a cone-beam volumetric scan. The basics of CBCT have been addressed in Chapter 5.

The detector on CBCT can be an image intensifier or amorphous silicon *flat panel detector* (FPD). FPD produces high spatial resolution and contrast in hard tissue. In preimplant assessment, it is more important to exam the hard-tissue anatomy. Therefore, it is better to use CBCT with an FPD. The size of the detector's *field of view* (FOV) also determines the volume of tissue included in each scan. The range is from 37 × 50 mm to 160 × 220 mm. This covers a region of from one or two teeth to the whole skull to suit individual clinical requirements. For complete implant planning, it is advisable to have a machine that covers at least one whole jaw and the occlusal table of the opposing arch. This enables implant planning to be done with the opposing occlusion in mind. The amount of occlusal clearance may determine the site for implant placement.

Unlike HCT, CBCT was designed and developed for dental practice. CBCT is compared with HCT in Table 15.3.

The scan time for CBCT is between 10 to 40 seconds depending on the required spatial resolution. The radiation dose can be less than linear or spiral tomography for multiple sites in the same arch.

The machine itself is relatively simple. Only one center of rotation is needed for CBCT compared to the complex rotational geometry required for the modern panoramic radiographic unit. CBCT basically consists of an X-ray generator on one end and FPD on the other end for a simple rotational movement. The need for repair and maintenance is minimized improving its reliability. CBCT usually operates at a higher kV than the panoramic radiographic unit. The range is from 75 to 120 kV. Some machines use pulsed X-ray for further dose reduction.

A major problem in any CT scanning of the jaws is image artifacts caused by metallic restoration in the dentition. Such artifacts in fan-beam systems (see Figure 5.1), as used in HCT, occur only in the axial (horizontal or parallel to the occlusal plane), whereas in cone-beam systems (see Figure 5.1), as used in CBCT, they also occur in the vertical plane.[9] Nevertheless, because the region affected by the beam hardening receives information from other angles, the streaks are shorter than they would be for HCT.[9]

High density alloys in crowns (the atomic number of gold is 79) and large amalgams (the atomic number of mercury is 80 and of silver is 47) significantly attenuate the primary X-ray beam.[9] This produces opaque radiating lines around the restorations or apparent radiolucencies on the adjacent teeth due to extensive beam hardening (see Figure 4.9). The patient should be positioned so that the occlusal plane is parallel to the central ray. This would limit the streak artifact to the coronal part of the dentition. The more important alveolar bone and adjacent vital structures can still

Table 15.3. Cone-beam computed tomography versus helical computed tomography

Advantages

1. Cost of CBCT is approximately 4–8 times less than HCT.
2. Because the CBCT is substantially lighter; no floor strengthening is required.
3. The CBCT's footprint is smaller.
4. CBCTs have better spatial resolution (i.e., smaller pixels) than the best HCT; 0.1 to 0.4 mm voxel size, respectively. The spatial resolution of CBCT is often higher than practically needed for implants, which is usually 0.2 mm voxel size.
5. CBCT, unlike HCT, uses isotropic cuberilles; therefore, the spatial resolution is just as good in the Z (long) axis as it is in the XY (axial) plane. See Chapters 4 and 5.
6. No special electrical requirements are needed for CBCT.
7. Unlike HCT, the room does not need to be cooled for CBCT.
8. CBCT is very easy to operate and to maintain; little technician training is required.
9. Radiation dose is considerably less than with a medical CT. Radiation dose can vary substantially between different CBCT makes.
10. CBCT exposes the patient in the upright position, the same as for a panoramic radiograph, and is associated with good patient tolerance
11. When the use of CBCT units with a field of view (FOV) of 8 cm × 8 cm or less and is confined to the jaws, they need to be read only by a specially trained general dental practitioner or specialist.

Disadvantages

1. Because the contrast resolution for CBCT is only 12 to 14 bits in contrast to 16 to 24 bits for HCT, differences between soft tissues can be appreciated only in the latter.
2. Both because of the preceding point and the fact that the patient is investigated in the upright position in CBCT, intravenous contrast cannot be used for CBCT.
3. CBCT units using a lower kilovoltage may experience spray artifacts from titanium implants. Titanium may cause less artifact with high kilovoltages because of its lower atomic mass (see discussion in text).
4. Related to Advantage Point 11, when the (FOV) of the dataset of HCT is greater than 8 cm × 8 cm and/or includes extragnathic structures, the images need to be reviewed and reported on by a radiologist.

be visualized without artifact. The use of higher kV helps reduce the streak artifact because of better penetrating power. A conclusion of Draenert et al.'s report is that a unit operating at a higher kilovoltage may minimize metallic artifacts.[9] They found that their 4-row *multidetector computed tomography* (MDCT), operating with a kV of 120, accurately reproduced an implant *in vitro* in contrast to the original NewTom (9000) operating at 85 kV. Other possible reasons could be fewer primary rays, software problems, X-ray beam geometry, and filtration. This important issue has not yet been fully addressed at the time of this writing (2010). Extensive bridgework or other restorations using metal can disrupt any CT image (Figure 15.6). In extreme cases, when the patient needs to have bridgework replaced and implant placement, the bridge can be removed prior to exposure by CBCT.

All implants are made of titanium, which has an atomic number of 22. Although it does not cause metallic artifacts when scanned by 120 kV MDCT,[9] Draenert et al. queried whether that would also hold for a 120 kV CBCT, because of other confounding factors already mentioned.[9] This is particularly important for CBCTs taken of a patient who already has implants (Figure 15.7). Recently Schulze et al. reviewed the beam hardening of two CBCT units, *in vitro*.[9] Schulze et al. compared the beam hardening at 80 kV and 110 kV. They found

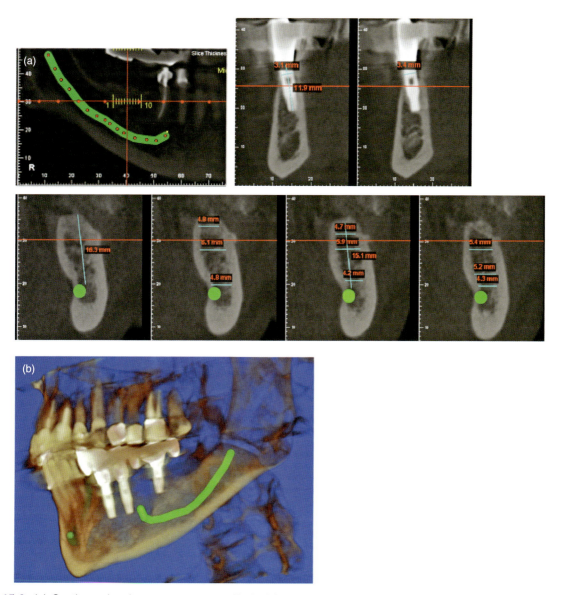

Figure 15.6. (a) Sections showing measurements (Dolphin) made on data from an iCAT scan. (b) 3D reconstruction of the mandible with the mandibular canal indicated by a green line. Figures courtesy of Dr. Babak Chehroudi, Faculty of Dentistry, University of British Columbia.

that the beam hardening was more pronounced for 80 kV.[10]

Currently, increasing numbers of dental practitioners have installed CBCTs in their clinics, mainly because of the unprecedented diagnostic power and convenience provided by CBCT. The use of CBCT for implant planning has become the generally accepted modality for pre- and postoperative assessments (Figure 15.7). Furthermore, CBCT (iCAT) imparts less radiation to the eye lens, all three major salivary glands, and the thyroid than HCT and the Scanora.[11]

Computer-Aided Planning

Because CBCT is routinely prescribed—particularly in North America and Asia—for preimplant assessment, the anatomy of the jaws and adjacent structures, such as the maxillary antrum and nasal

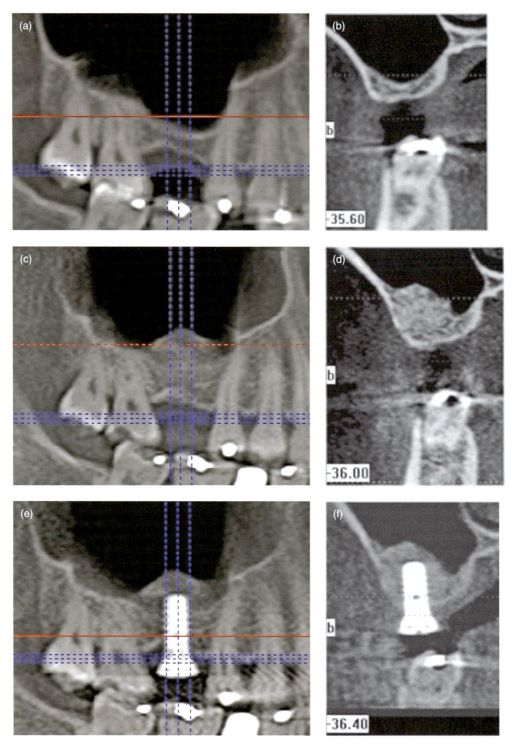

Figure 15.7. Cone-beam computed tomography of a case that has had maxillary antral lift. (a,b) The reduced height of the maxillary alveolus. (c,d) The increased height after maxillary antral lift. (e,f) Implant has osseointegrated. Figure 11.41 displays bilateral maxillary antral lifts. Although there is good osseointegration between the implant and the bone, it is deficient on the implant's distal aspect.

cavity, are demonstrated with a high spatial resolution and viewed in multiplanar reformatting or 3-D, which facilitates computer-aided planning. This also provides a simulation of the preoperative situation and the projected postoperative result in 3-D. This patient to virtual-patient simulation to patient is enormously facilitated by the use of surgical stents (discussed later). The resultant reconstructions (simulations) facilitate clinician-patient communication. Although this kind of reconstruction is made possible only by CT scanning and special computer software, it is still reliant upon the surgeon positioning the implant of the correct size precisely in the sites chosen by the clinician who will provide the definitive restoration. This not only presupposes that a surgical stent is supplied with the prescription to the surgeon, but that the entire dataset has been carefully reviewed to identify anatomical features that may complicate the desired outcome. These anatomical features are not just the mandibular canal, mental foramen, and maxillary antrum, but also other important structures such as accessory mental foramen, mandibular incisive canal, lingual foramen, concavities (particularly on the lingual mandible), floor of the mouth, nasopalatine (incisive) canal, and thin and inclined incisal alveolar ridges.

After scanning, all anatomical data is calculated in a 3-D matrix but stored as a series of axial slices (Figure 15.8). Depending on the resolution and slice thickness, the number of axial slices can range from 30 to 300. Common axial, coronal, and sagittal views are not the best way to demonstrate anatomy in the dentomaxillofacial complex. The maxilla is pyramidal and the mandible is U-shaped; hence they not easily interpreted in simple planar originations. The HCT scan is usually viewed using special software to reformat images in a curved or oblique plane. It is even better to view the jaws with 3-D reconstruction.

In 3-D reconstruction, clinicians can select different thresholds for tissues of different densities. These thresholds set the range of CT number (*Hounsfield unit*, HU) for pixels to be displayed. The reconstructed image can selectively display teeth, bone, or soft tissue. A combination of different tissues with different transparency and color can give very realistic reproduction of the patient's anatomy (see Figure 4.8). This type of multilayer 3-D display is also called volume rendering. Until quite recently, this sort of 3-D volume rendering processing could be done only on a medical CT workstation by very expensive hardware and software.

Some advanced software provides computer-aided diagnosis and planning. Representative software are Simplant (Materialise, Leuven, Belgium), NobelGuide (Nobel Biocare, Zurich, Switzerland) and Dolphin Implanner (Dolphin Imaging and Management Solutions, Chatsworth, USA). These have been especially developed to assist dental practitioners to perform preoperative planning. They contain the size and shape of commonly available implants for the clinician to select and place in the patient's alveolar bed as captured by CBCT. The clinician can appreciate the placement of the implant in a virtual computer simulation. This provides the clinician with a presurgical 3-D review of the desired postoperative result. The best position of the implant is often oblique. Therefore, the 3-D presentation of alveolar ridges is much better than any section simply orientated in axial, coronal, or sagittal planes—the standard reconstructions of the data capture by HCT. Furthermore, the spatial resolution of the 3-D of data captured by CBCT is superior to that of HCT, not simply because of its far superior spatial resolution in the axial plane, but also in the Z axis (patient's long axis) due to its isotropic cuberilles (see Figure 5.3) in contrast to the anisotropic cuberilles of HCT (see Figure 4.6).

Failure to osseointegrate is perhaps the most important reason for failure (Figure 15.9), but it is not the only one. In addition to pain and numbness, placing an implant into any neurovascular structure, both within and outside the bone (Figure 15.10), may result in life-threatening hemorrhage. Even if the implant avoids those structures and does successfully osseointegrate, it may still be considered a failure because its position would not allow an aesthetically and/or functionally acceptable result. In order to minimize all of these causes for failure a stent is recommended.

The rigid surgical guide stent with bur holes can be used to guide drilling and implant placement to a high degree of precision. Figures 15.11 and 15.12 display two different stents. Both have metal-lined holes to facilitate the precise placement of drilled holes. The stent in Figure 15.12 permits the direct screwing of the stent to the bone to ensure that the holes are cut precisely at the site required by the clinician who will provide the definitive prosthesis; this enhances the final aesthetic result of the restoration. This type of stent

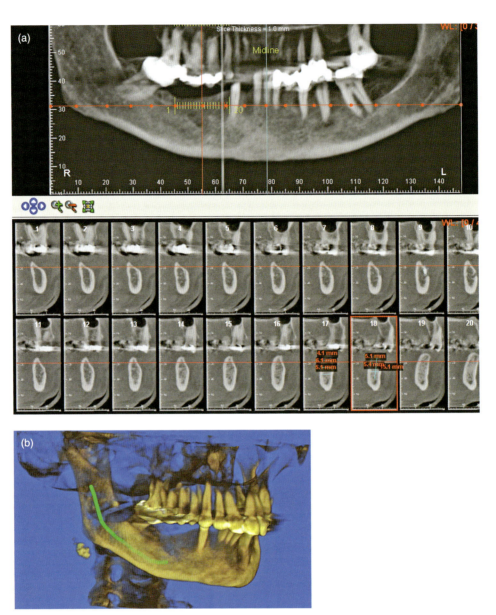

Figure 15.8. (a) Sections showing measurements (Dolphin) made on data from an iCAT scan. (b) 3D reconstruction of the mandible with the mandibular canal indicated by a green line. Figures courtesy of Dr. Babak Chehroudi, Faculty of Dentistry, University of British Columbia. **Note:** The radiopaque structure just below the angle of the mandible. This position is consistent with calcification within a lymph node.

is particularly useful in difficult cases or when multiple implants are needed.

In addition to the more common anatomical concerns to avoid the mandibular canal and maxillary antral cavity there are others of at least equal importance. These are the accessory mental foramen, mandibular incisive canal, lingual foramen, floor of the mouth, submandibular fossa, nasopalatine (incisive) foramen, buccal concavity of the maxilla, and thin and/or inclined alveolar ridges.

ACCESSORY MENTAL FORAMEN

Accessory mental foramina have been reported in 7% of CBCTs in one study.[12] Nine out of the fifteen cases observed ran distoinferiorly from the anterior

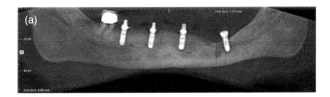

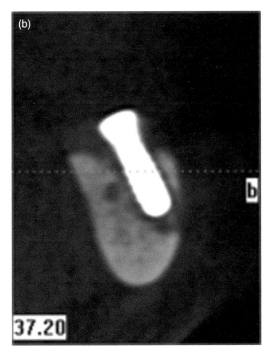

Figure 15.9. Cone-beam computed tomography (a,b) showing failure of osseointegration. The implant is completely surrounded by a radiolucent space.

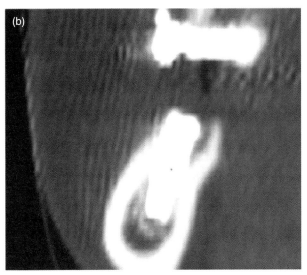

Figure 15.10. Helical computed tomography (a,b) displaying an implant inserted into the mandibular canal.

loop of the intraosseous mental nerve. Naitoh et al's Figure 6b displays a case course obliquely through the buccal cortex to reach the foramen.[12]

MANDIBULAR INCISIVE CANAL

Makris et al. reviewed 100 CBCTs.[13] They reported the mandibular incisive canal to be 1.5 cm from the mental foramen to the symphysis. The entire canal was visible in only 18% of cases. It displayed a marked tendency to be closer to the buccal cortex at the mental foramen (85% buccal to 10% lingual) to an increased tendency for the lingual cortext at the symphysis (60% buccal to 35% lingual).[13] Walton reported that 24% of patients with implants in the symphysis reported neurosensory disturbance.[14] This was temporary in almost all cases.

LINGUAL FORAMEN

The traditional anatomy textbook describes lingual foramen as a rare opening in the lingual side of mandible along the median sagittal plane. It sometimes shows up in periapical radiographs of mandibular incisors as a small well-defined, corticated foramen substantially apical to the apices of incisors. It is considered to contain only blood vessels but no nerves. Makris et al. reported that 81% of their 100 dried mandibles investigated by CBCT revealed one lingual vascular canal.[13] They reported that in those cases with only one foramen, it was above the genial spine, whereas the supplemental foramina were below the spines.[13] This is consistent with Rosano et al.'s *in vitro* findings.[15] They reported that 70% of cases have one foramen above the spines and one or two foramina below the spines.[15] Figure 15.13 shows five examples of lingual foramina and intraosseous canals. The fact

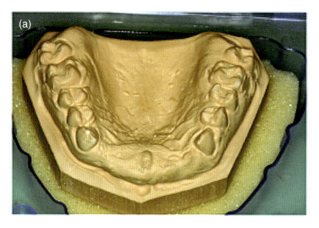
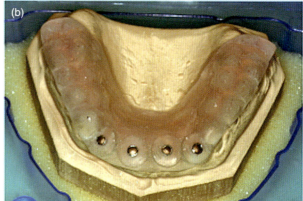

Figure 15.11. (a) Stone model of upper jaw exhibiting the edentulous anterior site that will receive implants. (b,c) The surgical stent made of acrylic with metal lined holes to guide the drill.

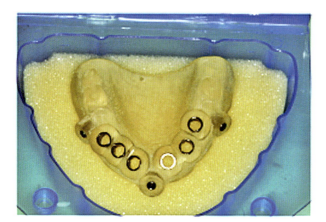

Figure 15.12. The surgical stent made of acrylic with metal lined holes to guide the drill. This stent differs from that of Figure 15.11 because it has additional holes so that the stent may be stablized by screwing it into the alveolus prior to drilling the holes for the implants.

that they convey blood vessels raises the possibility of uncontrolled bleeding into the floor of the mouth if damaged.

FLOOR OF THE MOUTH

The blood supply of the anterior sextant of the mandible is completely supplied by branches of the sublingual arteries, which penetrate the bone by way of the lingual foramina.[15] Furthermore, the close proximity of the sublingual arteries to the lingual cortex, particularly at the symphysis, makes them vulnerable to damage if that cortex is perforated.[15] Rosano et al. have identified 16 published cases of life-threatening hemorrhage associated with dental implantation.[15] There are three more recent reports of hemorrhage.[16-18] Both angulation and the size of the implant need to be carefully considered to avoid perforation of the lingual wall.[17]

THE SUBMANDIBULAR FOSSA

The submandibular fossa can be difficult to examine clinically. This lingual concavity of the posterior body of the mandible can be very deep and markedly limit the height of alveolar bone available for the implant; then it may need to be inserted at an angle. Parnia et al. classified the degree of concavity of the submandibular fossa, into three groups; Type I (less than 2 mm, 20%). Type II (2–3 mm, 52%) and Type III (more than 3 mm, 28%).[19] Although such a perforation into the very vascular floor of the mouth may cause a life-threatening hemorrhage, injury to salivary glands—although infrequently reported—has happened.[20] Furthermore, such injury may result in plunging ranulae.[21,22]

VARIATION IN NASOPALATINE (INCISIVE) CANAL

The nasopalatine canal in the maxilla has a highly variable diameter, path, and morphology. The diameter of the narrowest part of the canal can range from 3 to 8 mm. Figure 15.13 displays a range of normal nasopalatine canals. The diameter is often not uniform throughout its course. Because the position of the canal may not be exactly in the midline, it is important before any implant is placed in the maxillary central incisal region that CBCT is performed to examine this anatomy. An implant

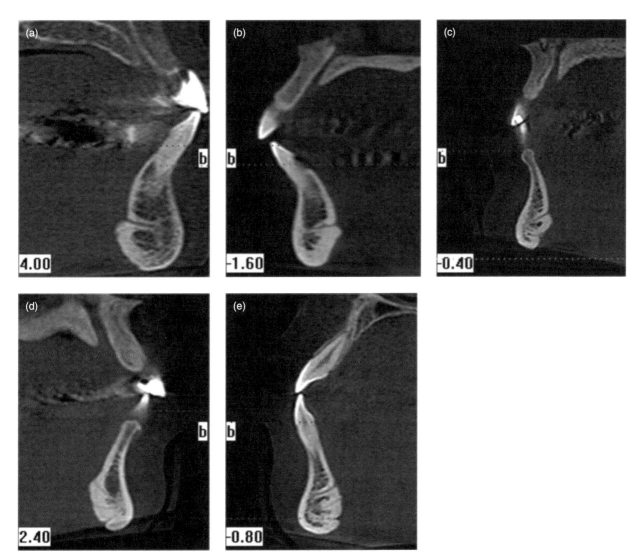

Figure 15.13. Lingual canals observed on cone-beam computed tomographic images. (a) 1 lingual canal and a narrow maxillary ridge. (b) 1 lingual canal and a narrow maxillary ridge. (c) 2 lingual canals and a narrow ridge in both jaws. (d) 3 lingual canals and a narrow mandibular ridge. (e) 3 lingual canals and a narrow ridge in both jaws. Note also the normal incisive (nasopalatine) canals.

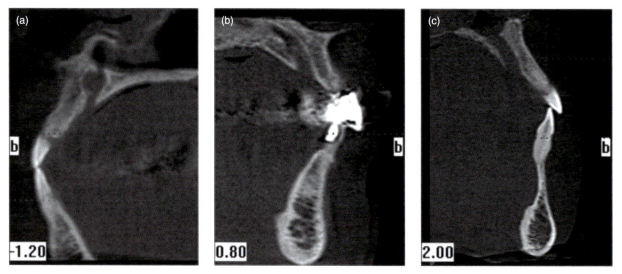

Figure 15.14. Variety in shape of incisive (nasopalatine) canals (a,b,c) observed on cone-beam computed tomographic images. Figure 15.14a displays an early nasopalatine duct cyst.

placed into the nasopalatine canal can lead to a failure in osseointegration. Furthermore, nasopalatine nerve damage will cause numbness of the incisive papilla. Figure 15.14 show examples of nasopalatine canals unusual to the point of abnormal. Figure 15.14a displays an early nasopalatine duct cyst (see Chapter 9) that should be recognized and treated at this stage.

BUCCAL CONCAVITY IN THE MAXILLA

It is common to have a concavity on the cortex buccal to an extraction site. The degree of this concavity would affect the position of the implant. Buccal fenestration can easily occur if no detailed preoperative radiographic assessment has been made (Figure 15.13). Failure to assess the buccal concavity can lead to fenestration of the alveolar bone, recession of the gingival margin, and exposure of the implant's thread.

THIN AND INCLINED ALVEOLAR RIDGE

The alveolar ridge in the incisal region of both jaws can be very thin (Figures 15.13, 15.14c). If the narrowest part of the ridge is less than 3 mm, implants are contraindicated, unless bone grafting has been successfully performed. Furthermore, the vertical angulation of this narrow alveolar ridge can form an acute angle to the standing teeth. This phenomenon is more common in the maxillary canine and premolar region. Implant planning in this region could consider inserting the implant at an acute angle to the suprastructure. An angled abutment may be needed.

Conclusions and Closing Remarks

The recent introduction of CBCT has completely revolutionized oral and maxillofacial radiology and implantology. In spite of the substantial body of literature that has emerged, only very little of it so far approximated the highest standard of clinical evidence. Therefore, its use, as with any other radiological investigation, needs to be clearly indicated for that particular patient. Nevertheless, there is a duty of care that all clinicians bear: to ensure that each potential implant patient has been properly evaluated so that each implant can be safely placed within the patient with the greatest chance of success. Due to an overall reduction in radiation dose, increased availability, facility of simulation, and an understanding of the risks of implant placement, it is becoming more difficult to justify not using 3-D imaging as part of preimplant planning.

Dental practitioners who use CBCT in their clinic for implant treatment often are not specially trained in oral and maxillofacial radiology. Should there be any other pathology, such as jaw cysts or neoplasm in the jaws, it can easily be missed because the most attention will be given to the implant sites. Failure to identify, diagnose and treat, review, or refer is considered at least unprofessional conduct in almost all jurisdictions. Therefore, it is very important to fully discharge this professional responsibility by carefully reviewing all regions of the scan in a systematic way. Such a review should also include the soft tissue outlined by air (see Point 1 Disadvantages in Table 15.3). Should there be doubt concerning any radiographic features, the case should be referred to an oral and maxillofacial radiologist for a full report. The EADMFR guidelines firmly advocate that the entire dataset of any CBCT investigations of areas outside the jaws and any CBCT investigation using an FOV greater than 8 cm × 8 cm should be reviewed by an oral and maxillofacial or medical radiologist in order to identify pathology so that it can be promptly investigated and appropriately treated.[2] The range of the more common and/or important extragnathic pathology that is likely to present on medium-to-large FOV CBCT are addressed in Chapters 16 through 18.

Those CBCTs with a FOV of 8 cm × 8 cm or less and whose use will be confined to the jaws (which may include the floor of the nose and the temporomandibular joint), can also be interpreted by an adequately trained nonradiologist dental practitioner.[3]

References

1. Tyndall DA, Brooks SL. Selection criteria for dental implant site imaging: a position paper of the American Academy of Oral and Maxillofacial radiology. *Oral Surg Oral Med Oral Pathol Oral Radiol Endod* 2000;89: 630–637.
2. Harris D, Buser D, Dula K, Grondahl K, Haris D, Jacobs R, Lekholm U, Nakielny R, van Steenberghe D, van der Stelt P. European Association for Osseointegration. E.A.O. guidelines for the use of diagnostic imaging in implant dentistry. A consensus workshop organized by the European Association for Osseointegration in Trinity College Dublin. *Clin Oral Implants Res* 2002;13: 566–570.
3. Horner K, Islam M, Flygare L, Tsiklakis K, Whaites E. Basic principles for use of dental cone beam computed tomography: consensus guidelines of the European Academy of Dental and Maxillofacial Radiology. *Dentomaxillofac Radiol* 2009;38:187–195.
4. Bryant SR, MacDonald-Jankowski D, Kim K. Does the type of implant prosthesis affect outcomes for the completely edentulous arch? *Int J Oral Maxillofac Implants* 2007;22:117–139. Erratum in *Int J Oral Maxillofac Implants* 2008;23:56.
5. Vazquez L, Saulacic N, Belser U, Bernard JP. Efficacy of panoramic radiographs in the preoperative planning of posterior mandibular implants: a prospective clinical study of 1527 consecutively treated patients. *Clin Oral Implants Res* 2008;19:81–85.
6. Butterfield KJ, Dagenais M, Clokie C. Linear tomography's clinical accuracy and validity for presurgical dental implant analysis. *Oral Surg Oral Med Oral Pathol Oral Radiol Endod* 1997;84:203–209.
7. Bou Serhal C, van Steenberghe D, Quirynen M, Jacobs R. Localisation of the mandibular canal using conventional spiral tomography: a human cadaver study. *Clin Oral Implants Res* 2001;12:230–236.
8. Diniz AF, Mendonça EF, Leles CR, Guilherme AS, Cavalcante MP, Silva MA. Changes in the pre-surgical treatment planning using conventional spiral tomography. *Clin Oral Implants Res* 2008;19:249–253.
9. Draenert FG, Coppenrath E, Herzog P, Müller S, Mueller-Lisse UG. Beam hardening artefacts occur in dental implant scans with the NewTom cone beam CT but not with the dental 4-row multidetector CT. *Dentomaxillofac Radiol* 2007;36:198–203.
10. Schulze RK, Berndt D, d'Hoedt B. On cone-beam computed tomography artifacts induced by titanium implants. *Clin Oral Implants Res* 2010;21:100–107.
11. Chau AC, Fung K. Comparison of radiation dose for implant imaging using conventional spiral tomography, computed tomography, and cone-beam computed tomography. *Oral Surg Oral Med Oral Pathol Oral Radiol Endod* 2009;107:559–565.
12. Naitoh M, Hiraiwa Y, Aimiya H, Gotoh K, Ariji E. Accessory mental foramen assessment using cone-beam computed tomography. *Oral Surg Oral Med Oral Pathol Oral Radiol Endod* 2009;107:289–994.
13. Makris N, Stamatakis H, Syriopoulos K, Tsiklakis K, van der Stelt PF. Evaluation of the visibility and the course of the mandibular incisive canal and the lingual foramen using cone-beam computed tomography. *Clin Oral Implants Res* 2010 (Apr 19) [Epub ahead of print].
14. Walton JN. Altered sensation associated with implants in the anterior mandible: a prospective study. *J Prosthet Dent* 2000;83:443–449.
15. Rosano G, Taschieri S, Gaudy JF, Testori T, Del Fabbro M. Anatomic assessment of the anterior mandible and relative hemorrhage risk in implant dentistry: a cadaveric study. *Clin Oral Implants Res* 2009;20:791–795.
16. Del Castillo-Pardo de Vera JL, López-Arcas Calleja JM, Burgueão-García M. Hematoma of the floor of the mouth and airway obstruction during mandibular dental implant placement: a case report. *Oral Maxillofac Surg* 2008;12:223–226.
17. Pigadas N, Simoes P, Tuffin JR. Massive sublingual haematoma following osseo-integrated implant placement in the anterior mandible. *Br Dent J* 2009;206: 67–68.

18. Dubois L, de Lange J, Baas E, Van Ingen J. Excessive bleeding in the floor of the mouth after endosseus implant placement: a report of two cases. *Int J Oral Maxillofac Surg* 2010;39:412–415.
19. Parnia F, Fard EM, Mahboub F, Hafezeqoran A, Gavgani FE. Tomographic volume evaluation of submandibular fossa in patients requiring dental implants. *Oral Surg Oral Med Oral Pathol Oral Radiol Endod* 2010;109: e32–36.
20. Nahlieli O, Droma EB, Eliav E, Zaguri A, Shacham R, Bar T. Salivary gland injury subsequent to implant surgery. *Int J Oral Maxillofac Implants* 2008;23:556–560.
21. Loney WW Jr, Termini S, Sisto J. Plunging ranula formation as a complication of dental implant surgery: a case report. *J Oral Maxillofac Surg* 2006;64:1204–1208.
22. Mandel L. Plunging ranula following placement of mandibular implants: case report. *J Oral Maxillofac Surg* 2008;66:1743–1747.

Part 4
Radiological pathology of the extragnathic head and neck regions

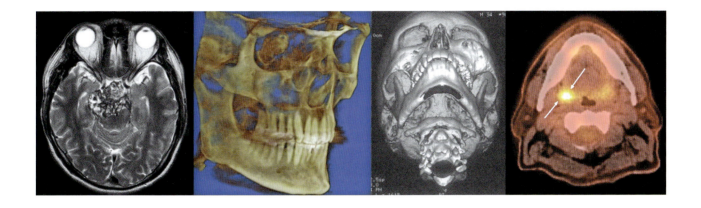

Chapter 16
Introduction

This chapter introduces some important anatomical and technical aspects of the radiology of those lesions that present outside the maxillofacial regions. The extragnathic head and neck will be considered as two separate anatomical regions, the skull and the neck. The benign and malignant lesions affecting either or both of these regions are addressed respectively in Chapters 17 and 18.

The Skull

This section places particular emphasis on the base of a skull, which appears increasingly within moderate-to-large *fields of view* (FOVs) for *cone-beam computed tomography* (CBCT) (Figure 16.1). Although it is generally held that this area is properly the purview of the medically trained head and neck or neuroradiologist, it is important that the oral and maxillofacial practitioner prescribing and interpreting these images be aware of some of the important lesions that may incidentally present (Figure 16.2). Knowledge of such lesions enable those practitioners to recognize the abnormality and refer it promptly and appropriately. It is expected that datasets of moderate-to-large FOVs would be reported by a radiologist (see Chapter 5).

BASIC ANATOMY OF THE SKULL BASE

The anatomy of the skull base is both complex and subject to much variation, such as asymmetrical pneumatization and the appearance of the foramina, which can present as pseudolesions.[1] The base of the skull is divided into three fossae: *anterior cranial fossa* (ACF), *middle cranial fossa* (MCF), and *posterior cranial fossa* (PCF).[2] The boundary between the ACF and MCF is the posterior margin of the lesser wing of the sphenoid. The boundary between the MCF and PCF is delimited by the posterior superior margin of the petrous temporal bone and the posterior clinoid process of the body of the sphenoid. The emphasis of this chapter is placed on the MCF and PCF because they are most likely to appear on the moderate-to-large FOV CBCT images. Of these, the portions of the base of the skull that are most likely to be captured are the clivus and petrous temporal bone (Figure 16.3). Some important lesions that may affect the orbit have been considered, because the orbit is just as likely to be captured as the MCF and PCF.

The *sella turcica* (ST), sometimes called the *pituitary fossa* is a salient point in cephalometry essential for the diagnosis and treatment planning of craniofacial abnormalities. Although significant differences in skeletal class of the face and jaws were found only between Skeletal Class II and Class III (the former has smaller ST diameters, the latter has larger diameters),[3] this opportunity should also be used for assessment of the size and shape of the ST that may arise from other causes. The dimensions of a normal ST are set out in Table 16.1.[4] The dimensions of the ST are larger in the older patient.[4]

CROSS-SECTIONAL IMAGING OF THE SKULL BASE–MRI AND CT

Current imaging of skull base relies heavily on *computed tomography* (CT) and *magnetic resonance imaging* (MRI). *Helical computed tomography* (HCT) is superior to MRI for evaluating bone erosion and destruction, particularly if the unit used has a high spatial resolution. A rudimentary knowledge of skull base structures and foramina is required for both interpreting the spread of cancer through the skull base and searching for certain neurogenic tumors (such as schwannomas). Some cases occur within foramina. In particular, the pterygopalatine fossa, sphenopalatine foramen, foramen rotundum, inferior orbital fissure, carotid canal, and jugular/hypoglossal foramina should be studied (see Figure 16.3).

Oral and Maxillofacial Radiology: A Diagnostic Approach, David MacDonald. © 2011 David MacDonald

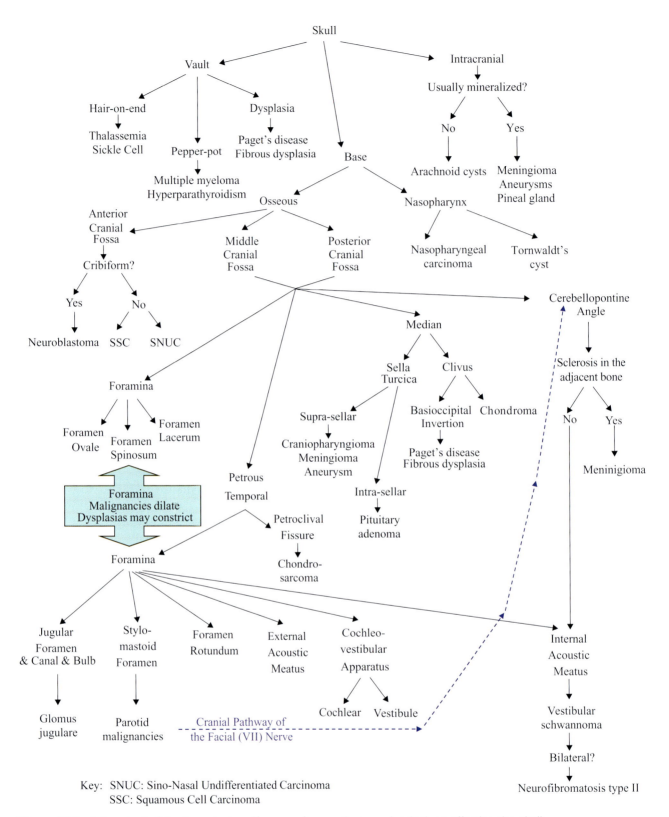

Figure 16.1. A flowchart of the important and/or more frequently occurring lesions affecting the skull.

Chapter 16: Introduction

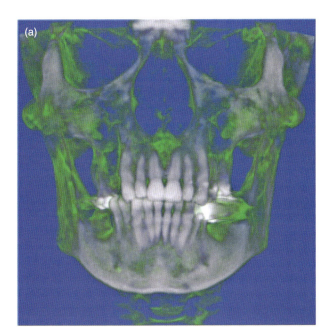

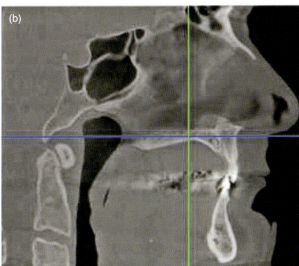

Figure 16.2. Cone-beam computed tomography (CBCT) as part of the assessment of a patient being considered for osseointegrated implants in both jaws. This was made on an iCAT type of CBCT unit with its large field of view. It can be seen that the base of the skull has been captured in (b). Figures courtesy of Dr. Babak Chehroudi, Faculty of Dentistry, University of British Columbia.

Table 16.1. Dimensions between 6-year-olds and 21-year-olds of both genders (derived from Axelsonn et al.[3]

Years	Dimension	Males	Females
6 years	Length	8.8 SD 1.5	8.5 SD 1.3
	Depth	6.3 SD 0.8	6.6 SD 0.7
	Diameter	10.0 SD 1.3	9.8 SD 1.3
21 years	Length	8.9 SD 0.9	8.4 SD 1.6
	Depth	7.3 SD 1.1	7.2 SD 1.2
	Diameter	11.3 SD 1.1	11.7 SD 1.1

MRI is introduced in Chapter 6. It is excellent for visualizing soft-tissue involvement, including the bone marrow. For the head and neck, precontrast MR images are usually first acquired in axial/sagittal planes with T1-weighting. T1-weighted images show anatomy well with good contrast between muscles and intervening fat planes (Figure 16.4a,b). This is because on T1-weighted images muscles are relatively dark (isointense) and fatty tissue is bright, or more accurately of high signal intensity (hyperintense). In adults, bone marrow is normally bright due to fatty infiltration. When tumor, infection, or inflammation displaces fatty marrow, it becomes "dark" on T1-weighted images, making MRI exquisitely sensitive for detecting these pathologies (e.g., Figure 18.19c). Be aware that in younger people the bone marrow is not yet fatty and may be normally dark on T1-weighted images. Fatty marrow replacement occurs at different times in different bones; in the mandible it generally happens between 10–15 years of age. T1-weighted scans are also called "anatomy scans" or "fat scans" (see Chapter 6).

MR images are then acquired with T2-weighting (Figure 16.4d). On these images, structures containing a high percentage of water are bright or hyperintense. Hence tissues that are neoplastic, edematous or inflamed tend to be bright on T2-weighted images. T2-weighted images are also called the "pathology scans" or "water scans" (see Chapter 6)).

By definition, muscles and brain tissue are considered isointense on T1- & T2-weighted images—their signal intensities are used as a reference to label other structures as hyper- or hypointense. T2-weighted images are always acquired before giving MRI contrast. MRI contrast has a significant effect on T1-weighted images.

After intravenous administration of gadolinium contrast, some normally T1-weighted isointense structures will become hyperintense or bright. The nasal mucosa will vividly enhance;

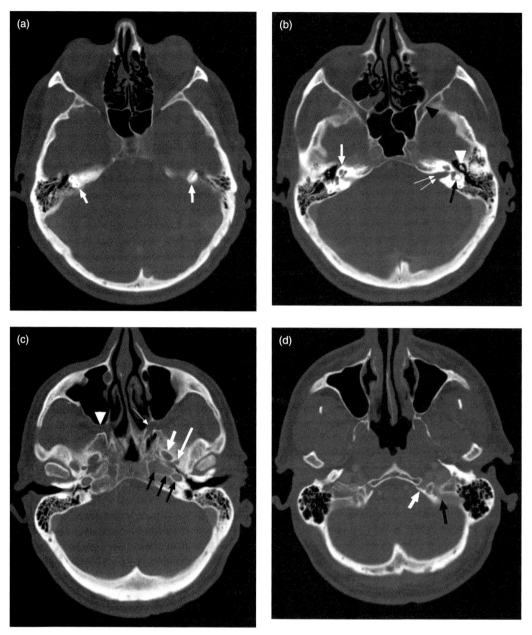

Figure 16.3. Computed tomography; normal axial sections (bone windows). (a) At the level of the base of the sella turcica, showing ethmoid and superior mastoid air cells. The vertical semicircular canals of the inner ear are visible within the petrous temporal bone (arrows). (b) At the level of the left internal auditory canal (double arrows). The cochlea of the inner ear are visible (single white arrow) and the "signet ring" shape of the left horizontal semicircular canal is seen (black arrow). The ossicles of the middle ears are visible, more so on the left (white arrowhead). The left inferior orbital fissure is marked with a black arrowhead. The ethmoid, sphenoid and mastoid air cells are visible. (c) At the level of the condylar head. The foramen ovale (short white arrow) and spinosum (long white arrow) are visible. The left carotid canal is seen ending medially at the foramen lacerum (black arrows). The pterygopalatine fossa is shown on the left (white arrowhead) and the sphenopalatine foramen on the right (thin white arrow). The conchae (turbinates) are partially visualized in the nasal cavity. (d) At midmaxillary sinus level. The nasal cavity and posterior nasopharynx are visible. The hypoglossal canal (white arrow) and jugular foramen (black arrow) are visible on the left. The mastoid air cells are extensively pneumatized.

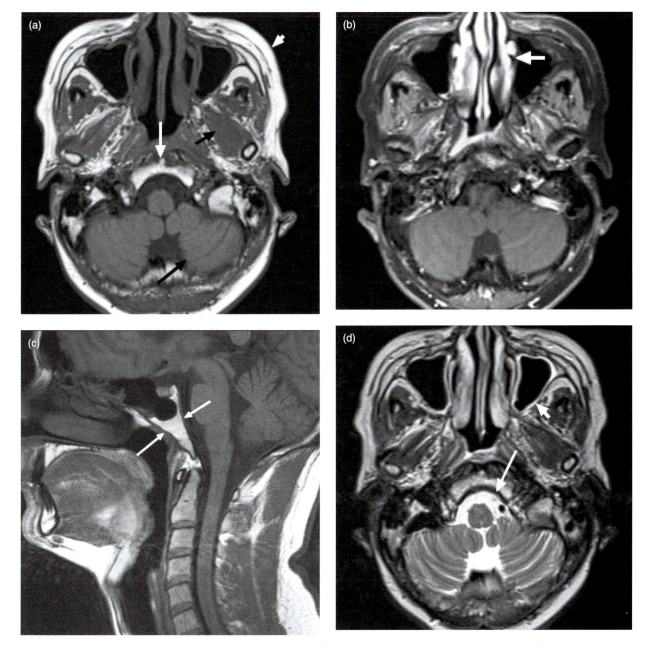

Figure 16.4. Normal magnetic resonance imaging (MRI) soft-tissue contrast pre- and postcontrast. (a) Axial non-contrast T1-weighted MRI. Note the bright subcutaneous fat (short white arrow) and the fatty marrow in the clivus (long white arrow). The lateral pterygoid muscle (short black arrow) and cerebellar hemisphere (long black arrow) both exhibit isointense T1-weighted signals. (b) Axial postgadolinium fat-saturated T1-weighted MRI. Compare this to noncontrast T1-weighted image (a) and see how brightly normal nasal mucosa enhances. Compare the subcutaneous fat signal with that from (a); it is now dark due to the fat saturation. (c) Sagittal non-contrast T1-weighted MRI. Note the high signal in the clivus corresponding to fatty marrow (arrows). (d) Axial non-contrast T2-weighted MRI. Note the normally bright cerebrospinal fluid surrounding the medulla (long arrow). The thin layer of inflamed mucosa in the left maxillary sinus also displays a hyperintense T2-weighted signal (short arrow). **Note:** Other good examples displaying the effect of fat saturation on postgadolininium T1-weighted MRI are Figures 18.15 and 18.25.

looking at the nasal septum and turbinates on T1-weighted images is a good way of determining whether MR contrast has been administered (compare Figures 16.4a,b). Many inflammatory, infectious and neoplastic lesions will also enhance with contrast. In order to see these optimally, a technique called "fat saturation" is usually used in the head and neck. This decreases the signal from normal fat, making it dark and allowing pathologic enhancement to stand out (Figure 16.4b). Postcontrast fat-saturated T1-weighted images are excellent for detecting malignant lesions (see Figures 18.2(e,f), 18.5(c), 18.6(b,c), 18.7, 18.8(b,c), 18.9, 18.11, 18.12(c), 18.15(a,b,c), 18.16, 18.17, 18.18(b), 18.20(c,d), 18.24(b,d), 18.25, 18.29, 18.30(b,c)) as well as many benign pathologies affecting the skull base (Figures 17.2(b), 17.5(a,b), 17.6, 17.9(c), 17.10(b,c), 17.12(a,b), 17.13(b), 17.14(c), 17.15, 17.19(c)). MRI is contraindicated in certain cases—for example, patients with either a pacemaker or an intraocular metallic foreign body (see Chapter 6). In these patients CT imaging will be done.

Swellings of the Neck

RELEVANT ANATOMY AND CLASSIFICATIONS

Imaging of swellings of the neck, in particularly those of the suprahyoid, form an important part of oral and maxillofacial practice. The clinician classically divides the neck into its anatomical triangles and locates the lesion in question accordingly (Figure 16.5). The *oral and maxillofacial* (OMF) clinician should be aware that head and neck radiologists commonly consider lesions according to alternate compartments and spaces (Table 16.2). Such awareness will enhance the quality of the OMF clinician's communication with his/her radiologist so as to optimize patient care.

Since there are these and other different ways to classify head and neck anatomy it is important to know something about each classification and how they differ before using them as a template to study pathology.

Classical anatomists divided the neck into simple triangles based on superficial muscular landmarks that could be easily appreciated via surface anatomy or superficial dissection.[5] This schema is useful for clinical examination and plan-

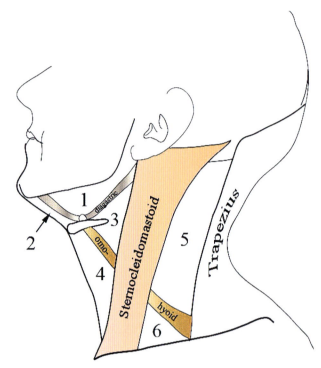

Figure 16.5. Classical triangles of the neck. Synonyms are listed: 1 = submaxillary or diagastric or submandibular triangle; 2 = submental or suprahyoid triangle; 3 = superior carotid triangle; 4 = inferior carotid triangle; 5 = occipital triangle; 6 = subclavian triangle. Collectively, 1 + 2 + 3 + 4 = anterior triangle and 5 + 6 = posterior triangle.

ning surgical approaches to the neck. After the advent of cross-sectional imaging techniques, a new system became popular based on cervical fascial planes that divide the neck into compartments.[6] This system is clinically relevant because these planes can limit and guide the spread of disease occurring within a compartment. Because lymph nodes occur in multiple compartments and are interconnected by lymphatics, nodal classification by fascial planes is less relevant. Different systems of cervical lymph node classification have evolved. Therefore, we have separate systems in use, each with its advantages and disadvantages.

CLASSICAL TRIANGLES OF THE NECK

The neck is divided initially into two triangles: anterior and posterior by the midplane of the sternocleidomastoid muscle. The anterior triangle is further divided by the omohyoid and diagastric muscles into the submental triangle (also termed *suprahyoid*), submaxillary triangle (also termed

Table 16.2. Deep spaces of the suprahyoid neck

Primarily Contains	Name	Major Contents/Comments
Muscles	Masticator space	Muscles of mastication
	Prevertebral space	Prevertebral and paraspinal muscles
Vessels and nerves	Carotid space	Carotid artery, internal jugular vein, cranial nerves IX → XII
Salivary glands	Parotid space	Parotid gland
	Submandibular space	Submandibular gland
	Sublingual space	Sublingual gland
Lymphoid tissue and mucosa	Pharyngeal mucosal space	Tonsils, pharyngeal mucosa
Potential spaces	Parapharyngeal space	Fat, vessels, nerves; located between masticator, carotid, and pharyngeal mucosal spaces
	Danger space	Potential space; a pathway for prevertebral abscess enlargement
	Retropharyngeal space	Fat and lymph nodes

diagastric or *submandibular*) and superior/inferior carotid triangles. The posterior triangle is split by the omohyoid muscle into a large occipital triangle and a smaller subclavian (also called the *supraclavicular* or *omoclavicular* triangle).

Submandibular triangle

The submandibular triangle is particularly relevant to oral pathology. Primary lesions are generally observed on clinical inspection, ideally while still small and symptom-free. On occasion the symptoms may not immediately suggest a malignancy; a floor-of-the-mouth *squamous cell carcinoma* (SCC) causing an obstruction of the submandibular duct may produce symptoms suggestive of a submandibular duct calculus. The need for imaging at this stage may not be immediately obvious. Nevertheless, advanced imaging such as MRI can reveal the degree of invasion (see Figure 18.22). Those SCCs, which appear more posteriorly in the oral cavity, are larger and/or already have invaded bone, so advanced imaging is now indicated. This emphasizes the importance of early and appropriate imaging—the SCC can be detected, diagnosed, and treated early so as to improve the prognosis.

Cysts such as ranulas and dermoid cysts are not life-threatening unless they achieve substantial dimensions to threaten the airway. Figure 6.8 shows a sublingual dermoid cyst.

Anterior triangle

The principle organs are the thyroid and parathyroid glands and larynx. Laryngeal SCC is the most common malignancy of the throat.[7] Figure 18.27 displays a case. Figure 18.4 demonstrates a case of chondrosarcoma affecting the larynx. The pathology of those lesions causing swelling of the thyroid are many and will not be considered.

Posterior triangle

The posterior triangle contains numerous lymph nodes that can become enlarged from metastatic, infectious, or inflammatory disease. This is its main relevance to the OMF clinician.

FASCIAL COMPARTMENTS OF THE NECK

The fascial planes of the neck form the boundaries of a compartmental anatomic classification scheme that is often used by head and neck radiologists. The fascia are initially divided into two layers: superficial and deep. The superficial cervical fascial compartment lies beneath skin, containing subcutaneous fat, the platysma muscle, and vessels and nerves. The deep cervical fascia divides the remainder of the neck into multiple anatomic compartments or "spaces" and defines the aerodigestive tract; a description of these is set out in Table 16.2. These spaces are most often used when describing pathology occurring above the hyoid bone. Because the suprahyoid region is generally most relevant to OMF pathology, the infrahyoid region, except for some laryngeal cancers, is not covered here.

PHARYNGEAL MUCOSAL SPACE

This contains the pharyngeal mucosa and the tonsils. *Nasopharyngeal carcinoma* (NPC) arises from squamous mucosa in this space and is an important lesion, particularly among those of East Asian origin. The incidence of NPC has declined in Chinese immigrant populations compared with natives in China.[8] Figure 18.15 exhibits an NPC that occupied the entire posterior pharyngeal wall and has spread to retropharyngeal nodes and the chain of lymph nodes under the cover of the sternocleidomastoid muscle.

PAROTID, SUBMANDIBULAR, AND SUBLINGUAL SPACES

The parotid submandibular and sublingual glands are encased in their own individual spaces. Though the parotid is the largest of the salivary glands, it tends to have a higher proportion of benign to malignant neoplasms than the submandibular, sublingual, and minor salivary glands. Examples of MRI of *pleomorphic salivary adenoma* (PSA) are seen in Figures 6.2–6.4, 6.11, and 17.17. Malignant neoplasms arising in this gland are displayed in Figures 18.12–18.14.

MASTICATOR SPACE

The masticator space contains the muscles of mastication, i.e., the pterygoids, temporalis, and masseter muscles as well as the inferior alveolar nerve (V3). Intrinsic lesions are uncommon; this space is most often invaded by pathology from outside its borders, such as SCC (see Figures 18.16, 18.17, 18.20).

CAROTID SPACE

The carotid space contains the carotid artery, jugular vein, and vagus nerve. Carotid body tumors are relatively rare tumors arising from neuroendocrine cells at the carotid bifurcation—hence, within the carotid space. These are also called *paragangliomas* and are described later in Chapter 17. Many of the deeper and larger lymph nodes lie adjacent to this space.

PREVERTEBRAL, DANGER, RETROPHARYNGEAL, AND PARAPHARYNGEAL SPACES

The prevertebral, danger, and retropharyngeal spaces are normally very thin, approximating collapsed tubes lying immediately anterior to the vertebral column. The prevertebral space is immediately anterior to the vertebral bodies and extends from the skull base to the coccyx. The danger space lies anterior to the prevertebral space and stops inferiorly at the diaphragm. If an abscess forms in this space, the infection can easily spread inferiorly into the chest—hence, the "danger." The retropharyngeal space lies between the danger and premucosal spaces. It contains lymph nodes that often become involved with metastatic disease in head and neck cancers. The parapharyngeal space resides between the masticator, carotid, and pharyngeal mucosal spaces and contains mainly fat. When tumors are nearby, the pattern of displacement of this space is useful for determining where the tumor originated.

There are more named fascial spaces than those listed above. The reader who wants to become more familiar with all the spaces is referred to excellent books by Harnsberger[9] and Som and Curtin.[10]

The new level-based systems, designed to be used with cross-sectional imaging are discussed in Chapter 18 (see Figure 18.31 and Table 18.1).

References

1. Schmalfuss IM, Camp M. Skull base: pseudolesion or true lesion? *Eur Radiol* 2008;18:1232–1243.
2. Pierot L, Boulin A, Guillaume A, Pombourcq F. Imaging of skull base tumours in adults. *J Radiol.* 2002;83:1719–1734. (In French)
3. Alkofide EA. The shape and size of the sella turcica in skeletal Class I, Class II, and Class III Saudi subjects. *Eur J Orthod* 2007;29:457–463.
4. Axelsson S, Storhaug K, Kjaer I. Post-natal size and morphology of the sella turcica. Longitudinal cephalometric standards for Norwegians between 6 and 21 years of age. *Eur J Orthod* 2004;26:597–604.
5. Drake RL, Pawlina W, Carmichael SW, Albertine KH. Anatomical Sciences Education, clinical anatomy, and

the anatomical record: take your pick. *Anatom Sci Ed* 2008;1:2.
6. Harnsberger HR. *Handbook of Head and Neck Imaging*, 2nd ed. Mosby, St. Louis 1995.
7. Li XY, Guo X, Feng S, Li XT, Wei HQ, Yang HA, et al. Relationship between a family history of malignancy and the incidence of laryngeal carcinoma in the Liaoning province of China. *Clin Otolaryngol* 2009;34:127–131.
8. Yu WM, Hussain SS. Incidence of nasopharyngeal carcinoma in Chinese immigrants, compared with Chinese in China and South East Asia: review. *J Laryngol Otol* 2009;2:1–8.
9. Harnsberger HR. *Diagnostic imaging. Head and neck*, 1st ed. Amirsys, Salt Lake City 2004.
10. Som PM, Curtin HD. *Head and neck imaging*, 4th ed. Mosby, St. Louis 2003.

Chapter 17
Benign lesions

The following are some common (and not so common) benign swellings affecting the skull base and suprahyoid region. All of these tumors can be seen on either CT or MRI, although as noted earlier, destructive bony pathology is usually better seen on CT.

Pituitary Adenoma

Pituitary adenomas (Figure 17.1) are benign tumors that arise from the adenohypophysis, or anterior lobe of the pituitary gland. They may be endocrinologically active and secrete the same hormones that the pituitary normally produces, only in excess amount. The most common "functioning" pituitary tumor is a prolactinoma[1]; the excess prolactin may produce amenorrhea and galactorrhea in women and impotence in men. Both prolactinomas and "nonfunctioning" adenomas occur at approximately the same frequency: 25% of total. Nonfunctioning tumors may present with decreased visual acuity when the tumor begins to compress the adjacent optic chiasm (Figure 17.2). Other lesions that may resemble pituitary adenomas include aneurysms, meningiomas, and schwannomas. Craniopharyngiomas and Rathke's pouch cysts also occur in the sella but are predominantly cystic instead of solid.

Craniopharyngioma

Craniopharyngiomas are complex cystic, predominantly suprasellar, mass lesions that are frequently partially calcified.[2] They are thought to arise from remnants of Rathke's pouch, an embryologic structure that migrates superiorly from the primitive oral cavity to eventually form the adenohypophysis or anterior lobe of the pituitary gland. Craniopharyngiomas comprise 6–13% of all pediatric brain tumors and have a bimodal age distribution, peaking between 5–14 years and after 65 years of age.[3,4] Three-quarters of these tumors are suprasellar, with most of the remaining lesions combined sellar-suprasellar. Rarely are they completely intrasellar. The craniopharyngioma usually presents as a small enhancing solid nodule within or adjacent to a lobulated cystic lesion with enhancing walls. The cystic part is often divided into multiple compartments, each with a slightly different though homogeneous T1- and T2-weighted MRI signal intensity. When they become large enough, craniopharyngiomas can produce obstructive hydrocephalus by blocking the foramen of Monroe (Figure 17.3).

Rathke's cleft cysts can sometimes mimic craniopharyngiomas.[5] These lesions also arise from Rathke's pouch and are seen in the same location. Unlike craniopharyngiomas they are generally purely cystic (no solid component), rarely calcify, and minimally enhance with *intravenous* (IV) contrast.[6,7] They also rarely produce symptoms.[8]

Intracranial Aneurysms

Intracranial aneurysms are focal dilatations of cerebral blood vessel walls with various shapes—some saccular, others fusiform. Arterial aneurysms are commonly caused by injury to the blood vessel walls from atherosclerosis and/or focal hemodynamic pressure. These aneurysms usually occur at vessel bifurcations, most often affecting the middle cerebral, supraclinoid internal carotid, posterior communicating, cavernous carotid, anterior communicating, and vertebrobasilar arteries in descending order of frequency.[9] The prevalence of *unruptured intracranial aneurysms* (UIA) is estimated to range between 0.4–6%, depending on the study design.[10] There is a 4% increased risk of having a UIA if a first-degree relative has a cerebral

Oral and Maxillofacial Radiology: A Diagnostic Approach, David MacDonald. © 2011 David MacDonald